Evolutionary Finance

EVOLUTIONARY FINANCE

Bartholomew Frederick Dowling

332.015195
D'14e

First published in 2005 by
PALGRAVE MACMILLAN
Houndmills, Basingstoke, Hampshire RG21 6XS and
175 Fifth Avenue, New York, N.Y. 10010
Companies and representatives throughout the world.

PALGRAVE MACMILLAN is the global academic imprint of the Palgrave Macmillan division of St. Martin's Press, LLC and of Palgrave Macmillan Ltd. Macmillan® is a registered trademark in the United States, United Kingdom and other countries. Palgrave is a registered trademark in the European Union and other countries.

ISBN 13: 978–1–4039–9665–7 hardback
ISBN 10: 1–4039–9665–2 hardback

This book is printed on paper suitable for recycling and made from fully managed and sustained forest sources.

A catalogue record for this book is available from the British Library.

Library of Congress Cataloging-in-Publication Data
Dowling, B. (Bartholomew)
    Evolutionary finance / Bartholomew Frederick Dowling.
        p. cm.
    Includes bibliographical references and index.
    ISBN 1–4039–9665–2 (cloth)
        1. Finance – Econometric models. 2. Econometrics. 3. Bioinformatics.   I. Title.

HG173.D65 2005
332'.01'5195—dc22   2005043354

10 9 8 7 6 5 4 3 2 1
14 13 12 11 10 09 08 07 06 05

Printed and bound in Great Britain by
Antony Rowe Ltd, Chippenham and Eastbourne

This book is dedicated to my three girls: Tracie-Louise, Breanna-Rose, Isabel-Jane

Contents

List of Figures and Tables

Figures

Tables

Acknowledgments

This book was written while I was a Research Associate at the Centre for Financial Research, Judge Institute of Management at Cambridge University. Without doubt, thanks must initially go to the support provided by this esteemed institution. That said, Oxford University too must claim its fair share of gratitude as the skeletal foundations of this book formed the basis of my Masters Thesis in Mathematical Finance at this equally impressive educational facility. My time at Oxford was particularly memorable thanks to the gifted (if not slightly unorthodox) lecturing by Paul Wilmott who proved to be a truly great educator, the erudite and thoroughly impressive Jeff Dewynne, the innovative and awe-inspiring Neil Johnson, and the wonderfully entertaining Sam Howison.

Still, as one will soon realize in perusing the following pages, my debt to academia is but a mere portion of the total bill of thanks this weighty tome owes to certain individuals. In particular, my time spent as a Chief Economist, Fixed Income & FX Strategist, Financial Engineer, and Global Asset Allocator based at first in the Antipodes, then to Wall Street, then to the City of London has provided me with a wonderful platform for some memorable exchanges as to how financial markets *really* work. To date, I have had the privilege of presenting my opinions in more than 18 countries in over 2000 various client meetings, conferences and seminars. This is not to mention the innumerable client conference calls, one-on-one exchanges and even "brainstorming" sessions with colleagues at, most recently, Merrill Lynch. To all these individuals – obviously too innumerable to mention here – I owe a sincere debt of thanks. It was through these interchanges that my views were molded and refined. Just as steel sharpens steel, I have been appreciative of the – at times near combative – forum in which these opinions were exchanged.

One particularly memorable experience for me was while I was in the position as a Desk Strategist to a team of proprietary traders and market-makers during the LTCM Crisis and (the associated) unwind of the Yen Carry trade. Watching markets gap down in a vain attempt to find a floor, experiencing first-hand the panic on the faces of traders, being physically shook by the sheer volume of noise erupting on the trading floor as individuals were – quite literally – screaming at each other (including at me, who was doing his own fair share of screaming back) was a truly exhilarating experience – both frightening and exciting at the same time. This was the blast furnace of capitalism in action. Unfortunately – somewhat ironically – given my academic background prior to this event, I was perplexed to find that there appeared to be no formal mechanism for describing precisely what was going on (much to the chagrin of my group of traders who were always looking for that informational edge). Sure, we all knew markets were governed by fear and greed – Keynes taught us as much and back in those days these basic concepts were in the process of being significantly embellished by the Behavioral Finance and information asymmetry theorists. Still, as each gut-wrenching day came to an end and the traders would gradually fade away to the various bars across town, I would hit the textbooks in a vain attempt to understand the apparently incomprehensible. Why had asset prices appeared to over-react to one particular innocuous information release and ignore other apparently more sentient information, why had the informational "tipping point" suddenly morphed trader concern into outright panic, why had the informational "seeds of destruction" of an event such as the LTCM Crisis (and associated Asian and Russian Debt Crisis) been written in data, well known and discussed at length up to a full 12 months earlier, but somehow suddenly the benign became malignant? I was severely disappointed by what finance theory had to offer during those lonely hours with only the cleaners and the socially challenged night traders to keep me company. So started my seven-year quest. Along the way I have managed to find scattered throughout the academic literature various snippets of useful insights but unfortunately, up until now, there was no unifying theory as to how the various microcomponents of information interact to "build" asset prices. After being on the frontline of applied finance for more than 16 years and dodging – and indeed, manufacturing and firing my own – informational "bullets," I decided to put pen to paper in an attempt to set the record straight – and I am thankful to Cambridge University for allowing me to pursue this ambition.

So what has been the response? Various drafts of various segments of this book have appeared on my company's – Evolutionary Finance™ Ltd – website. Thankfully, I have found practitioners within markets warmly endorsing my views. Indeed, many put to me that I was "mad" to even attempt to start outlining such a comprehensive investment framework without at first attempting to claim some proprietorship over its content. This put me directly into the capable hands of Barker Brettell Patent Attorney's in Cambridge to whom I owe an enormous debt of gratitude in assisting me with my patent application of the overall business method associated with my own "structured information" view on Evolutionary Finance. Kevin Upson at JellyBean Web at Cambridge did a wonderful job in designing my company's website while Tony Donnelly from UBS assisted enormously in providing worthwhile tips for the establishment of the Evolutionary Finance™ Ltd's absolute return strategy vehicles which have been designed specifically to leverage from the Evolutionary Finance principles espoused in this book. To all these individuals I am extremely thankful for allowing me to put this path breaking new finance theory into practice.

Dr Bartholomew Dowling
Cambridge, UK

Preface

This book is for all those who question assumption. Too often a Finance 101 lecture begins with the immortal words "let us assume markets are fully efficient" and that's the end of the discussion when it comes to market structure. Unfortunately, what follows from then on may look impressive but like any house – no matter how imposing – built upon a weak foundation, the cracks soon appear. We go back to Finance 101 and question that initial assumption. In particular, our focus is on the way in which market "efficiency" – in the sense of how information is imparted into price – can be better understood. We build the information generation mechanism from the "ground-up" using bioinformatics as a guide. The analyst's "idea" and the "gene" have a lot in common and we feel that some of the more exciting recent developments in the fields of genomics and bioinformatics have interesting implications for finance thinking – you just need to view the world through a different lens that's all.

It is the construction of such a lens that constitutes a large part of this book. The preliminaries of what we label as the "traditionalist approach" toward finance versus an emerging challenge of various "new" views will be discussed in Chapters 2 and 3. However, we have purposefully set aside Chapters 4, 5 and 6 to outline the theoretical foundations of our unique perspective on *Evolutionary Finance*. Here the reader will be made progressively more aware of the strong similarities between biomathematics and financial mathematics – via the conduit of informational analysis. While an excursion into the formal theoretical underpinnings of this exciting new field of thought is a necessary evil for methodically detailing our case, we stress that the onus will remain very much on what is necessary for the *applied* rather than pure theory for theory's sake.

Indeed, it is the pursuit of the applied aspects of our philosophy that form the foundations for Chapter 7 – which covers Evolutionary Finance principles in *practice*. As will be seen, virtually no aspect of applied finance will be left untouched – from asset selection, to strategy choices, to portfolio construction. In fact, as will be explained, we have now leveraged these applied principles to such an extent that we have developed our own proprietary Natural Selection™ software platform to assist us in making exceptional (risk adjusted) return decisions for asset and portfolio construction across an array of absolute return mandates within our firm – Evolutionary Finance™ Ltd. In short, Natural Selection™ provides a user-friendly interface for assessing an interesting array of Evolutionary Finance perspectives – from mapping the formation of the informational genome of an asset, to the developing of a step-by-step recommendation for the formation of an Evolutionary Stable

Strategy, to the construction of an optimal (time-variant weighted) evolutionary portfolio. These insights ultimately assist us to better manage drawdown risk while pursuing exceptional returns in our day-to-day running of funds and in the provision of research at Evolutionary Finance™ Ltd.

Finally, in Chapter 8 we will covey to the reader what we feel to be the overarching nature of our philosophy. Treating the evolution of information as the outcome of some biological-like molecular process is a powerful concept. Indeed, humans themselves can be perceived to be information generating "machines." Just as we cannot avoid shedding skin, hair etc we also cannot avoid creating information for each other – whether this is by hopping on a bus, buying a particular brand of coffee or buying a particular stock. In this book we focus purely on the financial consequences of the information generation "by-product" of human action. We do this by constructing a framework to better understand the information generation and agglomeration process itself – which ultimately enables us to form far more effective financial decisions. Subsequently, our tact will be of particular interest to those investors looking for innovative interpretations of market dynamics – especially active fund managers and (like ourselves) members of the hedge fund community. Indeed, up until now a notable omission from the finance literature has been a comprehensive theory explaining how the various disparate categories of financial information – technical, fundamental, economic and political – somehow come together to form a cohesive whole that, in turn, drives asset prices. This glaring oversight has been to the chagrin of many market practitioners who feel the gap between finance theory and finance practice has become virtually insurmountable. We attempt to unite these two worlds once more by illustrating how various financial information "bytes" come together to form memes, which in turn form themes and thus drive overall market bullish/bearish sentiment. We believe that a better understanding of the formation of the contextual fabric of this "informational genome" helps us to form a superior view as to what dictates asset price movements on an intra-day, inter-day, weekly, monthly and even across year basis. Indeed, we argue that by explicitly modeling the biological-like molecular formation of information in such a way that perhaps therein resides our greatest contribution to the field of finance – both from a theoretical and applied perspective alike.

On a presentational note, one of the first things you may have noticed already in reading the preface to this book is that the entire text – with the obvious exception of the Acknowledgments – will be written in either the second or third person. The writer adopts this style in deference to the collective wisdom of innumerable interactions with like-minded people at different points across the globe regarding the issue at hand and by way of tribute to the army of scholars that have preceded us. Without doubt, it is upon the shoulders of these innumerable giants that the author presently stands. Whether this has enabled us to see further we leave that for you – the reader – to judge. We hope you enjoy the book.

1
Introduction

It is the purpose of this book to introduce the field of bioinformatics to financial modeling. Why? Well, think of what you are doing. You are reading a book on the way information is imputed within financial markets. In the process you are absorbing information yourself. Whether you agree or disagree with what is about to be presented in the ensuing chapters you cannot help but be affected by it. The human mind displays a voracious appetite when it comes to seeking out and assimilating new information – even information that we don't necessarily agree with. Financial markets are no different. They display an insatiable appetite for new thoughts and ideas. In short, financial markets are huge information consuming entities. What is the most obvious manifestation of their ruminations regarding this vast information flow? Asset prices. So we ask the obvious question – why then, up until now, has a comprehensive theory of the actual *building blocks* of information been noticeably lacking from the finance literature?

Generally, most standard Economics and Finance texts still have little to say about "information" in its own right. Indeed, quite often the study of the role of information within financial markets is treated more as a "means to an ends" rather than an ends in itself. What discussion there is, generally revolves around studies of the Efficient Markets Hypothesis (EMH) or the sterile cost/benefit analysis espoused under the "Information Economics" banner. Thankfully, all is not lost. The worm is finally starting to turn and surreptitiously for the past 20 years a quiet revolution has been underway within the hallowed halls of academia. The reigning orthodoxy of what we label as the "traditionalist" school (a viewpoint we will elaborate upon in the Chapter 2) has come under fire from a new breed of thinking. Full rationality is out, bounded rationality is in. Representative agent models are being replaced with heterogeneous systems. Stochastic has been overridden by deterministic. Analytical solutions are being supplanted by a more computationally orientated approach. Sounds complex? It is. But the good news is that hand-in-glove with the rise of these "new views" has been a belated recognition of the centrality of information to asset price setting. Indeed, more and more studies are now being published under the collective subject headings of "incomplete markets," "information asymmetries" and "artificial models of the marketplace." This book aims to add to this emerging literature by developing a series of building blocks that assist investors by helping them to understand the actual texture and fabric of information itself. We do this by focusing on the analogous nature of *biological-like molecular replication* to the formation and transmission of ideas – in particular, those ideas expressed by analysts within the financial community.

So how best to introduce such a novel concept as biological-like molecular replication of analyst opinions within financial markets? Actually, it is quite easy. Think of it in these terms. The decoding of the human genome has been one of the principle mathematical achievements in recent history. The genome itself is a sequence of letters denoted C, A, G, T that runs over 600,000 A4 pages long – which in book form takes up 270 ft of shelving space. Encoding such information into binary form has allowed the development of sophisticated mathematical techniques to better understand this extremely long informational stream – particularly the detection of key sequences. So too can financial market information be encoded into a similar format so that we can better understand what is initially a dauntingly large information-set. The beauty of such an approach is that the encoding can take place irrespective of whether the original information is from a technical, fundamental, economic or political standpoint. In one fell swoop we thus have a unified theory of information inputs into investor decision-making. Indeed, just as it is possible to build a genomic sequence to describe each and every organism phenotype, we can just as easily build a "market information" genomic sequence to accompany each and every "asset price phenotype."

As will be demonstrated in the ensuing pages, we believe market "themes" are made up of component "memes" which in turn are made up of informational "bytes." This occurs in much the same fashion that DNA is made up of genes which in turn are made up of chemical bases (C, A, G, T). The mathematical building blocks of both approaches remain the same. Indeed the key objective within both approaches remains the same as well – the detection of those deterministic binary sequences that ultimately have a bearing upon phenotype. In the case of biology, it is the phenotype of an organism (its appearance). In the case of finance, it is the phenotype of an asset (its price). Consequently, it is the task of Chapters 2 and 3 of this book to initially lay the foundations for both traditionalist and "new" views of finance before moving on to Chapter 4 where we describe how the sequencing of byte, meme and theme informational building blocks actually determine asset price dynamics. In doing so, it is incumbent upon us to present a framework where an infinite stream of such information is generated willingly by the analyst community. We do this in Chapter 5 by developing a fully endogenized model of intertemporal information on "consumption and production" within financial markets. The net outcome of our labors is a structure that not only describes the insatiable demand for new themes and ideas by investors (which is supplied by a veritable army of analysts – as identified by a "representative" individual – who each manufacture information bytes) but our framework also identifies how information comes together in an evolutionary/biological-like fashion to form larger sequences – memes, themes and even overall market sentiment. It is these informational sequences which then, in turn, determine asset prices in a meaningful way. Ultimately, this has implications for the distributional form of asset prices as well – a point which will be discussed at some length in Chapter 6.

Ergo, a large portion of what is to follow is primarily orientated toward building a revolutionary new theoretical framework for understanding the role of information within financial markets. Certainly we do not abide by the traditionalist finance philosophy that information (and therefore asset prices) should be treated as though it is the outcome of some random number generation process. Indeed, the contrast between our highly structured evolutionary/biological-like approach toward information and the "random number generator machine" model espoused under the traditionalist finance framework could not be more different (Figure 1.1).

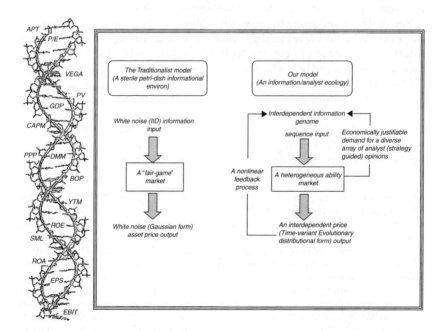

Figure 1.1   Their view versus our view on information within financial markets

So what to call our revolutionary evolutionary approach toward the treatment of information within financial markets? We label our style of analysis "Evolutionary Finance" in attempt to encapsulate the necessary elements of our cross-disciplinary "finance meets science" methodology.[1] The foundations of our Evolutionary Finance paradigm should be seen to be centered on the biological-like microfoundations of the way information is actually generated, interpreted and imparted within financial markets. It is this basic criterion that differentiates our Evolutionary Finance approach from models of a more traditionalist finance nature and even the "new" finance views that will be elaborated upon in greater detail in Chapter 3. Indeed, we would argue that the basic charter of Evolutionary Finance is to answer some fairly fundamental questions that as yet go unanswered within the finance sphere (or at least, are not explained by an all-encompassing model such as ours). How does information emerge and why do investors pay for it to emerge? How do the various components of what investors come to perceive to be "information" somehow bolt together to form memes, themes and overall market bullish/bearish sentiment? Are there any feedbacks to be considered in such a process? All are worthy questions for understanding the potentially complex nonlinear dynamics that can manifest between market information and asset prices. We argue that our own perspective on Evolutionary Finance develops a framework to better answer these fundamental questions and it is here that we feel that our greatest contribution resides.

But all this would come to naught had there not been an *applied* aspect to our work. When it comes to the earlier studies examining the link between information and asset prices, the typical approach has been to either undertake a piecemeal analysis via modeling specific interactions under various stylized frameworks, or alternately at the holistic level, treating information (and therefore asset price growth) as though it is the outcome of some stochastic random process. Our conjecture is to wholeheartedly rebut this claim. The reason why information in financial markets has up until now been treated as though it were the outcome of some random number generator is that the deterministic foundations of the production and

<hr/>

[1] We are not alone in coining the phrase "Evolutionary Finance." Others, principally Professor Thorsten Hens from the University of Zurich, Professor Blake LeBaron from the University of Wisconsin-Madison and Professor Michael Dempster from the University of Cambridge have somewhat interchangeably used the term to describe the use of "evolutionary dynamics" (mutation and selection) in the study of the evolution of trading strategies and the evolution of financial innovations. If anything, our definition of Evolutionary Finance is even more broad-ranging than that of our forebears given our focus on the centrality of (biologically like determined) information to most financial market decision-making – and not just the formation of trading strategies in a manner highly sympathetic to the concept of Genetic Algorithms/Genetic Programming. That said, without doubt, our work benefits immensely from the insights of these pioneers in this exciting new field of finance.

consumption of information – along with the way such information neces-sarily bolts together – have been poorly understood. Needless to say, the applied aspects of our methodology vis-à-vis those that have come before this book could not be more markedly different. In particular, we fully support the notion of active investing. Our founda-tion assumption is that we live in an informationally imperfect world where investors actively search for the "truth" about the informational determin-ants of a particular asset's price. How is this search conducted? For starters, a considerable portion of time and effort (along with a great deal of money) is devoted to the production and gathering of information itself. Active invest-ment institutions – whether they reside on the "buy-side" or "sell-side" – collectively employ tens of thousands of analysts, economists, strategists and market technicians whose primary goal is to garner that all important (ini-tially private) information "edge" – an "edge" that they believe is not yet reflected in asset prices. Unfortunately, up until now, an all-encompassing theory as to how these various elements of information – technical, funda-mental, economic and political – somehow bolt together to form a cohesive whole when percolating from private to public realms has been notable by its absence from the literature. It is this "informational building" process that drives asset prices – not some random number generator. Consequently, while most earlier theorists – especially those theorists of a traditionalist finance ilk – have tended to judge harshly alpha-seeking practitioners, our approach is the opposite – we actually condone active investing behavior wholeheartedly. By doing so we hope to bridge the veritable chasm that has arisen between a large portion of the theoretical and applied aspects of finance simply because of the harsh judgment allotted to active investing by these earlier (traditionalist based) theoretical studies. It is our intention to right this wrong primarily through a better understanding of the pro-cess of information building itself and importantly, how this interacts with asset prices.

One benefit of our approach is that we are thus able to form a better under-standing of the mechanics underlying the strategies of both analysts and investors alike. Seeing yourself not as a solitary entity but rather as an active agent who generates and absorbs information (either consciously or uncon-sciously) as part of a cohesive whole is a powerful concept. Recognizing one's role within this information "ecology" is therefore of crucial significance for *sustained* investment success.[2] Identifying how strategies *evolve* from one winning strategy to another as particular investment themes are being built, identifying the leading memes or even information bytes in this process so as to better forecast the emergence of winning strategies, or even recogniz-ing the power of one's actions should the market anoint you as a "dominant player" in the information hierarchy (i.e., as the likely initiation source of

---

[2] Farmer (1998) also refers to the "ecology" of financial markets.

new winning themes) are all crucial aspects of day-to-day investing covered by our Evolutionary Finance approach. Indeed, as will be discussed at length in Chapter 4, investors appear to interpret information via a mechanism that can be regarded as being highly analogous to the biological-like process of molecular attraction that we use as a modeling framework.[3] New information is initially encoded and then associated and assimilated within the investor's mind so that an interpretation can be formed from the sum of *all* information as to the future direction of asset prices. This algorithmic sequence of events actually mirrors the way information is "built" by the analyst community – thus opening the door for strategic behavior on behalf of enterprising analysts. Understanding this process is therefore crucial to any active investor looking to maintain their investment performance over the longer term.

While on the topic of the longer term, our Evolutionary Finance approach is particularly useful for identifying how information and institutions evolve symbiotically together through time. Looking at the world through an Evolutionary Finance lens quickly reveals just how vast a proportion of our present market institutions bear some form of informational role within our evolutionary/molecular-based information context. Investment banks and fund managers produce information, news services and salespeople distribute information, regulators monitor the dispersal of information and assess its incorporation into asset prices – it seems everybody has some form of an informational responsibility. Without doubt, with this responsibility also goes the necessary realization that one's informational role is itself symbiotic – the institutions that supply and nurture information *evolve* hand-in-glove with the evolution of information itself. There is little doubt that with the growing sophistication of financial instruments and analyst advice regarding these instruments, so too does our financial institutional infrastructure evolve into a more complex array of entities. The fact that market participants can better monitor, understand and (hopefully) predict such change under our Evolutionary Finance framework bodes well for the ongoing success of those who pursue our revolutionary viewpoint. Given this all-encompassing nature of information within financial markets we thus ask the obvious question – why hasn't someone previously attempted to form a better interpretation of this byte-to-meme, meme-to-theme information construction and interpretation process through the development of a comprehensive framework that actually models the building blocks of information itself? This appears to be a glaring oversight by finance theorists – at least up until now.

---

[3] This should not come as a surprise as both these fundamental aspects of information are the two sides of the same ledger – one is supply, one is demand. Both therefore have a reinforcing effect in shaping the outcome of the other – as the way we "think" about the world (informational demand) ultimately influences the format of the resources we use to interpret the world (informational supply).

But the benefits do not stop here. As will be explained in Chapters 6 and 7, our methodology also has implications for an array of applied financial engineering principles – including portfolio construction. Much of "Modern" Portfolio Theory (MPT) as it presently stands is nearly 50 years old. Over this course of time there has been little in the way of adaptation or change to the basic foundations of mean-variance optimization between risk and return for portfolio derivation.[4] Our Evolutionary Finance approach strikes at the very heart of MPT by questioning the distributional form of asset prices that implicitly underlies this procedure. Ironically, the Gaussian distributional form that constitutes the foundations of MPT assumes that information arrival is a "random" event and therefore asset prices too evolve in a random fashion – an assumption we will amplify in considerable detail in the next chapter.[5] Any role for the *active* investor under such a scenario is severely limited since it is assumed there can be no "informational edge" gained from the subsidization of private information "manufacture" by analysts. Rather, it is assumed active investors can only exploit short-term informational "inefficiencies" and their performance in this role is mixed at best.

As stated earlier, our Evolutionary Finance approach totally refutes this claim. Indeed, we argue that there is a crucial role for the creation of "alpha-generating" information by the analyst community and that the distributional form of asset prices – like information itself – *evolves* through time in a deterministic fashion as (analyst generated) new themes replace outmoded notions. This evolutionary (information-driven) process sometimes occurs gradually, sometimes violently, but always there will be implications for the way asset prices behave as they are effectively skewed from one end of the return spectrum to another. In fact, we use our all-encompassing evolutionary model of the marketplace in the *applied* sense to predict how key information bytes (at times, disseminated by "seer analysts") generate an information contagion/coagulation process that results in the generation of emergent memes, themes and even shifts in overall market sentiment. By doing so, we are thus able to better predict the evolution of the time-dependent distributional form of asset prices in response to this biological-like molecular formation of information. This informationally driven/deterministic micro-foundation approach toward the distributional form of asset prices represents

---

[4] The obvious exceptions here are resampling techniques and the introduction of alternate distributional forms, for example, truncated Levy – see Chapter 6 for more details.

[5] It is ironic because a plethora of active investors use MPT to construct "optimal" portfolios but in doing so they are using a procedure that recommends an investing stance that is in obvious contraction to what they are doing (in other words, in all but exceptional circumstances, passive investing). Indeed, it is on this issue of the degree of market efficiency underlying MPT that Markowitz and Sharpe still differ – eventhough both shared the Nobel prize for the concept.

a complete watershed in thinking as to how to best construct an optimal portfolio of assets. Indeed, as we will demonstrate, our approach affects many preexisting "best practice" financial engineering techniques.

So in summary, we are both supportive of the notion of active investing and we also provide some guide as to how one should do so in an information driven market. To develop our case, we devise an array of techniques that in many respects supercede the prevailing (traditionalist based) "received wisdom" financial engineering "best practice" methods. What's more, we actually identify the implications of our own advice by assessing how an army of active investors all generating and absorbing information affects both the performance and the distributional form of asset prices. It is through such a comprehensive framework that we are able to construct optimal time-dependent investment recommendations and portfolio structures across a diverse array of assets and informational environs – in essence, a new way of thinking for a new way of interpreting information within financial markets.

Finally in closing our introductory chapter, perhaps it is time to play devil's advocate and, ask is our evolutionary perspective within financial markets necessarily a "new view" after all? One field of great significance in "evolutionary" type thinking within the economics/finance sphere can be traced back to the early twentieth-century economist Joseph Schumpeter and the Austrian School of thought.[6] This field – now labeled "Evolutionary Economics" – concentrates on the way in which economies grow and evolve through time. For our purposes, apart from the notion of "equilibrium" itself being a time-dependent concept, what we find especially appealing from this body of thought is its notion of evolutionary institutional change – a topic we have briefly touched upon in this chapter and will elaborate upon in some detail in the final chapter of this book.[7] However, what differentiates our approach from this earlier analysis is our focus on the *informational* aspect of this evolutionary-type thinking. Our interpretations of the symbiotic evolution of information and institutions form a key aspect of our long-term thinking as to the development of successful strategies for market participants – whether they are in an information creation or interpretation role. What is particularly encouraging is that within the field of finance itself we are now able to develop this concept in much greater detail thanks to the recent advances within the field of bioinformatics. These mathematical advances allow us to better detect the emergence of those key sequences of information that are likely to have not only short-term price impacts but also long-term institutional change impacts as well. As we have briefly touched upon in this chapter, our results have remarkable implications for finance as it is practiced today – both from a theoretical and applied perspective.

[6] See Schumpeter (1939). For an excellent discussion of the earlier influences on Austrian thought, refer to Hodgson (1996).
[7] See Nelson and Winter (1982) for instance.

So while asset prices may seem to bob up and down like flotsam in a sea of information, we actually make sense of this information for the formation of more elaborate and efficacious strategies for active market participants – and for the commensurate construction of their evolutionary portfolios. To conclude, you may not agree with everything that is about to be presented in the following pages, but without doubt it will certainly shakeup one's notions as to how financial markets *really* work.

2
The "Old" View of Finance

Up until recently, one could look under "information" in the index section of any major financial textbook and be genuinely surprised at the paucity of references. Why? Quite often information was taken as a "given," an exogenous occurrence that undoubtedly affects market behavior but alas remained largely unexplained. More recently, information has been moved closer toward center stage with a plethora of studies now manifesting in the field of "informational asymmetries" and "incomplete markets."[1] Still, even in these studies, "information" itself is mostly treated as an intangible commodity with little or no attempt to understand its underlying attributes or the forces that govern them. It is our objective to put right this somewhat alarming oversight and bring information analysis firmly into the mainstream of finance thinking. However, before doing this let us at first examine what little information there is on "information" within the existing finance literature.

2.1 The efficient markets hypothesis: The traditional (albeit incomplete) standard-bearer for information assessment

The traditional benchmark for interpretations as to the way information is imparted within financial markets has been the Efficient Markets Hypothesis – or "EMH" as it is affectionately known by finance theorists. EMH comes in three separate guises as laid out by Fama (1970).[2]

Strong EMH. It specifies that asset prices fully reflect all information from public and private sources at each and every point in time. Obviously, a Strong EMH interpretation as to the way information is imparted assumes

---

[1] See, for example, Brunnermeier (2001) on the presence of asymmetric information and Hirshleifer and Riley (1992) for the link between information and uncertainty.
[2] The term "efficient market hypothesis" was coined by Harry Roberts (1967) but since Robert's paper was never published, it was Fama's (1970) discussion that subsequently became renown within finance circles.

no "insider" information is present, that no one investor has monopolistic access to price sensitive information or no one investor has "superior" ability. The obvious *applied* implication of such a purview is that investors cannot consistently make "above normal" profits from trading with market information – there is little point in trying to "time" the market or actively invest. Prices are assumed to adjust to new information (both private and public) instantaneously. Furthermore, many argue that the Strong EMH belief structure assumes that private information is effectively costless (e.g., Reilly and Brown (1997)) – a point we will elaborate upon in greater detail in Section 2.3.

Weak EMH. It assumes that current asset prices reflect only all relevant *asset market* information. What precisely does this mean? Asset market information includes historical prices, market-positioning data, rates of return, volatility and so on. Basically, the Weak EMH definition implies that asset prices reflect all information that has *already* been generated *within* the marketplace. The applied implication of this belief structure is that investors are able to generate excess return ("above normal" profit) by searching for private information that is not presently in the marketplace and acting upon this information by subsequently positioning for the ensuing excess return as the information percolates from the private to public arena. In other words, Weak EMH assumes that the speed of information dissemination is slow. Obviously this purview is more supportive of our biological-based evolutionary approach toward information generation and dispersal than Strong EMH (which, as will be explained later, is more supportive of what we label as the traditionalist finance viewpoint).[3]

Semi-Strong EMH. This is a sort of a "half-way house" between the Strong and Weak versions of EMH – a "compromise" so to speak. The Semi-Strong definition of EMH encapsulates the conditions of *Weak EMH* but adds a more onerous timing element (or "speed" condition) in the sense that it is assumed asset prices adjust rapidly to the release of all market sensitive information. Under this definition, investors can still make profits by engaging in the genetic-like production and then biological-like dispersal of (initially private) information and then adroitly positioning themselves for the (hopefully) enthused take-up and subsequent rapid dissemination of their insights into the wider marketplace – but in pursuing such active return it is very much a case of the "quick and the dead."

All this may seem like a bad case of semantics (most definitions are) but it is important for our purposes to note that the distinction between the

---

[3] Also it will be explained a little later, institutionally there appears to a reasonable amount of support for the *Weak EMH* approach toward information given the vast array of investment advisory institutions all with their associated army of analysts whose primary purpose is to produce private information.

various forms of EMH rests almost entirely upon perceptions regarding the *flow of information* – its cost, its speed and (somewhat related) the ability of investors to accurately interpret this information.[4] That is why we find the paucity of references within the greater body of finance literature as to a more in-depth study of the underlying characteristics of information so perplexing. As we will explain in much greater detail in the ensuing chapters, it is these "liquidity" aspects of information that are perhaps best explained via a biological-like molecular building block analogy – a line of thought we will pursue in great depth. In fact, it is our evolutionary information building blocks of "bytes," "memes" and "themes" and the way they affect the "liquidity" aspects of information that ultimately determines how close a particular market's structure replicates the perfection outlined by a "Strong" version of EMH. Should investors have perfect foresight as to the way such information is formed, then this purview would validate what we label as the "Strong EMH/traditionalist" viewpoint of the way financial markets operate – a view we might add that has contributed significantly to the virtual ignoring of the importance of information analysis in its own right (see Sections 2.2–2.5 for detailed explanations as to why). That said, it is hard to argue for such a pro-Strong EMH stance when even much of the existing finance literature does not appear to understand how the microfoundation component parts of information necessarily bind together to form a cohesive whole. But more on this later.

Subsequently, to conclude this section, up until now rightly or wrongly (but mostly wrongly) "information" and "market efficiency" have been used somewhat interchangeably within the greater body of (traditionalist dominated) finance literature. This is simply because (as stated earlier) the underlying characteristics of information itself were taken as a "given" – in effect, "information" in its own right was treated as an intangible commodity. What little attempt there has been at "information analysis" *per se* has been effectively buried under the collective weight of "market efficiency" analysis – part the way toward a greater understanding of the drivers between information and the markets, but by no means a complete model (and certainly incomplete when it comes to understanding the building block frictions that generate the divergences from the Strong EMH purview). It is our objective to better understand these "frictions" thanks to a better (evolutionary-based) framework for modeling the way financial market information actually comes together into a meaningful form. In doing so we hope to put right this shortcoming within the existing literature and remove the present annoying tendency for many finance texts to refer readers to "see Efficient Markets Hypothesis" or something of the like under index searches for "Information."[5]

---

[4] Such are the "frictions" of markets as Lo and MacKinlay (1999) describe them.

[5] For instance, the wide selling investment text Reilly and Brown (1997) falls into this trap in a considerable portion of its referencing under "Information."

2.2   A little more on the link between the theory and the applied

Before moving on, readers unfamiliar with the *applied* aspects of contemporary finance theory, should note that – as intimated in the various definitions of EMH in the previous section – the assumption as to what form of market efficiency necessarily prevails within the market at a given point in time is pivotal in the formation of one's viewpoint as to what constitutes "optimal" investment decision-making. For example, a belief in the "strong" version of EMH implies that there is little role for "active" investment management. Rather, such a mindset advocates a more "passive" investment philosophy of simply ensuring one's existing portfolio best replicates the present "market" portfolio – as commensurate with your risk preference.

Sounds a little extreme doesn't it? Still, we should ask – has such a belief structure had much of an influence in the applied world? The answer is – absolutely. In fact, one could go as far as to say that (surprisingly) Strong EMH has had a "big bang" impact in the inevitable conversion of finance theory into finance practice – especially from a financial engineering standpoint. Not only has the (somewhat limited) characterization of information flow under Strong EMH been responsible for advocating the "market" portfolio as *the* optimum solution for all investment decision-making – and subsequently given rise to the plethora of "portfolio indexing" funds – but it (via the conduit of Gaussian form) also forms the basis of many an elaborate extension of finance theory into the applied financial engineering realm as well. This includes such models as CAPM (Capital Asset Pricing Model), VaR (Value at Risk), Black–Scholes option pricing and even the various guises of fixed income factor models. Indeed, given this unerring association between Strong EMH and "standard" financial engineering methods within the greater body of existing (traditionalist dominated) finance literature, it can be argued that Strong EMH forms the very foundation (some would say a weak foundation) of what we witness as financial engineering's "best practice" today. Ergo, it is not surprising that our Evolutionary Finance purview has the potential to have significant ramifications for a wide field of finance theory thanks to its more comprehensive understanding of the actual building blocks of financial information itself (Figure 2.1).

Obviously given the above impressive edifice of (traditionalist) finance literature, we certainly have our work cut out for us. That said, apart from our approach, has there been any other form of rebuttal to date to the Strong EMH/traditional finance viewpoint? Is there any additional contrarian voice supporting our own perspective? Again, the answer is – absolutely. For starters, it is interesting to note that purely from a practitioner's standpoint that structurally much of the prevailing financial system sits institutionally at odds with the Strong EMH/traditionalist mindset. As pointed out in Chapter 1 and as practitioners within financial markets will readily testify, a large portion of the existing institutional fabric of what we have come

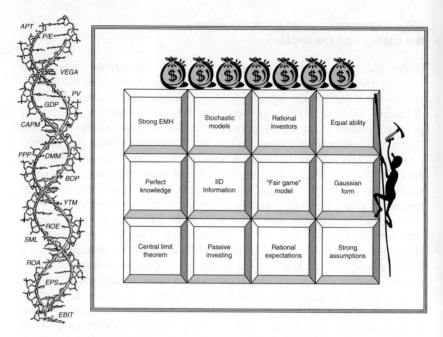

Figure 2.1   The strong EMH/traditionalist financial engineering edifice

to regard as the "financial system" is devoted to the production and dissemination of (initially) private information which then percolates into the public realm. This sits totally at odds with the Strong EMH/traditionalist belief structure. Investment research, financial news providers, active fund managers, hedge fund managers and a significant proportion of sales teams within the broker/dealer community all fall under this bailiwick and are thus guilty as charged when it comes to violating Strong EMH/traditionalist notions. Indeed, the implication of this unabashed willingness of erstwhile rational investors to fund such activities intimates a significant difference exists between what is effectively assumed by the Strong EMH/traditionalist viewpoint and what actually applies in the "real world." In short, it appears that it takes time for private information to percolate into the public arena simply because the biological-like molecular agglomeration of information is (by definition) a non-instantaneous process. Consequently the implications for "optimal" investment decision-making (and for financial engineering "best practice") are very much different in the "real world" to the theoretical purity of the Strong EMH/traditionalist utopia. We will dwell longer on the issue of rational investors funding private information manufacture in the following section but suffice to say at this stage that the obvious difference between theory and application in the finance world could explain a large portion of the present disillusionment expressed by practitioners with the endless reams of (traditionalist-based) research being published by the academic community.

Given that a dichotomy exists between private and public information thanks to the time it takes for financial information to agglomerate together in a biological-like molecular fashion (and the fact that investors cannot always perfectly predict the outcome of this process and so circumvent the evolutionary timeframe over which this process occurs) then one had better get on the "right side" of the information flow to prosper – or even, survive. Positioning to be an early recipient of new market sensitive information under such circumstances – or even better, to engage in the production and then dispersal of such information appears to be a sure route toward that all-elusive alpha that active investors so willingly covet. Certainly this provides the best mechanism via which to outperform one's peers. Hence, we again come back to the investor rationality argument. Why do so many Wall Street and City investment banks (and active fund managers for that matter) pay what appear to be huge sums of money to "star analysts" whose ultimate goal is to beat the market? Answer: it is the inherent belief that these star analysts have the ability to add "alpha" via the timely manufacture and release of (private) information into the public realm.[6] To summarize our arguments, we outline in Figure 2.2 (by way of diagrammatical analysis) a market efficiency spectrum that differentiates Strong/Semi-Strong/Weak EMH views from one another, the concurrent private/public information distinction that accompanies each of these market efficiency categories, and the "optimal" investment strategy that applies in each circumstance.[7]

---

[6] In the case of active fund managers, this "informational release" is generally through the price mechanism as their positioning ultimately will have a market impact – indeed, some may even encourage rumor in the market once they are advantageously positioned. We will elaborate upon such principles in more detail in Chapter 7.

[7] Note, it is important to stress here that we are not effectively espousing insider trading – in the sense that individuals who have privileged access to information purposefully exclude this information from the public until they can trade on it – as an embodiment of "private" information. Rather, what we are identifying is the more "law-abiding" pursuit of private information "manufacture." As Kirzner (1979) in his analysis of the reward for entrepreneurial behavior would no doubt approve, those who have the skills and the models to make use of what is effectively public information and convert this into much prized private information justly deserve the spoils afforded to them in terms of economic rent. Do analysts actually earn an "economic rent" (or "alpha" in investing parlance) for investors who use their recommendations? Some of the cynics would argue no – especially post the chastened investor experience of the early 2000s bear market in equities – but recent research by Barber *et al.* (2001), Krische and Lee (2001) and Jha *et al.* (2003) demonstrates that analyst recommendations do indeed add value to the portfolios of those investors who care to take on board the advice.

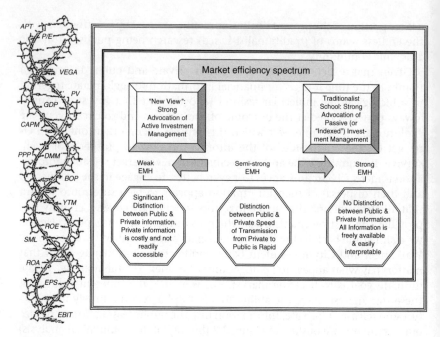

Figure 2.2   The market efficiency spectrum and optimal investor behavior

Needless to say, it is important to highlight that it should be necessary for all investors to make a judicial assessment of information characteristics – and by default, market efficiency – *before* they decide upon their investment style and enter the market. Unfortunately, as mentioned earlier, up until now, the amount of analysis on "information" in its own right within the field of finance has mostly been confined to studies of EMH and not in understanding better the actual fabric and texture of information itself. In part, the emerging field of information asymmetries goes some way toward addressing some of these problems but it must be said that really up until now there has been no true formal framework for a thorough analysis of the actual determinants of what investors come to perceive as "information" and in particular, the biological-like microfoundation elements that come to govern its creation, agglomeration and speed of dissemination. This is where our evolutionary framework for information analysis comes to the fore. But before exploring our thoughts in greater detail, it is necessary to do a little more justice to the EMH framework by highlighting the additional structures that have been put in place to better understand the principles behind the various definitions outlined by Fama (1970) and summarized earlier in Section 2.1.

2.3   Cost, ability and speed: Important information determinants

Apart from a three-part series of definitions, is there anything more to the formalized framework of EMH and its link to information analysis? To help clarify his thoughts, Fama (1970) found it best to augment his definitions with the aid of a simple model. Under the principles of a "fair-game" Fama argued that, market participants of similar financial ability (provided they have equal access to information) cause the market price and the expected price of any asset to converge. In a sense, everyone comes to roughly the same conclusion as to the correct price of an asset when presented with the same information. As a consequence – consistent with our discussion in the previous section – under the principles of a true "fair-game" there is no opportunity for consistently "above normal" or excess profit from embarking on a career of (private) information production and then positioning for its biological-like agglomeration and dissemination within the wider investment community in the hope of gaining some "informational edge" in market pricing. Under the auspices of Fama's "fair-game" all information is assumed to be free – both public and private – thus negating the concept of private information altogether. To be sure, Fama argued that on occasion *slight* differences in the *ability* between investors in interpreting information can at times result in excess profit to information producers – a ray of hope for being slightly supportive of our approach – but then he snuffed out our inspirational flame by arguing that these returns would be soon arbitraged away as any superior skill-set immediately becomes "public" information and thus is "learned" by the entire financial market – including one's competitors. We would counterargue by stating that to date, since there has been no formal attempt to understand that actual fabric of information itself, such an assumption of instantaneous information transferal by Fama can at best be described as "heroic." Still, "sticking to his guns" Fama characterized any sporadic excess returns to (private) information producers resulting out of the investor learning process as profiting from "noise trading" – an often referred to insight that highlights that private information manufacture cannot be a *consistent* profit generator.[8] In sum, Fama was hardly supportive of the notion of active

---

[8] Assuming the population is finite and capital is costly, long-term accounting implies noise traders would soon cease to exist due to a lack of financial backing. In short, under "Traditional Finance" assumptions, noise trading has an economic return of less than zero – see Friedman (1953). This promotes the question – why then do markets continue to gyrate on "non-information" days? Our answer is that "Traditional Finance" assumptions are too restrictive and in actuality information is continuously being generated, categorized and associated in a biological-like fashion. Indeed, Delong *et al.* (1990) illustrated that "noise trading" can indeed be a profitable pursuit under less restrictive "quasi-rational" assumptions regarding investor behavior. Finally, note that there is an important academic literature highlighting the link between "noise trading" and the

investing and by default is hardly supportive of our evolutionary purview on information within financial markets. The principles of Fama's "fair-game" model are illustrated in the following box.

Fama's "Fair-Game" Model

Fama's (1970) "fair-game" model is built around the following central equation

$$\ddot{E}(\overline{P}_{a,t+1}|\ddot{I}_t) = [1 + \ddot{E}(\overline{r}_{a,t+1}|\ddot{I}_t)]P_{a,t}, \tag{2.1}$$

where
$\ddot{E}$ is the expected value operator;
$P_{a,t}$ the price of security a at time t;
$P_{a,t+1}$ the price of security a at time $t + 1$;
$r_{a,t+1}$ the one period percentage rate of return for security a during period $t + 1$ and
$\ddot{I}_t$ is the set of information that is assumed to be fully reflected in the security price at time t.

Assume the expected price of the asset a fully reflects the discounted value of all information (public and private) that has a material affect upon the asset's value.[9] Under such conditions, the excess market value between actual price at $t + 1$ and expected price at t

$$ex_{a,t+1} = P_{a,t+1} - \ddot{E}(P_{a,t}\phi_t) \tag{2.2}$$

is zero:

$$\ddot{E}(\tilde{ex}_{a,t+1}\ddot{I}_t) = 0. \tag{2.3}$$

That is, information transferal from private to public realm is instantaneous and this makes the return from holding private information effectively zero. If expected excess return $\ddot{E}(\tilde{ex}_{a,t+1})$ is in any way affected by $\ddot{I}_t$ then Strong EMH does not apply.[10]

That said, there is a supportive cavalry on the horizon riding to our rescue. Consistent with our investor rationality arguments presented earlier, Grossman and Stiglitz (1980) have pointed out that in the "real world" private information is not costless and, as a consequence, a dichotomy must exist between what can be categorized as "public" and "private" information. This empirical fact implies there are profits for private information producers to

presence of "uncertainty." Minor amounts of "uncertainty" are generally argued to be consistent with the concept of "noise trading" – see McKenna (1986).

[9] Ultimately for such an outcome to exist, investors must be fully rational – as consistent with Muth's (1961) original exposition of the rational expectations hypothesis.

[10] Note: we are referring to expected *excess* return. Fama's "fair-game" model does not imply that the *expected* return on any asset is effectively zero.

reap as investors regularly pay for the privilege of accessing the best analysts to garner such private information insights *before* they are disseminated into the wider public arena.[11] Again, as partly alluded to in the previous section, Grossman and Stiglitz argued that assuming such investors are economically *rational* (and not sequentially replicating the mistakes of perpetually loss-making noise traders) then one must assume that embodied in each market price is at least some amount of monopolistic reward for those individuals who produce private information.[12] And herein resides a poetic justice for the Strong EMH/traditionalist protagonists. By pointing out the obvious incongruity between the real world experience and the traditionalists' assumption of full economic rationality on behalf of investors – a necessary condition for Fama's "fair-game" principles to hold – Grossman and Stiglitz (1980) had the effect of "turning the worm" so to speak back on the traditionalists. In short, it was precisely the traditionalist group of academics who had been such arch proponents of economic rationalism to bolster *their* arguments for Strong EMH. This is because a fully economic rational individual will assiduously seek out new information as it arrives and correctly interpret this information – including its full biological-like molecular formation potential. However, according to the Grossman and Stiglitz argument, rationality (along with a casual observation of the real world "costly information" experience) is precisely the reason why it can be interpolated that there are differences in the abilities of analysts and even more importantly investors cannot possibly fully project the full consequences of information as it binds together in a biological-like fashion and thus investors are willing to pay a price to those analysts with superior ability who can assist them in forming a better investment judgment. Subsequently, it is implied there must be an "informational component" embodied in every asset's price to cover the cost of private information manufacture by the analyst community.[13] This is an important premise we will build upon extensively in our evolutionary model of intertemporal information production and consumption presented in Chapter 5.

Trying not to labor the point too much, we contrast the two diametrically opposed views on the link between analyst ability and the actual cost of information embodied in each asset's market price with the aid of the following diagrams (Figure 2.3). In essence, Strong EMH implies a perfect competition ("fair-game") world for information. Each individual is effectively a price-taker for that component of the market price that reflects the

---

[11] Empirically this argument has some merit – as evidenced by the exhaustive study by Asquith *et al.* (2002).

[12] For an empirical study illustrating the link between private information and asset returns under a rational expectations framework, see Easley *et al.* (2002).

[13] Indeed, Elton and Gruber (1995) point out that a preferable definition for market "efficiency" under such circumstances would be to state that the marginal cost of information equates to the marginal benefit.

cost of information. This is because it is assumed there is no distinction between the investor's accessibility to information that is either in the "public" or "private" realm – all market relevant information is effectively free and therefore the public/private distinction does not apply. Importantly, this includes a complete understanding of the full biological-like molecular formation consequences for the full information-set of each information byte's arrival.[14] In short, everybody knows the full consequences of a certain information byte's price impact (both in contemporary and latent form) thus asset prices adjust instantaneously and correctly to reflect this knowledge. Contrast this with the Semi-Strong and Weak versions of EMH demonstrated in the opposite diagram where a gap opens up between what is effectively "public" and "private" information simply because it is assumed that there is some form of superior (monopolistic) ability on behalf of talented analysts engaged in the production of private information.[15]

If one wished to move away from static analysis for a moment and take the element of time into account, then implicit in such a framework as presented in Figure 2.3 is the assumption that the biological-like molecular agglomeration of information is slow enough that worthy analysts can expect a certain stream of returns from their (initially, private) information insights as they percolate into the public domain. Ability and speed are inexorably intertwined under such a framework. This ultimately brings us to our next point in the information debate in the sense that cost and ability are not the only factors one needs to consider in forming a view of market efficiency. As Fama's definitions outlined in Section 2.1 testify, speed too is an important criterion. Why? The speed of information transferal as dictated by the biological-like molecular principles governing information categorization, assimilation and aggregation will determine the *stream* of excess profits any analyst can expect to earn as a reward from their monopolistic skill-set in producing what is initially coveted private information and then allowing this to percolate into the public domain.

---

[14] A lesser assumption would be that information is sufficiently "noisy" that nobody can predict what is about to happen – so again, everybody is effectively playing on a level playing field. Ironically, both arguments are consistent with the Random Walk argument as popularized by Malkiel (1996) and discussed in detail in Section 2.5. We thoroughly refute the Random Walk claim as our modeling in Chapter 5 will demonstrate that just because aggregate information (and the ensuing price action) may on the surface give the appearance of a random/chaotic pattern of behavior, this does not mean that certain elements of this information set are not predictable should one have the correct methodology for filtering this "noise" – which we believe we have via our evolutionary approach toward information in financial markets. Therefore we feel investors should not be assumed to be either "super smart" or completely ignorant, but something more in-between.

[15] Such "ability gaps" are clearly demonstrated by the perplexing phenomenon of the "irrational" early exercise of exchange traded options by the customers of discount brokers as opposed to the traders at large investment houses – see Poteshman and Serbin (2003).

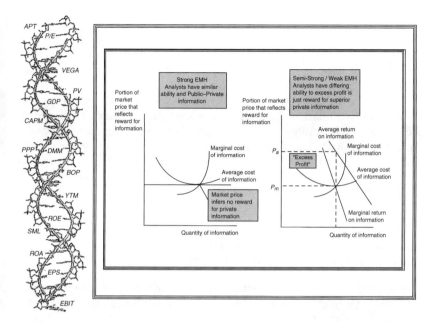

Figure 2.3   The link between ability and cost in determining market efficiency[16]

As any student of Finance 101 will testify, present valuation techniques highlight the importance of time as a factor in discerning the economic viability of any enterprise. Information generation by the analyst community is no different. If a talented analyst is to actively invest considerable time and effort to engage in the production of a certain information byte, then it stands to reason that they – being "economically rational" – will also have some expectation as to the time over which they will reap a reward for this information. The longer the time, the greater the overall reward. It is this "scalar" property of time that makes it so piquant to the various definitions of market efficiency espoused in Section 2.1. Indeed, it is a focus on time that forms a key principle of our evolutionary model of intertemporal information production and consumption presented in Chapter 5.

However, before moving on to such analysis let us now briefly review the empirical evidence as to the "real world" validity of each alternate form of EMH to see if this sheds any extra light on the actual speed of information

---

[16] Morrison and Vulkan (2003) illustrate the existence of excess profit from mining "imperfectly distributed public information" (or in other words, "private information") under Game Theoretic conditions. They illustrated that this remains the case even when the entry by would-be information manufacturers is free simply because of the latent belief structure held by these speculative agents. We will elaborate more on the finer aspects of Game Theory in Chapter 3.

dispersal. This potentially stands to be useful in the ensuing chapters simply because our evolutionary model of biological-like information formation is (by definition) a non-instantaneous event. An empirical jury coming down in support of a Strong EMH type paradigm as typifying most financial markets at most points in time would effectively strike a killer blow against our hypothesis of an army of economically rational analysts all producing new ideas and disseminating them into the marketplace in search of monopolistic reward. Thankfully the empirical jury is by no means so unequivocal in its verdict. Rather, any feedback as to the validity of any of the various alternate forms of EMH in the "real world" seems to be at best characterized as being in a permanent state of quandary – not able to make up its mind if Strong, Semi-Strong or even Weak EMH reigns supreme. This is the best of all possible results for our hypothesis – the reason for which will become clear in the next section.

2.4 Do empirical studies of the EMH shed any light on the actual speed of information transferal?

Do the plethora of empirical studies on the issue of EMH provide any answers as to the likely degree of market efficiency that prevails in a particular market and (concurrently) the most likely "real world" scenario for the actual speed of information dispersal? Unfortunately, it seems the only clear conclusion that can be drawn from the virtually innumerable number of studies conducted on EMH is that there is in fact no clear conclusion. Rather, it appears that various forms of EMH seem to typify different markets at different points in time – an interesting insight that we will revisit toward the end of this section. But back to the actual empirical studies themselves, by way of background such tests have generally fallen into five stylized forms.[17] Some of the studies under these various banners are documented in a Table 2.1.

---

[17] Unfortunately, as will be discussed a little later in this section, from a methodological standpoint, many of these cited studies purview the EMH spectrum as binary. In short, premise "0" – or the "null hypothesis" - is that all information is public information (which the market immediately factors-in) and that the only private information is illegal "insider" information. While premise "1" – the "alternative hypothesis" – is that both public and private information coexists but unfortunately there is no attempt to explain (as is our purposes) the *evolution* of this information from private to public realm. We believe this to be an extremely limiting testing methodology. The thrust of our Evolutionary Finance technique is to explicitly model the "evolution" of information from private to public realms using a biological-like molecular formation analogy as a general framework for analysis. This, under certain assumptions, allows for a more *graded* view of the market efficiency spectrum. In short, we believe a better understanding of the true microfoundations of information generation and transmission would provide a far better starting-point as a methodological test-bed in assessing the relative efficiency within markets from an empirical standpoint. We cite this as a potential avenue of fruitful future research.

Table 2.1 The various categorizations of empirical EMH analysis

Accuracy of forecasts in generating "excess" returns	Information transmission over market "events"	Power of technical analysis	Seasonal patterns in time series	Firm characteristics and excess return
Latane et al. (1970); Black (1973); Holloway (1981); Copeland and Mayers (1982); Dimson and Marsh (1984); Rozeff (1984); Stickel (1985); Elton et al. (1986); Kiem and Stambaugh (1986); Campbell (1991); Pesaran and Timmermann (1995); Womack (1996); Barber et al. (2001).	Reilly and Hatfield (1969); Kraus and Stoll (1972); Pettit (1972); Grier and Albin (1973); Watts (1973); Firth (1975); Ibbotson (1975); Dodd and Ruback (1977); Joy et al. (1977); Charest (1978); Watts (1978); Aharony and Swary (1980); Rendleman et al. (1982); Foster et al. (1984); Pierce and Roley (1985); Jain (1988).	Fama (1965); Fama and Blume (1966); Levy (1967); Jensen and Bennington (1970); Pinches (1970); Praetz (1972); Fama and MacBeth (1973); Cootner (1974); Fama and French (1988); Brush (1986); Conrad and Kaul (1988); Poterba and Summers (1988); Pruitt and White (1988); Glosten (1989); Fama (1991); Campbell et al. (1993); Ball et al. (1995); Benik and Bossaerts (2001).	Granger (1975); Rozeff and Kinney (1976); Branch (1977); Dyl (1977); French (1980); Gibbons and Hess (1981); Brown et al. (1983); Gultekin and Gultekin (1983); Reinganum (1983a); Keim (1983); Berges et al. (1984); Lakonishok and Smidt (1984); Keim and Stambaugh (1984); Tinic and West (1984); Kato and Shallheim (1985); Keim (1985); Keim (1986); Chang and Pinegar (1986); Harris (1986); Ariel (1987); Jones et al. (1987).	Dimson (1979); Banz (1981); Reinganum (1981); Roll (1981); Basu (1983); Brown et al. (1983); Stoll and Whaley (1983); Reinganum (1983b); Shiller (1984); Balvers et al. (1990); Chan et al. (1991); Reinganum (1992).

Don't be overcome by the extensive categorization and citation in the Table 2.1. The point readers should focus on simply stated is that all these tests have a common denominator – they search for associations between market information and market returns. Systematic excess return from private information (such as analyst forecasts) or under-utilized public information (such as technical rules, seasonal patterns or firm characteristics) implies a refutation of Strong EMH. Why? Primarily because under the Strong EMH conditions – as outlined in the previous sections – such information should be already reflected in an asset's price. Thus a *systematic* excess return to any one investor from having an "informational-edge" is implausible under the austere Strong EMH world. What these studies attempt to do is to categorize this "informational-edge" into a variety of forms. In terms of specifics, forms (1) and (2) of our categorized EMH testing methodologies in Table 2.1 – testing the accuracy of analyst forecasts in generating excess returns and measuring the speed of information transmission during market "events" – are obvious in their rationale for empirical EMH testing in the respect that they explicitly seek to clarify the existence and transmission of profit-making private information as it percolates into the public realm. That said, the latter three forms of EMH testing are much more subtle in their empirical approach since they attempt to identify systematic excess return from specific trading rules rather than attempting to measure informational impact directly.

For instance, form (3) of the EMH testing methodology highlighted in Table 2.1 – estimating the profit generating capacity of technical analysis – is especially interesting as it attempts to test the hypothesis that price trends are an informational variable to market participants (an idea seized upon in some of the "new" views within finance described in Chapter 3 and indeed, technical analysis comprises one of the four major information categories for our information byte components).[18] If price trends themselves were to exhibit predictive patterns in a fashion consistent with the precepts of technical analysis, then this would imply a refutation of the principle of Strong EMH. Why? Discerning future prices from patterns exhibited in past prices under the guise of technical analysis is totally at odds with the "instantaneous adjustment" principles espoused under Strong EMH. Indeed, the traditionalists are particularly harsh in their judgment of the efficacy of technical analysis. To quote Malkiel (1996):

Technical analysis is anathema to the academic world. We love to pick on it. Our bullying tactics are prompted by two considerations: (1) after paying transaction costs, the method does not do better than a buy-and-hold

---

[18] The other major informational categories being – fundamental, economic and political information.

strategy for investors, and (2) it's easy to pick on. And while it may seem a bit unfair to pick on such a sorry target, just remember: It's your money we are trying to save. (Malkiel 1996, p. 139)

As stated earlier, Strong EMH argues that all information (both public and private) that is relevant to determining the future price of an asset is instantaneously incorporated into an asset's price upon its production – this includes latent price information. Subsequently there is no incentive for any analyst (including technical analysts) to systematically produce information in an attempt to position themselves for reward once this information is dispersed to the wider investment community as effectively there is no opportunity to do so. Under Strong EMH principles, the production and dispersal of information are inexorably intertwined and are instantaneous events. Indeed, as also mentioned earlier, even the skill-set used to produce information is effectively in the public domain the moment it has been developed. In one fell swoop this totally removes the incentive for skilled analysts to devote themselves to a career of producing information – especially when it comes to enhancing memes and themes which are already in place in the hope of an even larger price impact and therefore subsequent reward. Under Strong EMH it is assumed investors already know the full consequences of a given information byte's price impact-that we will demonstrate in Chapter 4 has the potential to be governed by Complex nonlinear principles when various information bytes come together to form memes. We find this assumption of perfect foresight a little "heroic" to say the least. Indeed, in Chapters 4 and 5 we will formalize the concept of "latency" in *all* information bytes that when unlocked generates the price movements that technical analysts feed upon for their particular style of analysis. In short, in contrast to the traditionalist's viewpoint, Evolutionary Finance principles are highly supportive of the pursuit of technical analysis as a mechanism for alpha-generation.

As for form (4) of the empirical EMH testing methodologies identified in Table 2.1 – determining if asset prices follow a reliable seasonal pattern – this particular methodology follows in a similar vein to the conceptual basis for the inclusion of technical analysis as a separate category for EMH empirical analysis. That is, estimating the degree of efficacy of seasonal patterns as reliable alpha-generators tests yet again the foundation for a degree of latency in information. Any statistically significant correlation between particular trading "seasons" illustrates a link between past and present prices and the Strong EMH null hypothesis is effectively rejected. Why? Simply because investors could use such information to formulate a reliable trading rule. Rather, investors should already know of such phenomena (assuming it exists) and this should be reflected in prices – thus denying any return from trading "seasonals." In the tests cited within this category in Table 2.1, various correlations were tested over alternate time horizons with different lags but the same principles applied – if any associative tendencies were identified between trading season and prices the Strong EMH hypothesis was rejected. The results were mixed.

Finally, form (5) of our empirical EMH testing methodologies – the link between firm characteristics and the presence of excess return – provides perhaps the most esoteric of all EMH empirical test-beds. Attempting to explicitly link firm characteristics such as size to the presence of excess returns implies a failure of market efficiency. Why? If a market were truly efficient then such descriptive characteristics of firms and its relationship with price should already be known by all investors – thus denying any form of excess return.[19] Rather, under Strong EMH principles characteristics such as firm size that do on occasion generate excess return can only do so under the somewhat random practice of "noise trading."

So, these are the various critiques, what to make of this extensive – albeit conflicting – array of empirical evidence as to which form of EMH best typifies markets in the "real world"? It is often quoted that it is impossible to step into the same river twice. Apparently, the same principle applies to markets. We believe the academic stalemate in failing to conclusively come down in favor of a particular form of EMH can perhaps best be described by the time-dependent nature of market efficiency itself – and, by default, the biological-like determined speed of information production and dissemination. As we will argue in Chapters 4 and 5, the forces governing the biological-like molecular formation of information are difficult to predict and have the potential to generate complex nonlinearities in pricing behavior. Without doubt, this has direct implications for any attempt at measurement of market efficiency simply because at times the market may take on a somewhat random appearance in response to these nonlinearities, while at other times, the market follows a more predictable pattern. Indeed, we would favor a spectrum rather than binary – all or nothing – approach to EMH testing where the market *evolves* from fulfilling the conditions from one form of EMH to another as information, price and indeed the supporting infrastructure of the market itself adjusts through time in response to the genetic-like forces governing emergent information. Obviously, to undertake such analysis one would require a better understanding of the true microfoundations of information generation and transmission – a charter which our Evolutionary Finance approach aims to fulfill and which is indeed achieved in our extensively documented Game Theoretic microfoundations of analyst behavior detailed in Chapter 6. But more on this later, for now let us finalize our critique of the "traditionalist" elements of information assessment within the existing finance literature by outlining the "old" view on the actual mechanics of information arrival.

---

[19] Note we are referring to *excess* returns here. Obviously smaller firms would command a larger risk premium given their higher potential to fail.

2.5   Is "Strong EMH" all there is to the "Traditional" view of markets and information?

To build their veritable battery of defenses in the financial engineering sphere the traditionalists required more in their armory than just a firm conviction of strong market efficiency and the principles of Fama's "Fair-Game." In particular, they needed to somehow interpolate their beliefs on the instantaneous speed of information dispersal into a model of how asset prices actually incrementally move through time – after all, it is this underlying premise that drives many a financial engineering technique. The answer to the traditionalist prayers came in the form of the Random Walk Hypothesis which – along with Fama's "Fair-Game" principles and some fairly heroic assumptions as to the actual pattern of information arrival – resulted in the much-vaunted Gaussian (Normal) distribution of asset returns.[20] As briefly touched upon in Section 2.2, it is this Gaussian form that permeates risk calculation across an entire spectrum of "Traditional Finance" analytical tools – from Black–Scholes option pricing, to fixed-income factor models, to mean-variance optimization.

The Random Walk[21]

Under a "Fair-Game" and Independent and Identically Distributed (IID) "random" information arrival conditions, asset price movements are argued to reflect a Markov process. As a consequence, the only relevant variable in predicting the future value of an asset is its present value (past history has no relevance). Such is the hypothesis of the "Random walk."[22]

Appealing to a Wiener process (or somewhat interchangeably, Brownian motion), the Random Walk asset price model can thus be expressed in the continuous time limit as

$$dP = \mu Pt + \sigma PdX, \tag{2.4}$$

where dP is evolution of asset price; μPt the drift and σPdX is the volatility of asset price as conditioned by a normal distribution (dX is effectively random).

---

[20] So named after Carl Friedrich Gauss (1777–1855) even though the first $N(\mu, \sigma^2)$ proof dates back to 1733 by A. de Moivre (1667–1745) as a special case for Bernoulli variables (with $p = 1/2$).

[21] For a widely discursive overview of the "pro" Random Walk argument refer to Malkiel (1996). To contrast, an interesting rebuttal – but more quantitative approach – is available in Lo and MacKinlay (1999).

[22] The formal foundations of the Random Walk approach in a financial context are traced back to Osborne (1964) and Samuelson (1965) but it should be noted that as far back as 1900 Bachelier proposed that asset prices follow a Random Walk type of intertemporal pattern. Readers should note that much of the mathematics in modeling Random Walk movements with respect to time dates back to sixteenth-century games of chance – see Hald (1990).

Mathematically, this is a very convenient form via which to model asset price movements. Ito's Lemma provides solution to Equation (2.4) and, more importantly, lays the foundation for the Black–Scholes model of option pricing. However, for our purposes it is sufficient to concentrate on the last term – volatility (σ) – and how this is specifically modeled by a Gaussian distribution. Critically, it is the assumption of IID information arrival that gives (σ) its definite "Gaussian flavor." This *imposed* attribute of volatility in asset prices in response to a random error component generated by the assumed random arrival of information manifests into a belief structure as to how asset prices evolve across an entire spectrum of "Traditional Finance" analytical tools – from Black–Scholes option pricing, to fixed-income factor models, to mean-variance optimization. Indeed, as mentioned earlier much of what we witness today as financial engineering "best practice" is premised on this (and, by association, the Strong EMH) outcome. In rebuttal, we argue that while the benefits of Gaussian form (σ) may come in terms of mathematical tractability (see later for more details) this comes at a considerable cost in terms of the relevance of "Traditional Finance" assumptions in explaining the prevailing "real world" orthodoxy of analyst and informational behavior. In particular, we argue that while information (and by definition, asset prices) may at times give the semblance of being a random occurrence, the foundations of "real world" information production and its biological-like molecular formation imply this is merely a surface interpretation of events – the "real world" drivers of information are deterministic, not stochastic, in form.

Finally, it is important to note that while the Random Walk Hypothesis should be seen as an extension of Strong EMH in aiding and abetting the traditionalist cause, the direction of causality cannot be inverted. LeRoy (1973) and Lucas (1978) have both demonstrated a somewhat "random" pattern in price movement is neither a necessary nor sufficient condition for strong market efficiency – this is an important point we will develop a little further in Chapter 3.

One crucial point to note in the above shaded box is that the assumption of Fama's "Fair-Game" is by itself not sufficient to ensure normalcy in the distribution of asset returns – in particular, the Gaussian format of (σ). It is only by an additional assumption that the arrival of information is sufficiently *random* in nature (IID) that the principles of a "Fair-Game" result in the characteristic Gaussian distribution of asset returns (as illustrated in Figure 2.4).

If one wanted to be harsh, one could accuse the traditionalists of "manufacturing consent" in assuming information arrival is IID and therefore "random." The obvious criticism here is that the arrival of some market information necessarily does not always come as a "surprise." What about regularly scheduled economic and company reports which, in some cases, are reliably and accurately forecast? The traditionalist retort is that – as mentioned earlier – either investors are adroit enough to price the full implications of such a release across all assets (including the biological-like molecular clustering impact that a particular information byte will have upon the existing information-set) or investors are collectively ignorant of such "knock-on"

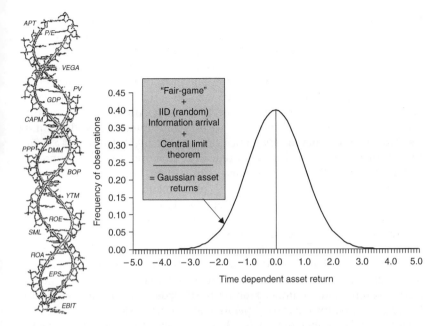

Figure 2.4 The "traditional finance" model of risk/return distribution – the Gaussian form[23]

potential. Either way, asset prices still react in a random fashion once (even well forecast scheduled) information releases are embedded into the existing information-set and a complex array of nonlinear feedback effects manifest in response to the agglomeration of this new information in a biological-like molecular fashion into the existing information-set.[24] Indeed, we respond by yet again saying that just because the outcome of this process may look random doesn't necessarily mean that its determining conditions are truly stochastic.

As we will display in Chapter 3, at times deterministic (rather than stochastic) forms can generate chaotic sequences that fool many a statistical test for "randomness."[25] Indeed, we would go further in arguing that the

---

[23] Note: we have assumed zero drift in asset returns for exposition purposes only.

[24] One solution to this overwhelming "noise" problem in the release of information has been suggested by Tay and Linn (2001) who argue that traders can adopt a "fuzzy logic" methodology to condense the plethora of possible alternatives into a few simple generalized notions. Such aggregative rules of thumb sometimes work well in the modeled environment – as will be explained in Chapter 3.

[25] Indeed, such an outcome could be construed as turning the argument back on those who have espoused the virtues of the Strong EMH mindset by arguing investors are

true biological-like deterministic foundations of the way financial information actually comes together, and importantly how investors interpret this information, can explain the varying empirical results of the tests of EMH presented in Section 2.4. As mentioned earlier, at times, this clustering of information bytes into memes, themes and even overall market sentiment occurs in a reasonably predictable fashion and is thus supportive of a Weak EMH outcome. At other points in time, the complex nonlinear relationships that necessarily go "hand-in-glove" with such a framework underpinning the drivers of information dynamics generate pricing outcomes of a definite chaotic nature – which is consistent with an empirical result supportive of Strong EMH. We argue that what is important to understand here are the true microfoundation drivers of this phenomenon – as will be explained in detail in Chapter 6 – and not to opt for simplistic notions of information "randomness" as indicative of the market's informational state simply because on occasion price outcomes seem to (erroneously) intimate as much. Instead, it is far more important to understand the *evolution* of pricing behavior from predictable to "random-like" states. To do this, one should believe in a spectrum (rather than binary outcome) for market efficiency and indeed, form a comprehensive understanding of the true biological-like molecular forces governing information generation and transmission.

Finally, before concluding, why all the fuss about the need for Gaussian form in the first place? Obviously the traditionalists went to a lot of trouble to make the necessary assumptions to generate this particular density structure for asset price returns but that still doesn't explain why one would necessarily indulge in such behavior? Well, as we briefly intimated in the shaded box earlier, the reason why this particular distribution was "aimed for" via assumption in preference to other distributional forms is simply because of its mathematical tractability. Gaussian forms are additive – in short, summate a sequence of Gaussian forms and you end up with a Gaussian form. This made the algebra easier – especially when it is considered that much of the traditionalist thought paradigm was built in the 1950s and 1960s when most analytical solutions were sought by putting pen to paper and not computationally derived (as is the case in many instances today). Furthermore, the Gaussian distributional form has the appealing attribute of finite variance which made the calculus easier when seeking intertemporal solutions. Finally, as we briefly highlighted earlier, in many cases the distribution of returns for any given asset does come to resemble through time what one would come to expect had prices (and therefore information) been truly a random sequence – provided one chooses the correct timeframe. But we

"fooled by randomness" – see Taleb (2004). It may be that such "randomness" validating tests bolstering such arguments have themselves been fooled by the deterministic complex/chaotic informational foundations that drive markets. We will discuss this in greater detail in Chapter 3.

reiterate, just because something does on occasion may look to be random, doesn't necessarily mean that it is truly stochastic in nature.

So to conclude, we would argue that the mathematical tractability of the Gaussian form appears to have been put ahead of the relevance of traditionalist finance assumptions about market (and in particular, information behavior) relative to what we observe in the real world. In particular, traditionalist beliefs regarding true market structure (Strong EMH/Fama's "Fair-Game") and the forces governing the generation and dispersal of information (IID) appear to be extremely limiting. The fact that these questionable assumptions underpin the entire edifice of the traditionalist approach toward risk and return assessment certainly raises concern as to the applicability of these mechanics as a one-stop solution for all financial engineering requirements. Many argue that it is time to investigate an alternative approach or at the very least, remove some of the more stringent behavioral/institutional assumptions underlying the traditionalist platform. We now turn to some of these "new view" challenges before presenting our own Evolutionary Finance purview of the way in which asset prices and information symbiotically interact through time in a manner that is analogous to a biological-like molecular pattern of informational clustering.

3

The "New" View of Finance

In this chapter we discuss the "new" view challenges to the traditionalist finance thought paradigm. Thankfully, there have been a variety of alternatives put forward to supercede the stringent set of assumptions imposed under the "old" view/traditionalist finance interpretation of the way markets operate. Even better, many of these "new" view challenges – as we label them – are starting to enter the mainstream of "received wisdom" in finance. Still, regrettably for us many of these "new" view challenges continue to regard information as a uniform intangible concept. There is no information spectrum, no texture as to how information is actually manufactured and disseminated, no extensive microfoundation structure of how the actual components of what investors come to regard as "information" actually bolt together. This is where our Evolutionary Finance perspective comes in – we aim to identify what actually constitutes a certain stream of financial information's "DNA."

But that is the topic for the next chapter. For now, what of these "new" view challenges? Are there any useful insights among these rebuttals of the traditionalist "old" view thought paradigm that might prove useful for our Evolutionary Finance purposes? Let us start by saying that even though these "new" view challenges treat information as though it were a generic commoditized concept, it by no means implies that we cannot adopt at least some of their key principles within our Evolutionary Finance framework. Indeed, in true spirit with Sir Isaac Newton's maxim "If I have seen further it is by standing on the shoulders of giants" we must stress here that we would be lost without being able to leverage at least some of the key insights put forward by these "new" view protagonists.[1]

To guide the reader along our eclectic path of selecting certain building blocks from within the "new" view way of thinking, we have decided to break the various challenges to the traditionalist finance line of reasoning into three separate categories.

[1] Quote from Sir Isaac Newton in his letter to Robert Hooke – dated February 5, 1676.

- Challenge no. 1: Determinism, Complexity Theory and the Nonlinear Dynamics School.
- Challenge no. 2: Bounded Rationality, Heterogeneous Agents and the Behavioral Finance School.
- Challenge no. 3: Trading Rules, Evolutionary Games and Artificial Markets.

We conclude this chapter by highlighting the mix of components that we have chosen from within each of these three categories to assist us in the formulation of our Evolutionary Finance framework. But let us start by giving a brief appraisal of each "new" view challenge in turn.

3.1   "New" view challenge no. 1: Determinism, complexity theory and the nonlinear dynamics school

Nonlinear dynamic systems come in many forms but two categories of special interest in finance "are Complex" systems and another popular class of models known as "Chaotic" systems.[2]

- Complex systems are nonlinear dynamic systems that display self-organizing behavior and are primarily *deterministic* rather than *stochastic* in foundation.
- Chaotic systems are nonlinear dynamic systems that are again, primarily *deterministic* rather than *stochastic* in foundation, but display non-replicating/unstable behavior dynamically.

Let us deal with Chaotic systems first since they are but a subset of the much larger class of Complex systems. Why are Chaos models of interest to "new" view clergy as a challenge to the traditionalist paradigm? Well, a lot hinges on the terminology of "deterministic" rather than "stochastic" in the earlier brief definitions. Deterministic systems are nonrandom in nature – their foundations are mathematical rather than statistical.[3] This stands in

---

[2] Typically such systems are dynamic in nature. Sterman (2000) categorizes six generalized classes of dynamic system – "exponential growth" (e.g., $y = b^t$), "goal seeking" (e.g., $y = \log b^x$), "S-shaped growth," "oscillation" (e.g., $y = 1/(x^2 + 1)$), "growth with overshoot" and "overshoot and collapse," for example,

$$\dot{x} = (a - cy)$$
$$\dot{y} = y(dx - b)$$

Chaos and Complexity models fall mostly within the oscillation category.

[3] The stochastic versus deterministic debate has a long lineage. Some argue there are few truly random sequences in life except perhaps in quantum mechanics where Einstein's proof of the random behavior of elementary particles near the Planck scale still reigns supreme. Still, even this is under threat thanks to recent advances in String Theory. Indeed, Duncan Watts in his book *Small Worlds* (1999) identifies that even the most

direct contrast to the principles of IID driven "random walk" espoused in the previous chapter. But again, as we stressed repeatedly in Chapter 2, just because something may actually look random on the surface doesn't necessarily mean that its true microfoundations are actually stochastic – as the following application illustrates.

A simple deterministic model of price discovery

Let us introduce a very simple rule of thumb via which prices are set.[4] To start with, let us assume price determination at time t_n is driven by the following logistic equation

$$P(t_n) = \phi P(t_{n-1}) - \phi P^2(t_{n-1}) = \phi P(t_{n-1})[1 - P(t_{n-1})], \tag{3.1}$$

where $\phi P(t_{n-1})$ implies demand created by buyers causes price to *rise* at rate ϕ; and $\phi P^2(t_{n-1})$ implies supply created by sellers causes price to *fall* at rate $\phi P(t_{n-1})$.

Obviously, this is a *deterministic* model in the sense that investor price setting is not random but conditioned by the rules specified in Equation (3.1).[5] But just because our model is deterministic in foundation does not imply that its dynamics are not "random" in appearance. For example, calibrating our model by setting ϕ at 4 in Equation (3.1) and for simplicity constraining $0 \leq P_n \leq 1$, it can be illustrated (see Figure 3.1) that the time-series of price movements generated by this (chaotic) deterministic process are almost indistinguishable from a series of hypothesized price movements generated by a purely stochastic random number generator.[6]

---

innocuous supposedly "random" event such as a chance meeting of a friend at a remote location generally has deterministic foundations. For instance, you may both have the same hobbies, vocation, be generally of the same age, and the same cultural/educational background. This narrows the odds of such a "chance" meeting considerably. Further, readers should note that within this book we take the purest notion of "determinism" rather than the "fine graining/coarse graining" outcome of Ergodic Theory which argues (in sympathy with the earlier cited uncertainty principle in physics) that statistical "fine grains" can be regarded as sufficiently deterministic at a "coarse-grain" level provided the right statistical mechanic rules are applied. For an erudite overview of the links between supposedly "random" systems and systems of a more deterministic nature refer to Beltrami (1999).

[4] More will be said on the use of heuristic rules of thumb in the next section.

[5] Note: our focus on a deterministic approach does not constrain us to only technical (ex post price) models of price evolution. Price setting in Equation (3.1) could just as easily incorporate additional "fundamental" factors such as the forces which govern future supply and demand. These were omitted for simplicity here but will be built upon extensively in Chapter 5.

[6] Note that the calibration values of our model were not chosen arbitrarily but rather in accordance with typical logistic equation principles where $0 < \phi < 3$ results in a regular period cycle, $3 < \phi < 3.5699$ results in the system becoming increasingly unstable (with the appearance of new strange attractors), and for $3.5699 \leq \phi < \infty$ the system becoming chaotic (and uniformly chaotic at $\phi = 4$). For more on the properties of logistic equations, see Devaney (1989).

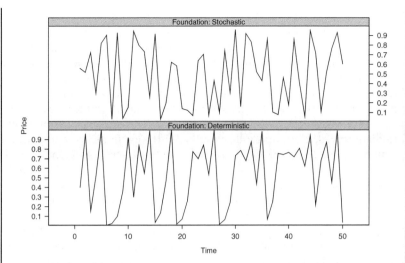

Figure 3.1   Stochastic or deterministic? Can you spot the difference?

Indeed, true to form, our deterministic logistic equation was able to fool a variety of tests for randomness such as scatter plot diagrams of the data, auto-correlation tests and nonparametric Wilcoxon Rank Sum testing (that indicates no significant difference between a purely stochastic form and our deterministic distribution). That said, plotting the first difference of price movement as specified in Equation (3.1) or testing for Normalcy using chi-square or Kolmogorov–Smirnov tests did indicate that our deterministic model failed in hypothesized "Gaussian" form (see first difference distribution and QQ plots in Figure 3.2) – thus illustrating that some Normalcy tests are obviously better than others.[7]

The key element for readers to focus on in Equation (3.1) is the specification of feedback – that is, prices today are related (in nonlinear fashion) to prices yesterday.[8] It is this feedback loop that gives the logistic equation its chaotic appearance for certain calibrations of the inputs. Feedback of this type is integral to this class of dynamic system. Note that this form of time-related interdependence is in direct contravention to the EMH testing methodologies identified in Chapter 2. That said, there is little doubt that if the chaotic outcome of such feedback was able to fool the variety of specific purpose tests we performed, it is also likely that some of the econometric tests cited in Table 2.1 are just as easily in question – especially those that came down in favor of a Strong EMH outcome. We will use both the concept of feedback and time-related interdependence in our modeling of the biological-like molecular formation of information in finance in Chapters 5 and 6.

---

[7] Note: this but a small subset of an entire battery of tests that have been constructed specifically to discern what is truly "random" and what is actually deterministic in foundation – for examples of such tests see Scheinkman and LeBaron (1989).

[8] For an examination of the link between price feedback and financial bubbles, refer to Cohen (1997).

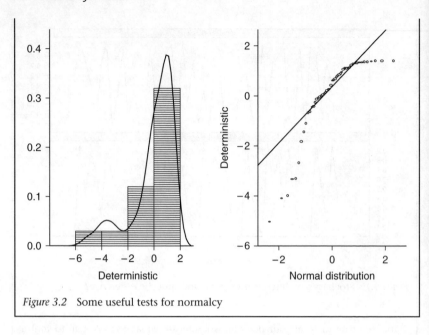

Figure 3.2 Some useful tests for normalcy

So what purpose does the above serve to assist the practitioner rather than the theorist? By emphasizing the deceptively random appearance of Chaotic sequences these "new" view determinists were able to discredit a key argument within the traditionalists battery of defenses. In particular, in sympathy with the precepts of technical analysis, a number of "scaling" experiments have been conducted examining the correlation (fractal) dimensions between successive timescales.[9] What is the purpose of such studies? To quote Peters (1996) "A fractal is an object in which the parts are in some way related to the whole."[10] In other words, fractal studies typically look for interdependencies (correlation) between datasets in different timescales in an attempt to identify the true deterministic foundations that make up what we perceive to be overall market dynamics. Obviously, any form of interdependency is in direct contravention to Strong EMH principles. Indeed, it is interesting that in many cases these studies illustrated a marked "leveling off" in correlation dimension as time scaled – thus highlighting underlying deterministic rather than stochastic drivers of movements in asset prices. The difficulty remains

---

[9] This is typically done by constructing a k-dimensional vector $X(t) = [P(t), P(t+\tau), P(t+2\tau)...P(t+(k-1))\tau]$ and plotting its trajectory in k-dimensions as t varies. Indeed, one can further formalize this approach by examining Hurst exponents. A Hurst exponent $\neq 0.5$ implies intertemporal correlations and thus fractal dimensions.
[10] We will present more on this holistic element to nonlinear studies of the market later in this section.

however in ascertaining the actual formulation of the underlying deterministic microfoundations. Unfortunately from the practitioner's perspective, this is where Chaos Theory has had a tendency to disappoint. In fact, the experience with the implementation of Chaos Theory into the applied finance field has been mixed at best. Vaga (1994) is a strong supporter of using Chaos Theory from a practitioner's perspective but Scheinkman and Le Baron (1989) found little evidence of nonlinear deterministic forces at work in their analysis of the US equity market.[11] That said, Peters (1996) pointed out that the Scheinkman and Le Baron study may have been biased against a deterministic outcome thanks to insufficient observations. Still, what this study did definitely demonstrate is the immense difficulty in modeling dynamic systems with multiple variables – in the case of Scheinkman and Le Baron (1989) it was a six-variable system. Indeed, while fractals may give us some insight as to the interdependent nature of various timescales in price movements, one cannot be truly sure that they understand the full workings of a Chaotic system without the specification of its functional form and initial conditions. The more variables in the system, the more difficult this becomes. Ergo, it becomes very problematic to make forward projections of price action based on only partial (fractal) glimpses of the dynamics of a truly Chaotic system.[12]

In short, it appears that the very "chaotic" nature of Chaos systems has proven problematic from the practitioner's standpoint. Urbach (2000) illustrated that even systems that are truly deterministic in origin are not immune from the presence of additional statistical "noise" that obfuscates the underlying drivers. So while identifying fractal dimensions may be useful for questioning the traditionalist finance evangelical-like belief in Strong EMH, it still leaves the question – if there are deterministic forces at work, how best to identify them and what is the rationale for their existence? This is where our Evolutionary Finance approach aims to make a major contribution by identifying precisely why fractal interdependencies exist – this is done via modeling the evolutionary-like formation of price sensitive

---

[11] That said, our own empirical work has also demonstrated the US market as a particularly difficult "nut to crack" in formulating reliable alpha generating strategies – simply because there are so many analysts picking over this market 24/7 and subsequently "bidding" the alpha out of any nascent exceptional return strategy. It appears somewhat ironic therefore that many tests of market efficiency etc. use the US market as a base-case scenario (mostly because of the quality of its time-series data) when practitioners realize that the US example is more the "exception to the rule" rather than the "rule" itself.

[12] Such fractal dimensions are typically measured as $\phi = \log \kappa / \log(1/2r)$ where κ is in effect the required number of circles to encompass a "noisy" price pattern and r is the standard circle radius. As $\phi \Rightarrow 2$ the more "noisy" the data. An alternate approach to this is to use "boxes" as opposed to "circles." Indeed such a tact was proposed by Mandelbrot (1997).

information using nonlinear processes such as evolutionary algorithms (part of the neural network "family" of nonlinear modeling techniques). Importantly we incorporate not just price data into our analysis but the entire information-set – technical, fundamental, economic and political – in building our "informational genotype" to match each asset price phenotype. By doing so, we overcome some of the earlier implementation problems of nonlinear dynamic modeling from the practitioners standpoint.

But more on this later. One question to answer right now is that if interdependencies between price ticks in high frequency data do seem to occur in a deterministic manner, it still does not explain why under most conditions bid/offer spreads do not fluctuate wildly from minute to minute – as implied under a pure Chaos regimen. This "self-organizing" nature of the market cuts directly across the principles of Chaos Theory in its purest sense. Thankfully all is not lost. Attention has now turned to a particular form of nonlinear dynamic system that contains many of the deterministic benefits of Chaotic systems – in that they look deceptively "random" when calibrated – but they do not suffer from as many of the infinite instability problems as pure Chaos. This new breed of nonlinear dynamic system are known as "Complex" systems and represent a wider class of nonlinear deterministic models – indeed, they umbrella Chaos models.

As highlighted at the start of this section, Complex systems are nonlinear dynamic systems that display *self-organizing* behavior and (like Chaotic systems) are primarily deterministic rather than stochastic in foundation.[13] They are particularly relevant to modeling biological-like systems due to the natural "checks and balances" endogenized within their mechanics and are thus utilized in our formal modeling presented in Chapter 5. To quote the eminent physicist Erwin Schrödinger – a living organism has the "astonishing gift of concentrating a 'stream of order' on itself and thus escaping the decay into atomic chaos."[14] As a consequence, as we will amply illustrate in Chapter 5, such systems – at the holistic level at least – are particularly useful for modeling the evolution of information in a biological-like molecular manner.

How do Complex nonlinear systems actually "self-organize"? Well, referring back to our much-loved logistic equation cited earlier, Complex systems cover all scales for $\phi \geq 3$. Therefore, they encompass what is generally referred to as Chaotic systems (especially uniformly Chaotic models where $\phi = 4$) but also describe nonlinear models that are intermittently "random" in appearance at certain times while at other times they are more regular in their

[13] Such systems are sometimes known as "self-organizing criticality" after Bak and Chen (1991). Indeed, the underlying characteristics of such systems may be configured in a manner that promotes the emergence of what Waldrop (1992) labeled as "spontaneous self-organization."
[14] See Schrödinger (1967).

dynamics – perfect for our purposes given the discussion in Chapter 2 on the empirical inconsistencies between the various EMH empirical studies. The "schizophrenic" nature of Complex systems is a characteristic of the "self-organizing" feedback driving their time path – as evidenced by the presence of "strange attractors" in phase portrait depictions of this class of nonlinear deterministic models.

Increasing levels of Complexity

To illustrate how increasing levels of Complexity arise from incremental movements in the nonlinear feedback mechanism ϕ cited in our logistic Equation (3.1), we illustrate the following charts. Figure 3.3 demonstrates that for values of $0 \le \phi \le 3.0$ (in the following example $\phi = 2.9$) the nonlinear dynamic system represented by our logistic equation is convergent to a singular steady-state.

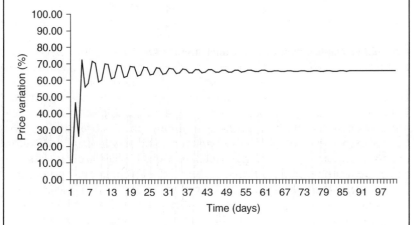

Figure 3.3   Our logistic equation calibrated at $\phi = 2.9$

Figure 3.4 illustrates that for $\phi = 3.5$ the system becomes increasingly unstable with the appearance of multiple strange attractors.

Finally, for $\phi = 4.0$ our system becomes uniformly Chaotic with no discernable oscillatory pattern (see Figure 3.5).

Complex systems encompass all calibrations where $\phi \ge 3$. To quote Mainzer (1997) "the degrees of increasing Complexity are defined by the increasing bifurcations which lead to Chaos as the most complex and fractal scenario."[15] Complexity arises primarily because the system jumps from one "strange attractor" to another. The characteristics of these strange attractors are explained by Lyapunov exponents. It is only in the range $3.5699 \le \phi < \infty$ that the system can be described as

---

[15] For a detailed discussion of the "robust yet fragile" nature of Complex systems, see Doyle (2001).

Chaotic (and uniformly so for $\phi = 4$). Hence, it can be surmised that Chaos itself falls within the larger family of Complex models.

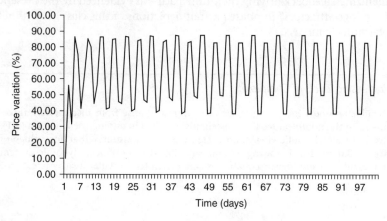

Figure 3.4 Our logistic equation calibrated at $\phi = 3.5$

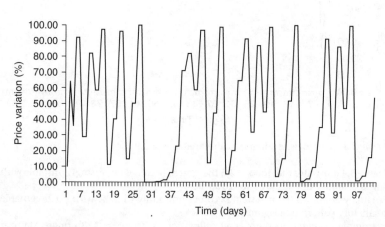

Figure 3.5 Our logistic equation calibrated at $\phi = 4.0$

It is not just within the field of finance that Complexity Theory has made its presence felt. The smallest microfoundation deterministic element in such systems can range from an economic agent, to a gene, to an organism, to a particle – thus making the analysis of self-organizing nonlinear Complex systems applicable to a wide range of academic fields, including our deterministic information-based Evolutionary Finance approach.[16] Indeed, the higher order characteristic of these nonlinear systems – which implies that they cannot be broken down into parts and solved piecemeal (as is the case with linear systems) – makes the "holistic" nature of Complexity Theory a key selling point in explaining behavior in everything from traffic flows, to ant colonies, to financial markets.[17]

In fact, the principle of the "sum is greater than the individual parts" in Complex systems has now been popularized to such an extent that it has even made its way into generalist management literature.[18] Long-standing proponents of this over-arching holistic perspective toward systems thinking such as Arthur (1994), Bak (1997) and Farmer (1998) fervently argue against the "reductionist" notions of traditional science – and finance for that matter.[19] In short, reductionists have typically argued that a system is best understood by examining the role of each component part – sometimes in minute detail and mostly in glorious isolation. To contrast, Complexity buffs argue that it is the *externalities* that stem from the interaction of the combined deterministic components that constitute the microfoundations of the system that make Complex systems so interesting.[20] Ergo, it is thus (by definition) impossible to examine such externalities via reductionist method.[21] In the ensuing chapters we take an eclectic approach in stressing that the externality potential locked up in each piece of information – which

---

[16] For a comprehensive assessment of the plethora of fields into which Complexity Theory extends, see Favre *et al.* (1995) and Mainzer (1997).

[17] For an example of the application of Complexity Theory to traffic flows see Benjamin *et al.* (1996). For an analysis of the Complexity principle across biology, economics and physics in general, see Auyang (1998).

[18] See Pascale *et al.*(2000).

[19] Such arguments have not always been easily won – Arthur (1989) for instance had to wait six years for his seminal paper on increasing returns in economics to be published. Obviously the traditionalist/reductionist camp was vigorous in its defense.

[20] Readers should note that we adopt the term "externalities" here in accordance with the way the phraseology is used in economics – in particular, models of endogenous growth – where "spillovers" from the interaction of various components result in a classic nonlinear "sum is greater than the component parts" outcome. See Romer (1986), (1989) and (1990).

[21] For example, observance of a single ant rarely provides insight into the overall operation of the colony as a whole. Likewise, observance of a single player in the market is unlikely to provide insight as to the complete market's macrodynamics. This holistic perspective contrasts the traditionalist notion of "reductionism" in the sense that all market dynamics can be interpreted through the observance of each agent in isolation. A fascinating book on this topic of reductionism versus a holistic approach is *Swarm Intelligence: From Natural to Artificial Systems* by Bonabeau *et al.* (1999). For an outline of how this school of thought can be applied to a variety of phenomena, see Johnson (2001).

we label as an informational "byte" can be best understood by examining the microfoundations of how this information is produced and actually binds together (in particular its sequencing). This is entirely sympathetic to the holistic principles underlying the Complexity theorists' cause. It is from these microfoundations that we attempt to better interpret the nonlinear dynamics observed in asset prices.

However, before concluding this section, one cannot stress enough that one of the true beauties of Complex systems from a finance perspective is that, as mentioned earlier, at times they display intermittent somewhat "noisy" (read "random-like") behavior, while at other times they display behavior that is of a more predictable nature.[22] In particular, Complex systems have the appealing attribute of displaying periods of both low and high volatility as they jump from one steady-state strange attractor to another. These bursts of extreme volatility followed by periods of relative volatility quiescence have an immediate analogous inference with observed market movements. Such "volatility clustering" as it is known to those in the profession, is an endemic feature of the time-series analysis of most asset prices.[23] The fact that such volatility (albeit at extremes at certain points in time) is bounded within a finite range thanks to the "self-organizing" nature of Complex systems makes this class of models particularly appealing to both the theorist and the practitioner in attempting to explain this erstwhile perplexing phenomenon. Indeed, if one is to further investigate the time-variant attributes of volatility within Complex systems, one can identify epochs where substantial departures from near-Gaussian like distributional forms occur simply because one's frame of reference contains a relatively high number of "extreme events" as the system jumps from one steady-state strange attractor to another. But at other times a near-perfect Gaussian-like distributional form of returns exists. Again, this Gaussian-like departure in distributional form for certain periodicities (where there can be a relatively high number of extreme events) but more Normal-like distributional forms for other epochs (where the "noisy" nature of Complex systems takes on more precedence) is an

---

[22] The fact that Complex systems display such behavior has resulted in analogous referencing between hydrodynamic systems and (Complex directed) price systems. See Ghashghaie et al. (1996) for an introduction.

[23] See Sornette (2001) for a fascinating comparison between the predictability of extreme volatility "clusters" in physical phenomena – such as earthquakes – and financial phenomena. Sornette was able to identify that significant market events – such as the 1929 and 1987 stock market crashes – were preceded by a sequence of "intermediate" events years in advance. Using the correct scaling techniques, Sornette was able to detect the presence of associative nonlinearities in the financial system via intertemporal correlations in various dimensions (fractals). Indeed, the recent popularity of event risk assessment programs such as *Crashmetrics* stands as testimony to the desire of practitioners to better understand time-variant volatility – especially at the extremes, see Wilmott (1998).

appealing attribute of this class of "self-organizing" nonlinear models and fits in nicely with our earlier observations of the empirical evidence on the various forms of EMH.

Naturally, we have been lead into a discussion of distributional form under Complex modeling. By way of background, a number of distributional forms have been suggested by Complexity theorists as an alternative to the Gaussian form in an attempt to better capture the short-term kurtotic behavior of Complex systems but also their tendency to approximate to $N(\mu, \sigma)$ as $t \to \infty$ and the impact of the various jumps to alternate strange attractors gets effectively washed out by the high number of observations. One interesting contribution comes from Mandelbrot (1963) who is perhaps best known for his work on Chaos rather than Complexity *per se*. Mandelbrot suggested a stable Levy distribution as an alternative to the Gaussian form given that – as desired – it displays discernable kurtosis.[24] One downside of this particular distribution however is that it also suffers from infinite variance – which makes the intertemporal math unwieldy. Thankfully Mantegna and Stanley (1996, 2000) – some of the founding fathers of the *Econophysics* school of thought – came to the rescue in advocating a "truncated" Levy distribution which had most of the appealing scaling attributes of the stable Levy distribution but had finite variance as well.[25]

The stable and truncated levy distribution

These distributional forms were originally described by Levy (1925) but their roots reside in the workings of Pareto (1897) in his approximation of the distribution of income. The characteristic "fat-tails" of such distributions arise in response to the presence of an inverse power law (in Pareto's case, the relationship between income and the ability to generate wealth – in much the same fashion as Zipf distributions have an inverse power law in describing the distributional frequency of words in the English language). In essence, it is the presence of feedback between the underlying variables that contributes to the manifestation of such "fat-tails" – hence the immediate compatibility of this form of distribution with Complex nonlinear models. Levy characterized the generalized version of this particular distributional

---

[24] It is interesting that an arch advocate of the traditionalist (Gaussian distribution) line of reasoning – Eugene Fama – also suggested such a distributional form. See Fama (1965).

[25] Econophysics is the application in finance of empirical techniques refined in Physics. Its foundations go back as far as the 1940s with the publication of a paper by Majorana (1942) highlighting the link between physics and the social sciences. Still, it can be argued that the generalized approach of using tools developed in Physics to divine the deterministic forces at work in financial data did not become "popularized" until the 1990s. Readers should note however that the inspiration flow has not all been one-way. For instance, notable economist Vilfredo Pareto's (1848–1923) use of power laws predates the use of such empirical analysis in Physics.

form (known as the stable Levy distribution) with the following formula

$$\ln[f(p)] = i\Im p - \Im|p|^{\text{Щ}}\left[1 - i\text{Ш}\left(\frac{p}{|p|}\right)\tan\left(\text{Ш}\left\{\frac{\pi}{2}\right\}\right)\right], \quad \{\text{Щ} \neq 1\},$$

$$\ln[f(p)] = i\Im p - \Im|p|^{\text{Щ}}\left[1 + i\text{Ш}\left(\frac{p}{|p|}\right)\frac{2}{\pi}\ln|p|\right], \quad \{\text{Щ} = 1\}, \tag{3.2}$$

where

3 is the scale parameter (for instance, the difference between daily and weekly data);
Ш the skewness of distribution (note $-1 \leq \text{Ш} \leq 1$);
Щ the peakness of distribution (note $0 \leq \text{Щ} \leq 2$), and
Э is the location parameter.

Altering the values of 3, Ш, Щ and Э in effect alters the characteristics of the distribution. For example, when Ш = 1 the distribution is "fat-tailed" to the right (the converse applies when Ш < 0). Under the special case of Ш = 0 and Э = 0 the distribution is both symmetric and stable with a probability density function

$$\Pr(p) = \frac{1}{\pi}\int_0^\infty e^{-3|p|^{\text{Щ}}}\cos(px)dp. \tag{3.3}$$

But the true beauty of this distributional form is that in truncated form it displays discernible kurtosis for small timescales, but it converges to Gaussian-like form for larger timescales.[26] This is because, Levy distributions emerge as a generalization from the Central Limit Theorem. In short, Levy distributions are characterized by the parameter Щ. So in the special case of Щ = 2, Ш = 0, 3 = 1 and Э = 1 the Levy distribution corresponds to the normal distribution with a probability density function

$$\Pr(p) = \frac{1}{2\sqrt{\pi}\sigma}e^{-p^2/2\sigma^2}. \tag{3.4}$$

In terms of calibration, Mandelbrot (1963) and Fama (1965) effectively measured Щ ≈ 1.7. This began an intense empirical debate centered on power law measurement.[27] The reason for the lively debate? The Strong EMH/traditionalist viewpoint effectively argues that Щ must always equal 2. The Complexity Theory/Chaos Theory viewpoint is that $1 \leq \text{Щ} \leq 2$ (since for 0 < Щ < 1 there is

---

[26] An alternate distributional form – Student's t distribution – also displays this appealing "convergence to Gaussian" characteristic as $n \to \infty$. One potential problem with this distributional form however, is that it is not stable for $n \neq 1$ (even though it is finite).

[27] Put simply, power laws state that some quantity ($\acute{\text{K}}$) can be expressed as a power of another quantity ($\acute{\text{к}}$). In effect, $\acute{\text{K}}(\acute{\text{к}}) = \acute{\text{к}}^{-c}$ where c is the "power." Expressed in log terms – $\log\acute{\text{K}}(\acute{\text{K}} = -c\log\acute{\text{к}}$ where c is the gradient of a straight line. In the earlier section, it must be said that the transformation issues become a little more complex as c is filtered through a number of additional variables but the basic concept remains the same.

no stable mean to the distribution). The outcome? Unfortunately (as was the case with other attempts at empirically validating the various forms of EMH identified in Section 2.4) it appears the empirical jury is still out – see Farmer (1999).

That said, one should note a practical problem of the stable Levy distribution (even with $1 \leq Щ \leq 2$) is that it displays infinite variance – the Fourier transform of Equation (3.3) only results in a *stable* and *symmetric* distribution, it says nothing about being *finite*.[28] Variance is only *stable* and *finite* for $Щ = 2$. As mentioned earlier, a solution has been suggested – and, again as mentioned earlier, popularized by Mantegna and Stanley (1996, 2000) – in the form of a *truncated* Levy distribution. However, in doing so a compromise one has to make is that as is the case with the Student's t distribution, the truncated Levy distribution is finite but not stable (both distributions however converge to Gaussian as the number of price observations approaches infinity). The benefits remain however that given the truncation to the Levy distribution there is not the infinite variance problem that plagues the intertemporal calculus of the stable Levy distributional form.

So what to make of these alternate distributional forms? Without doubt, they are certainly a step in the right direction as they do not suffer from the same fallibility of the traditionalist finance/Gaussian framework in the sense that they are *imposed* within the system via stringent assumption – the search for fractal dimension is in a sense a search for the power law determinants of these particular distributional forms. But it must be stated that this simplicity comes at a cost. While it can be divined that the financial system is necessarily deterministic in form (at least according to the empirical results of some of the power law tests) there is little explanation as to why determinism should be the case. In particular, there is little discussion of the underlying microfoundations as to how investors necessarily come to make the decisions in deterministic fashion. This is where our Evolutionary Finance approach aims to fill a considerable void in the literature. Not only is the atomistic creation of information and its biological-like molecular formation and dispersal perfectly suited to the microfoundation determinism of Complexity Theory, but it also provides a framework for the derivation of superior empirical techniques from the practitioner's standpoint. This is simply because we consider the entire information-set – technical, fundamental, economic, political – and not just asset prices alone. We will discuss this issue in much more depth in Chapter 5 where our all-encompassing evolutionary model of the marketplace is presented. Likewise, in Chapter 6 we will develop our own comprehensive approach toward distributional form modeling under our Evolutionary Finance auspices. For the moment however, let us examine in greater detail some additional "new" views as to the way investors actually interpret information in the "real world." Here, the concepts of bounded

[28] Infinite variance is a major pitfall for any distributional form chosen for modeling return distribution as it implies infinite risk against which one is required to attempt to form a price. This makes for an intractable problem in option pricing as risk no longer cancels in the Black–Scholes equation via Delta hedging.

rationality, heterogeneous agents and Behavioral Finance have made great strides in challenging the traditionalist finance/*homo economics* (fully rational agent) orthodoxy and it is toward this purview that we turn to next.

3.2 "New" view challenge no. 2: Bounded rationality, heterogeneous agents and the Behavioral Finance school

What we have categorized as "new" view challenge no. 2 is all about the demand-side – the investor's experience – of information. It effectively represents an attempt to address at least "one-side" of the coin in dismantling the edifice of stringent micro foundation assumptions that constitute the basis of the traditionalist thought platform – the other side of the coin being the supply-side of how information is actually generated (which in part, was addressed in the previous section – at least in the sense that information arrival is non-IID).[29]

Let us assume for the moment that:

(i)   not all individuals have access to the same information; or

(ii) that all the information that individuals require to make a "rational" judgment (with the benefit of hindsight) is not always available when it should be; or

(iii) that individuals have different risk preferences and/or abilities to interpret information.

All the above open up a veritable Pandora's Box of interesting human – and financial – behavior for both the theorist and the practitioner alike. Mining this rich vein of research has been a new breed of academic thinkers (or in some cases, reformed members of the traditionalist finance school) who have benefited from the appearance of purpose-orientated journals in which to publish – such as *The Journal of Psychology and Financial Markets*, the *Journal of Behavioral Finance*, the *Journal of Economic Psychology* and the *Journal of Behavioral Decision Making*. Indeed, their efforts in such worthwhile endeavors are now starting to be acknowledged via the process of peer review. Cases in point are the award of the 2001 Nobel Prize in Economics to Messrs Akerlof, Spence and Stiglitz for their work on information asymmetries and the award of the 2002 Prize to Messrs Smith and Kahneman for

---

[29] Readers should note that studies of information interpretation go part way toward developing a better understanding of the forces that govern the speed of information dispersal – an integral part of our Evolutionary Finance approach which will be expanded upon in greater detail in Chapter 4. Indeed, we argue that one of the beauties of our biological-like molecular approach toward information formation is that it is actually thoroughly consistent with the way neuroscientists have described the human mind's processing of information – a serendipitous outcome which we will elaborate upon in greater detail in Chapter 4.

their work on Human Judgment and Experimental Economics.[30] Unfortunately, as stated in the introduction to this chapter, much of this interesting work still characterizes information as some intangible commodity with little understanding of the true microfoundation building blocks of what investors come to perceive as "information" in the first place – a shortfall which we aim to address in Chapter 4 of this book.[31] Still, it is an interesting field of research and we thought we might start-off by highlighting one area that has been receiving a considerable degree of attention in recent years – the field of Behavioral Finance.

What exactly is "Behavioral Finance"? Let us begin with some definitions. Some opt for a reasonably broad definition – such as Shefrin (2000) who argues Behavioral Finance "is the application of psychology to financial behavior."[32] We opt for a more precise definition in the sense that we believe:

> Behavioral Finance is all about the fact that investors make mistakes. They do not possess complete information, they are not equal in financial ability, and their wants and desires alter. In other words, it treats investors as though they are human.

The foundations of this field of thought are generally regarded to reside in Kahneman and Tversky's (1979) seminal paper on Prospect Theory. Prospect Theory espoused that consumers specify their value function in terms of wealth rather than utility. This may not sound like much but the implication of Kahneman and Tversky's work was profound in the sense that it allowed individuals to have a greater aversion to loss and a diminishing sensitivity to return (either gain or loss) than would be implied under a pure "rational expectations" framework. Needless to say, this formalization of a definite skew within investors' preferences had a pervasive impact upon the direction of research in Behavioral Finance.[33]

---

[30] For a link between these two fields of thought, refer to references on "Experimental Game Theory" – for example, see Gintis (2000).

[31] The obvious exception here is McFadden (1998) who went part of the way by studying the link between the psychological concept of "perception" and information. He even went as far as to categorize "information" into different forms. Alas, this analysis still falls a long way short of an understanding of the true biological-like microfoundation elements of what comes to constitute "information" to investors – as will be discussed in Chapter 4.

[32] See H. Shefrin (2000) *Beyond Greed and Fear: Understanding Behavioral Finance and the Psychology of Investing*, p. 3.

[33] In short, Kahneman and Tversky's (1979) work highlighted the potential for investors to attach nonlinear decision weights to certain outcomes. For example, investors may attach a larger decision weight to a large loss than a small loss and such weights may not reflect the true probabilities of specific events occurring (i.e., investors display a nonlinear preference set illustrating an "irrational" fear toward large

But were Kahneman and Tversky truly the founding fathers of the field of Behavioral Finance? This may be a polemical question but it should be recognized that the applicability of full economic "rationality" as an appropriate assumption for all market participants at all points in time (and the subsequent potential for market failure) has been an age-old debate in the economics field.[34] Indeed, we would argue that John Maynard Keynes was in fact one of the first Behavioral Finance theorists – at least Keynes was one of the first economists to popularize the analysis of investor misjudgment.[35] Keynes' interpretations of herd behavior, investor myopia and the "greater fool" theory of investing are now legendary. One frequently cited quote in modern Behavioral Finance texts is Keynes' beauty pageant analogy

> professional investment may be likened to those newspaper competitions in which competitors have to pick out the six prettiest faces from a hundred photographs. . . . It is not the case of choosing those which, to the best of one's judgment, are really the prettiest, nor even those which average opinion genuinely thinks the prettiest. We have reached the third degree where we devote our intelligences to anticipating what average opinion expects average opinion to be. (Keynes, 1936, p. 156)[36]

As will be explained in the following section, it is possible to formalize this type of iterative decision-making process (if only at the metaphysical level) using the principles of Game Theory.[37] Indeed, our Evolutionary Finance purview takes such a formalization of the investor's time dependent assessment of information (including the implications of competing investors actions) to an even higher level by examining all information inputs – technical, fundamental, economic and political – along with this information's strategic behavior inference. Certainly, we would favor Keynes' gradual/adaptive approach toward information processing rather than the

losses). This insight has permeated an entire range of research from explaining volatility smiles to illustrating various deficiencies in standard Markowitz mean-variance portfolio theory.

[34] See Caldwell (1986) for an historical overview.

[35] Other greats need to be mentioned here as well such as Marx and Schumpeter in their analysis of the links between the shortfalls of capitalism (in particular, the painful consequences of the downside of investor's, "irrational exuberance") and business cycles. See Marx – English translation – (1954) and Schumpeter (1939).

[36] That said, it was Descartes not Keynes who coined the phrase "animal spirits." Keynes simply popularized the notion within the field of economics and finance. Rene Descartes used the term to describe how humans come to interpret information in an emotional fashion – in much the same manner that Keynes used the term.

[37] Yet another field of thought awarded the prestigious Nobel Laureate in Economics – this time in 1994 to Messrs Harsanyi, Nash and Selten for their work on non-cooperative games.

instantaneous adjustment process by omnipotent investors championed by the traditionalist finance school of thought.

That said, before progressing onto the theory of games (especially Evolutionary Games) to explain such iterative decision-taking procedures by investors we should at first ask the question – is there more we can say about the "new" view challenge represented by the field of Behavioral Finance as a rebuttal to the perfectly rational, homogenous investor norm championed by the traditionalist finance school of thought?[38] The good news is that – yes we can. In particular, Behavioral Finance models come under a variety of interesting names and describe a variety of interesting behaviors but basically they can be broken down into either:

- the fallibility on behalf of investors since they adopt too simplistic a – deterministic – rule of thumb for governing their investment decisions given the presence of limited information (and, somewhat in a related manner we would argue, the limited ability of particular individuals to interpret information);[39] or
- the existence of nonprofit maximizing "irrational" behavior thanks to the presence of heterogeneous utility functions. For example, Camerer *et al.* (1997) illustrated New York taxi cab drivers proclivity to "clock-off" on days of high activity – especially rainy days – which may appear "irrational" to the outside observer under profit maximization principles but was perfectly "rational" from a work/leisure preference standpoint for the taxi drivers.[40]

Certainly the first category of Behavioral Finance critique is of interest to us given our biological-like molecular "information building" approach. Interestingly from a Behavioral Finance perspective, this all-too-human investor fallibility of using a too simple heuristic rule of thumb falls under the

---

[38] This perfectly rational, homogenous utility function null hypothesis is sometimes defined as Expected Utility Theory (EUT) as popularized by von Neuman and Morgenstern (1944). See Levy *et al.* (2000) for a critique of EUT along with a comprehensive discussion of deterministic departures from EUT principles.
[39] Note the presence of the terminology "deterministic" in this category. It is the deterministic microfoundations of such heuristic rules of thumb that effectively forms an intellectual bridge between Behavioral Finance type thinking and the Complexity Theory models presented in the previous section – a framework built upon in the "new" view challenge no. 3 models of artificial markets presented in the next section.
[40] However, the identification of heterogeneous utility functions is not without its pitfalls. Mertens and Zamir (1985) have illustrated differences in prior to what can be argued is "public information" requires the inclusion of as many "agents" as there are heterogeneous utility sets in a private information game. This increases the complexity of equilibrium search considerably.

"bounded rationality" umbrella.[41] Bounded rationality implies that even if one were a truly "rational" investor, one can only make decisions on the basis of what one already knows – or in modeling jargon, the information-set available at time t. Hence, the "bounded" component to bounded rationality.[42] With the benefit of 20/20 hindsight, such decisions often prove to be "irrational" given the advantage of the full (historic) information-set, but this is the advantage of looking at decision-making in the *ex post* rather than the *ex ante*. Indeed, this is why we insist upon placing inverted commas when using the phraseology "rational" or "irrational" in this book simply because such subjective judgments do not always reflect an investor's mental health but rather the limitations of human knowledge. In Chapter 4 we explicitly model this upper threshold on a human being's information absorption capabilities in response to the ever-expanding stock of information that is continually thrust upon them. From a purely human point of view, how do investors cope with this "information overload"? Behavioral Finance tells us that in a world where time and cognitive resources are limited, human beings aim to simplify information by adopting heuristic "rules of thumb" – see Simon (1956).[43] It is when these information management devices known as heuristic "rules of thumb" break down that many of the observed "irrational" behaviors under the Behavioral Finance umbrella actually manifest. But as stated earlier, making such judgments often relies upon a good dose of "told you so" hindsight – which we might add, does nothing to endear the academic cause to the practitioner community who frustratingly must make such difficult (limited information) investment decisions on a regular basis.

In essence, what we are referring to here is a lack of "perfect foresight" on behalf of investors. That said, we would extend the categorization to include not only the mistakes investors make from failing to have all the necessary information (including future information) to produce the correct investment decision, but also the fallibility on behalf of investors stemming from their inability to correctly *understand* the complexity of the existing information-set as well. As we will demonstrate in Chapters 5 and 6, each and every information-set possesses Complex nonlinear interrelationships between its various components. It is well nigh impossible for the average

---

[41] Sometimes "bounded rationality" is referred to as "quasi-rationality" – see Black (1986) and Thaler (1994).
[42] For a detailed exposition of the concept of bounded rationality, refer to Simon (1982). But note, there is a difference between "bounded rationality" and "bounded willpower" as outlined by Mullainathan and Thaler (2000) – they also add a third condition of "bounded self-interest" to describe the lack of altruism sometimes displayed by economic agents.
[43] Kahneman and Tversky (1979) further developed the key concept of "heuristics" – simple rules of thumb that investors use to filter information. They identified "availability" heuristics as probabilities pulled out of recent memory and "representative" heuristics as analogous inferences from the past.

investor to understand (and more importantly predict) these dynamics at each and every point in time given that up until now, a formal model of the actual building blocks of information (and more importantly, how these building blocks interrelate) has been notable by its absence from finance theory. Ergo, it stands to reason that investors should be fallible ("bounded" if you like) in their interpretation of the type of Complex nonlinear dynamics presented in the previous sections when it comes to each and every information byte's arrival simply because they do not possess the correct framework to actually understand the full consequences of this information. The best investors can reasonably hope for under such circumstances is to form reasonably limited projections as to asset prices based upon relatively simple heuristic "rules of thumb" – a tact supported by the Behavioral Finance approach. Hopefully our Evolutionary Finance framework will leave investors a little bit better informed as to the true inter-linkages between various information signals which should (as a consequence) enable a higher degree of sophistication within their decision-making. Fingers crossed, this is the charter of our book.

But what of the more quirky/interesting behavior under the "Behavioral Finance" banner that typically captures the imagination of the popular media? What are some examples of more "aberrant" investor behavior that has made the Behavioral Finance school of thought the darling of the weekend financial press? Our survey from the literature in the box is far from exhaustive but hopefully it will provide readers with at least some guidelines as to what are the main investor fallibilities presented under the Behavioral Finance banner.[44]

A Behavioral Finance "Investor Fallibility" list

Regret and cognitive dissonance. This type of investor behavior represents a classic case of "living in denial." The psychology of "regret" implies that investors may be reticent to crystalize a loss due to the emotional pain of being wrong. As a consequence, they have a tendency to defer selling assets that have depreciated significantly in value or alternatively, sell assets early so as not to run the risk of encountering such a loss – thereby avoiding "regret."[45] If such behavior is taken to the extreme, "cognitive dissonance" occurs in the sense that investors will purposefully avoid all news on a loss-making asset simply because of the emotional pain of regret.[46]

---

[44] For a nice overview of some of the more widely recognized Behavioral Finance models of investor behavior, see Shiller (1999) and Wärneryd (2001). For a practitioners standpoint, see Montier (2002).

[45] See Shefrin and Statman (1985), Ritov (1996) and Josephs *et al.* (1996).

[46] See for instance the classic Erlich *et al.* (1957) study of new car purchasers selectively avoiding reading car advertisements post purchase for fear of information of an erroneous decision on such a large expenditure item.

Anchoring. This facet of Behavioral Finance concerns the power of "suggestion."[47] Consistent with our analysis presented earlier, it concerns the use of deterministic "rules of thumb" but also highlights that in an environ where true uncertainty is pervasive (in the sense that history does not provide a reasonable guide as to probabilistic outcomes) then investors are prone to broad association in forming their investing "rules of thumb." It is this tendency for broad association that can generate errors from "anchoring" to the wrong relativity. For example, if all stocks in a given market appear expensive on a price/earnings (P/E) basis, then an investor may be willing to (erroneously) accept a higher P/E on a given stock than would otherwise be the case in forming their heuristic rule of thumb for investing.[48] This is a classic symptom of a "bubble" type mentality in equity markets.[49]

Mental compartmentalization. This refers to the tendency for investors to compartmentalize assets into specific groups based upon superficial attributes. The danger in this behavior is that investors fail to see the associative tendencies between the assets in the different groups – a common fallibility our Evolutionary Finance approach intends to overcome through our assessment of the informational impact of technical, fundamental, economic and political information upon *all* assets within a given mandate (thus highlighting the potential for inter-linkages between assets in response to given informational input). Hopefully, this should raise investor awareness as to the true complexity of the informational fabric but it is interesting that under the "mental compartmentalization" banner it is also argued that investors appear to display surprising reluctance to alter their overly simplistic classifications – even in the face of irrefutable evidence of the need for reclassification.[50] So while we may raise overall investor awareness, it does not necessarily imply that all investors will suddenly drop their information filtering mental compartmentalization rules. Hence the old maxim "you can take a horse to water, but can't necessarily make him drink." This potential to "live in denial" for those investors who choose to ignore the evidence that we provide as to the Complex nonlinear inter-linkages that sometimes exist between information and various asset performances will ultimately be to the benefit of those who do take on board our sophisticated approach. This is simply because the presence of "denial" investors implies there is plenty of opportunity for alpha-seeking for those who do take on board our sophisticated informational analysis. Hence yet another old maxim "to the lambs go the slaughter."

---

[47] See Tversky and Kahneman (1974).

[48] Another example of this is forward FX market prices. Gruen and Gizycki (1993) have illustrated that forward FX rates display considerable anchoring to present rates. This remains the case even though present prices have been a poor predictor of the future. While protagonists of the traditionalist school would argue this is a classic case of random walk principles at work, the fact that investors continue to use forward prices even though their forecasting ability record is appalling is indicative of the power of anchoring in highly uncertain markets.

[49] At times such erroneous anchoring may be driven by "new era" type thinking as identified by Kindleberger (1996) where, in deference the Minsky (1982), Kindleberger points out that the lead-up to a crisis is generally started "with a 'displacement', some exogenous shock to the macroeconomic system."

[50] See Edwards (1968).

Overconfidence, representative heuristic and over-/under-reaction. Though they are often confronted with failure, investors as a class have a tendency to express excessive confidence in their own judgments. This may be a Darwinian attribute of financial markets in that "only the confident (or foolhardy) survive."[51] However, one consequence of this phenomenon is that investors have a tendency to attach too high an emphasis on the probabilistic projections of their own heuristic "rules of thumb." This can create over-reaction or under-reaction to the release of seemingly innocuous financial information – again, a common investor response that our Evolutionary Finance approach intends to better understand and possibly even exploit in its applied mechanics demonstrated in Chapter 7. In fact, investor overconfidence is so pervasive that the "invincible syndrome" can apply to seemingly random information as well. For example, a coin-tossing experiment will generally reveal 6 heads in succession or 6 tails in succession over 100 flips. It is possible that a similar sequence of random informational events in the market can trigger overconfidence on behalf of investors when aligned with investors priors. It then comes as a rude shock when investors ultimately find out that their models are fallible – thus provoking over-reaction or under-reaction depending upon each investor's initial position.[52]

Magical thinking. Under conditions of uncertainty investors have a tendency to form associations between nondependent outcomes somewhat related to the above. As if by "magic" a particular outcome may become associated with a market event and then it is possible that the two outcomes can become (erroneously) associated within investor's minds and therefore their deterministic rules – even though there is not an economic or financial rationale for doing so. In many senses, investors are vulnerable to "conditioning" much like Pavlov's dogs. Here a classic experiment was conducted by B.F. Skinner (1948) who starved pigeons and then fed them small quantities of food at regular 15-second intervals.[53] The poor birds subsequently began to develop abnormal behavior since they interpreted whatever they were doing before the food arrived as the cause of the food's delivery. Obviously some people get a kick out of doing this sort of thing (although one doubts whether an eminent psychologist like Skinner did) but it does describe why some sports people are very superstitious about a certain club, racket or even routine they follow before they play. Investors too are vulnerable to such superstition-based conditioning – which explains why certain proprietary traders can become very aggressive if the cleaners have inadvertently moved items around on their desk.

As can be seen from the earlier discussion, once one makes a break from the omnipotent/perfectly rational investor championed by the traditionalist finance school of thought, then all sorts of interesting investor behavior

---

[51] We are not alone in espousing such a Darwinian "solution" to the market – see Mullainathan and Thaler (2000). Indeed, to some extent, "overconfidence" and "regret" are somewhat linked. The need to protect one's self-esteem via avoidance of "regret" can be seen as a necessary precursor to overconfidence – see Joseph *et al.* (1996).

[52] See Kroll *et al.* (1988) and Barberis *et al.* (1997) for examples of this.

[53] See Skinner, B.F. (1948) "Superstition in the Pigeon" *Journal of Experimental Psychology*, 38, 168–72.

begins to present itself. We intend to leverage some of these principles in our own Evolutionary Finance approach – especially in our applied analysis presented in Chapter 7 of this book. But be warned, the traditionalists are not taking the challenge lying down – as Fama's (1998) rebuttal of Behavioral Finance insights from the *applied* sense testifies. Obviously, upon reading the various traditionalist finance criticisms of the Behavioral Finance literature, one soon realizes that a common misconception is that Behavioral Finance simply describes the principles underlying the earlier identified pursuit of "noise trading" (see Chapter 2 for more details). However, one fact to note is that Behavioral Finance theorists invariably argue that the principles of human fallibility described within their field of thought are *systematic* in nature. This implies that one can systematically make profits by capitalizing upon such errors – thus negating the "noise trader" criticism by the traditionalist finance vigilantes.

In concluding, since Behavioral Finance effectively argues that investors are error prone because they adopt the wrong deterministic rules (they effectively are informationally constrained), then the best Behavioral Finance type of analysis will be that which identifies the rules that are currently enforced by the investment community. If these rules can be identified, then greater insight can be ascertained as to the potential for such rules to go wrong. Indeed, this is a key principle of the strategy selection component within our applied Evolutionary Finance approach (as identified in Chapter 7). In this particular chapter we identify how winning strategies "evolve" in a manner consistent with our biological-like molecular approach toward information as a whole. The optimum strategy is therefore one that identifies an overarching sequence of strategies that actually takes into account the failure of investor's deterministic trading rules. In essence, these deterministic trading rules themselves can be regarded as a competitor's "strategy" and, like in a game of chess, our objective is to develop a set of winning strategies that capitalize on the fallibilities of our competitor's strategies. Such over-arching time-variant "Evolutionary Stable Strategies" as they are known form an integral component of the Game Theoretic analysis we will present in Chapter 6. However, before we can even attempt to outline in greater detail the microfoundations of our biological-like informational approach toward the selection of Evolutionary Stable Strategies we must at first do justice to the "new" view encapsulated by "artificial" models of the markets. As we said in the introduction to this chapter, it is upon the shoulders of giants that we have been able to potentially see farther and certainly it has been the marriage of both the Complex nonlinear microfoundations and the deterministic/heuristic rules of thumb presented in Sections 3.1 and 3.2 respectively, that has enabled the study of artificial markets to flourish. Indeed, some of the recent modeling of artificial markets has assisted us in our own Evolutionary Finance interpretation of the way markets operate. So now, over to the recent studies on the principles of artificial markets – especially those focused on trading rule based, evolutionary principles.

3.3 "New" view challenge no. 3: Trading rules, evolutionary games and artificial markets

The beauty of "new" view challenge no. 3 is that basically it is a summation of the previous two sections. Effectively put "new" view challenge no. 1 and no. 2 together and you have the antecedents of a deterministic/trading rule based model of an artificial market. It's as simple as (challenge) $1 + 2 = 3$. For our purposes however, what we find particularly appealing about this categorization of "new" view challenge models is their ability to provide a framework for how information is both generated and interpreted within "real world" financial markets – in obvious contradiction to the traditionalist finance thought paradigm. Still, as mentioned in the introduction, it should be emphasized that these studies continue to suffer from the fallibility of treating "information" as a generic homogenous concept – a shortfall our Evolutionary Finance technique aims to address in our own undertaking of artificial market modeling presented in Chapter 5. But before we can even begin to outline the broad attributes of such a model – and indeed, of artificial models of the market in general – we must at first lay some foundations by providing at least a cursory overview of Game Theory. This is because the principles of Game Theory – and in particular, a certain class of Game Theoretic models known as "Evolutionary Games" – reside at the very heart of this final (artificial market) categorization of "new" view challenges.[54]

So just what is "Game Theory" *per se*? Put simply, Game Theory is a mathematical representation of the strategic interaction between two or more "agents" – each attempting to maximize a specific utility function and each imparting information to the other through their actions. In many respects, as our analogy earlier has already intimated, Game Theory is like a mathematical formalization of the game of chess. Instrumental in Game Theory is the extrapolation of not only the actions of oneself but also the actions of others. In short, one needs to "forecast" an opponent's reactions to various contingencies. Here a military perspective comes to mind and indeed many elements of Game Theory have found their way onto various battle plans

---

[54] Furthermore, we aim to leverage some of the Game Theoretic principles presented here in considerably more detail in Chapter 6 – where our deterministic informationally driven foundations of the marketplace are put to work in generalizing an Evolutionary Distributional Form (EDF) for asset prices. Since the prerequisite knowledge required in Chapter 6 with regard to Game Theoretic principles is quite high, we will spend a little longer outlining the "basics" in this section than we have in outlining the theoretical underpinnings of the other "new" view challenges that have been presented in the earlier sections of Chapter 3.

over the course of the latter stages of the twentieth century – and those of the early twenty-first century as well.

As for a brief history of Game Theory, it is generally acknowledged that the popularization of Game Theory within academia arose with publication of von Neumann's and Morgenstern's *Theory of Games and Economic Behavior* (1944). This seminal publication represented the first true *comprehensive* formalization of the subject.[55] Since then, there have been innumerable additional developments within the field itself – thereby deepening and widening the applicability of the overall framework across a variety of alternate academic realms of thought. Indeed, the uptake of Game Theory as an academic pursuit in its own right has now risen to such a level where most postgraduate students in the fields of Finance, Economics or even Mathematics inevitably find a course in pure Game Theory a recommended (if not a compulsory) component of their curriculum – much to the chagrin of some of those within our younger generation. While by no means pertaining to be an exhaustive overview, some of the notable developments within this field since the publication of the *Theory of Games and Economic Behavior* include:

- 1950: the first formal experiment of the game known as "Prisoner's Dilemma" – explored in detail a little later in this section – by Melvin Dresher and Merrill Flood of the Rand Corporation;[56]
- 1950–3: John Nash – a leading figure in the field – publishes four seminal papers that form the basis of the now famous "Nash Equilibrium" – again, explored in detail a little later in this section – and the lesser known axiomatic bargaining theory and Nash program solution to cooperative games;[57]
- 1953: the publication of Kuhn's "Extensive Games and the Problem of Information" sets out a framework that allows a more comprehensive assessment of true Game Theoretic dynamics by highlighting the order

---

[55] That said, it is noteworthy that there have been a number of scattered accounts of the mathematical solution to various games throughout mankind's written history. For example, the Babylonian Talmud (a compilation of Jewish ancient law and tradition set down during the first five centuries AD) sets out the preconditions for a particular "marriage contract problem" where the division of a man's estate among three wives is discussed in considerable detail and a Game Theoretic solution is recommended – see Aumann and Maschler (1985). Indeed, as early as 1713, James Waldegrave provided the first known solution to a particular type of two-person game known as a "minimax mixed strategy" – see Kuhn (1968). In 1838 the brilliant mathematician Cournot discussed preconditions for the solution of a particular duopoly problem that represented a restricted form of Nash Equilibrium – see Cournot (1838). And Zermelo (1913) presented a mathematical formalization of the game of chess.

[56] For the original exposition of *Prisoner's Dilemma*, refer to Tucker, A.W. (1950) "On Jargon: The Prisoner's Dilemma" *UMAP Journal*, 1, 101.

[57] See Nash (1950a, b), (1951) and (1953).

in which decisions are made by various players and the information they possess when they make their decision;[58]

- 1954: Game Theory is applied to the field of political science by Shapley and Shubik – indeed, the Shapley value is ultimately used to determine the relative power of the members of the UN Security Council.[59]
- 1959: the notion of a Strong Equilibrium introduced by Aumann;[60]
- 1960: cooperative Game Theory is made more widely applicable thanks to the development of nontransferable utility games;[61]
- 1961: Lewontin applies Game Theory as a means of describing evolutionary biology;[62]
- 1966: Aumann and Maschler develop the theory of infinitely repeated games with incomplete information;[63]
- 1966: Harsanyi develops the now, most commonly used definition of cooperative versus non-cooperative games – a game is *cooperative* if commitments are fully binding and enforceable; a game is *non-cooperative* if commitments are not enforceable;[64]
- 1967–8: over the course of three papers, Harsanyi constructs the theory of games of incomplete information that ultimately lays the foundation for information economics to become one of the major bridges between "traditional" economics and Game Theory;[65]
- 1972: John Maynard-Smith develops the concept of Evolutionarily Stable Strategy which presents the foundations for Evolutionary Game Theory;[66] and
- 1994: Nobel Prize in Economic Science awarded to John Nash, John C. Harsanyi and Reinhard Selten for their contributions to Game Theory.

The areas that we are most interested in from this list given our evolutionary approach toward the way in which information binds together in molecular form within financial markets is the applicability of Game Theory to evolution itself – especially Lewontin's and John Maynard-Smith's seminal work. It is here that the earlier mentioned principle of Evolutionary Stable Strategies were developed – a concept we will draw upon heavily in Chapter 6. However, before embarking on a discussion of these rather complex principles (and an explanation as to why such games should provide part of the foundations

---

[58] See Kuhn (1953).
[59] See Shapley and Shubik (1954).
[60] See Aumann (1959).
[61] See Aumann and Peleg (1960).
[62] See Lewontin (1961).
[63] See Aumann and Maschler (1966).
[64] See Harsanyi (1966).
[65] See Harsanyi (1967–8).
[66] See Maynard-Smith (1972).

of our own "new" view challenge) we must at first provide a primer on the mechanics of Game Theory in its most simple form.

Game Theory: The Basics

There is no means by which we can effectively summarize the full gamut of Game Theoretic principles in an entire book – let alone a singular section within a chapter. Given this obvious limitation, for a more comprehensive overview of some of the principles of Game Theory, readers are referred to such excellent texts as Osborne and Rubinstein (1994), Friedman (1991), or even the seminal publication on Game Theory itself, von Neumann's and Morgenstern's *Theory of Games and Economic Behavior* (1944).

Instead, the purpose of this book is to focus on the primacy of the fabric of information – its biological-like molecular production, agglomeration and interpretation by investing agents as information *evolves* from private to public realms. As a consequence, we will only provide a cursory overview of what we feel is necessary from Game Theory to explain the basic principles of the artificial market stratum of this particular "new" view challenge and our own evolutionary (artificial) model of the marketplace presented in Chapter 5 – and in particular, the microfoundations of this model presented in Chapter 6. Principally therefore, our focus will remain on the criteria that we feel are important in understanding some of the deterministic microfoundations of market transactions as it is this "core" element of Game Theory that we will explore in detail in this section and which forms an integral part of our own artificial model of the marketplace presented in Chapter 5 and its microfoundations which are presented in Chapter 6.

To start with, it should be pointed out that in terms of a formal framework, Game Theory is as simple as it is elegant as a tool of analysis. In essence, most games can be represented concisely with the aid of a simple payoff table. The payoff table (in general) represents the actions of a given player along the vertical axis and an opposing player across the horizontal axis. Different cells of the table represent different choices made on behalf of the respective players. The numbers occupying each cell represent the "payoff" that each player receives – as implied by the combined choice of both participants.

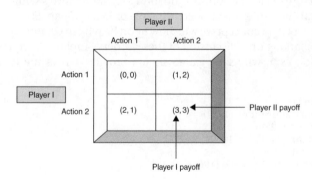

Figure 3.6 An example of a typical two-person game

In Figure 3.6, Player I's payoff appears as the first number of each pair, Player II's as the second. Measuring the "payoff" sometimes is difficult in a monetary sense and this makes it necessary in some instances to invoke the concept of utility. "Utils" represent the amount of satisfaction each player receives from the outcome of a particular game when a dollar amount (or pounds, shekels, infantry or whatever the chosen numeraire may be) is not an appropriate unit of measurement. Utility theory was yet another fruitful product of von Neumann's and Morgenstern's endeavors – at least in its *cardinal* form with the *ordinal* concept earlier developed by Paul Samuelson (1938) in his explanation of Revealed Preference Theory.[67] Indeed the formalization of the utility function constitutes a substantial first stage component in many studies within both the microeconomics and macroeconomics field – and indeed, forms the basis of the initial equation in our own evolutionary model of the marketplace presented in Chapter 5.[68]

So now we know how the game is played and what the game actually looks like but the next question to ask is when does the game actually stop? The game effectively stops when both respective players in essence decide (generally, but not always, independently of each other) that they have achieved the optimum outcome. This occurs when no player would be advantaged by making an alternate decision based upon the given array of payoffs – remembering that the magnitude of such payoffs relates in part to the other player's actions. This equilibrium may not represent the *maximum* possible payoff that a given player could potentially achieve from the array of choices, but it is the highest risk/reward payoff that they can achieve – with "risk" effectively constituting leaving oneself open to an opponent choosing a course of action that would make you considerably worse-off had you not chosen the equilibrium strategy.[69] This "logic exhaustion" type of outcome for both players is known as "Nash Equilibrium" after the solution by the "beautiful minded" John Nash – and espoused across his four seminal papers released over the period 1950–3.[70]

In our example, the Nash Equilibrium is represented by (3,3) – the lower right hand cell. Why? Assuming Player I moves first, their optimal strategy is to pursue Action 2 with a payoff of 3. As it happens, Player II also receives their maximal payoff by pursuing Action 2 in such instances – a payoff of 3. Since Action 2 is still the most preferred action for Player II even if Player I made a mistake and opted for Action 1, we can say Action 1 is *strictly dominated* by Action 2 for Player II.

[67] See Samuelson, P. (1938) "A Note on the Pure Theory of Consumers Behavior" Economica, 5, 61–71. For an exposition on cardinal utility – particularly with reference to its applications for games – refer to von Neumann, J. and Morgenstern, O. (1944), *Theory of Games and Economic Behavior*, Princeton University Press, Princeton, NJ.
[68] For a comprehensive overview of utility theory (both past and present) refer to Levy *et al.* (2000).
[69] As intimated earlier, at times the calculation of one's competitors' actions may require the formalization of what are known as "beliefs" – or "forecasts" as to how competitors would respond to particular information. Such "beliefs"/"forecasts" are at times conditioned by previous experience in "repeated games."
[70] The term "beautiful mind" refers to biography by S. Nasur on John Nash – see Nasur, S. (1994) *A Beautiful Mind: A Biography of John Forbes Nash Jr., Winner of the Nobel Prize for Economics 1994*, Simon & Schuster, New York.

A strictly dominated action never represents a Nash equilibrium since a Player could do better by switching to the dominating strategy. If the payoff for Action 2 were to be *at least* as great as that of Action 1 – irrespective of the choice of Player I – then Action 1 is said to be *weakly dominated* by Action 2 for Player II. It is the iterated elimination of weakly dominated strategies for both Players that results in the identification of a unique Nash equilibrium in our example.[71]

So what are the "basics" when it comes to mathematically formalizing a game? In general, the following three specifications are made – either explicitly or implicitly – before any mathematical Game Theoretic model is extended into a specific task.

- The game is played by a set of players $Z = 1, 2, \ldots, z$.
- Much of the actual "game" itself is represented by a set of St strategies where players get to choose from a strategy range $St = (st_1, \ldots, st_n)$ where $st \in St$. This is known as the strategy profile of the game.
- The key factor dictating the choice of strategy is the payoff function π – so $\pi_Ƀ$ represents a player's payoff for choosing strategy Ƀ. In aggregate, this choice of strategies ultimately results in the gravitation toward a (hopefully) optimum solution for the players involved is represented by $St \to T$ (where T represents the terminal node of the game).

The benefit from this considerable investment in notation is that concepts such as Nash equilibrium become more straightforward – at least to the mathematically inclined. For example, a Nash Equilibrium strategy st^* can be defined in the relatively simple terms that for each player $Z = 1, 2, \ldots, z$ and every strategy $st \in St$, a payoff condition for Nash equilibrium is

$$\pi_Ƀ(st^*) \geq \pi_Ƀ(st_Ч^*, st_Ƀ), \tag{3.5}$$

given that we assume our stylized Player I opts for strategy choice Ƀ while our stylized Player II opts for strategy choice Ч. In other words, neither Player can profitably deviate from their chosen strategy given the actions of the other Player. Simple and concise.[72]

So why do the principles of Game Theory necessarily form the foundations for the artificial market "new" view challenge? Well, the short answer is that they don't – at least in their purest sense. Provided both players are fully "rational" and fully informed of each other's past and *future* actions then Game Theoretic models are generally supportive of the traditionalist finance

---

[71] Indeed in our example, the iterated elimination of strictly *dominated strategies* allows for the identification of a unique *dominant strategy equilibrium*. Note: not all Nash equilibria are necessarily unique.

[72] Note: it was the formal proof of the existence of a (not necessarily unique) "Nash equilibrium" for a mixed strategy that helped bring John Nash into notoriety – see Nash (1950).

thought paradigm that all market participants instantaneously and rationally respond to the arrival of (random) information. In a sense, "rationality" dictates some simple and predictable premises for individual action within a Game Theoretic framework that can be used to augment the traditionalist cause. For example, we generally prefer more relative to less (at least up to a point), we (hopefully) are consistent in our choices both through time and across a wide array of possible alternatives, and we generally seek a balance between the cost of an action and the reward for that action. When married with perfect foresight, these rationality dictums have been used extensively by the traditionalists in furthering their particular purview as to how markets operate – and in so doing, using Game Theory as a convenient framework to formalize their views.

But hang-on, isn't Game Theory supposed to also provide a framework to *challenge* the reigning Strong EMH/traditionalist finance orthodoxy as well? In short, it does both. Just as Game Theory provides a nice theoretical framework to support the traditionalist thought paradigm, so too can the sword of logic be turned back on the traditionalists. By dismantling the edifice of rationalist/fully informed agent assumptions, "new" view protagonists were able to wrap the Game Theoretic framework around the "bounded rationality" behaviorist/psychological based studies identified in the previous section – such was the birthplace of the earlier cited relatively new field of experimental economics and (somewhat synonymously) the artificial models of markets that will be discussed in this section. However, before describing this new view challenge in detail we must at first pause for a moment to explore some of the more interesting *applied* implications of assuming full rationality within a Game Theoretic framework (the box below) before then going on to outline some of the principles underlying intertemporal models of Game Theoretic behavior – particularly as they apply to Evolutionary Games.

The applied pitfalls of "Rationality" within a Game Theoretic framework

The extremes of so-called "rationality" caused great consternation during the Cold War. It was argued that the likely incapacity of a victim of a nuclear holocaust to launch a retaliatory strike (and indeed, the fact that such a strike would not be in their interest to do so as it would limit the habitable living space for the human race to survive) resulted in the "rational" solution to the ensuing armaments build-up being a "first strike" against one's opponent – knowing full well that a "rational" opponent would not retaliate.

Such a "rational" recommendation from the Game Theoretic models of the nuclear arms race produced some rather bizarre results. For example, it has been argued that President Nixon had the CIA try to convince the Russians that he was insane, so that they'd believe that he'd launch a retaliatory strike even when it was no longer in his interest to do so. Similarly, it has been argued that the Soviet KGB leaked fabricated medical reports regarding Brezhnev's senility with the same end in mind. Indeed, the infamous "doomsday device" was developed to break this deadlock. This strategic solution equipped a worldwide fleet of submarines with enough missiles to destroy the USSR, and arranged their communications technology in such a way that the US President could not be *sure* he would be able to reach the submarines and cancel their orders to attack should any Soviet missile cross the radar "trigger line" – thereby removing a first strike advantage. This unfortunate outcome had the potential to occur even though a Soviet missile could have been launched by "accident."

While this may have eased the concerns of those familiar with the "rational" Game Theoretic solution to the initial armaments build-up problem, it certainly did not appease the concerns of the general public. Hence, Stanley Kubrick's classic film *Dr. Strangelove*, where the world is destroyed because the Russians build a similar doomsday machine and keep its incapacitation details totally secret from the Kremlin. So when a mad American colonel launches missiles at Russia there is little the respective heads of State can do but watch in dismay as the world is blown up. Such are the logical pitfalls of assuming "rationality" – at least in the applied sense anyway.

Now that we have covered the basics of Game Theory and of rationality itself, we next turn our attention to Game Theoretic behavior from an intertemporal standpoint. Adding such a "time dimension" is crucial to forming an understanding of the *evolution* of information (and its concurrent relationship to this particular artificial market "new" view challenge). In doing so, our attention is drawn not so much to each players *decision* but instead, their *strategy*. In short, how do players eventually gravitate toward a Nash Equilibrium position – assuming one exists? What are the informational criteria necessary for this to take place? Divining this is the secret as to how the market actually does evolve through time toward (hopefully) some steady state solution. To examine this intertemporal aspect of Game Theory in more detail we need to break away from the concise representation of games – or "normal" form games as they are known – and focus instead on a framework that allows each sequential decision by the respective players to be displayed.[73] Such a framework is known as the "extensive" form of Game Theory and the preferred tool of analysis for its representation is a "decision tree" structure (see Figure 3.7).

In the Figure 3.7 decision tree each node represents a different stage along the sequential decision-making process by the respective players as they (hopefully) converge toward a steady-state Game Theoretic solution. Players can converge to such a solution either via a "pure strategy" – seeking a *single*

---

[73] This does not imply that normal form games cannot be dynamic in the sense that they are unable to fall under the umbrella of "repeated games," but rather normal form games represent a concise summary of the end point of a game rather than giving a complete description of how this end point is actually achieved.

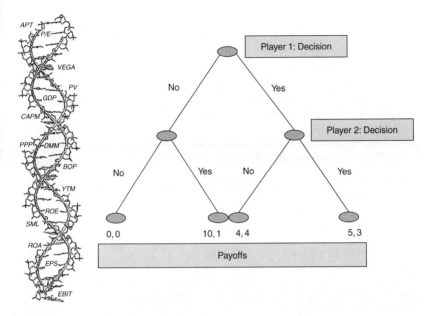

Figure 3.7   An example of an extensive form game

course of action to maximize risk/reward – or a "mixed strategy" – choosing from an array of potential risk/reward optimal strategies (each dependent upon some pre-assigned probability).[74] Under such a schemata, the respective payoff functions for each player are displayed at each stage of the game (below each node). Rational players choose the course of action that will result in the highest possible risk/reward payoff over the life of the game. This (at times unique) solution represents Nash Equilibrium.[75] The beauty for our purposes of such a structure is that it allows for the visual representation of the dynamic sequence of steps players can take toward the (hopefully concisely summarized) solution – in other words, it has the potential to display

[74] At this stage it is important to highlight to readers the clear distinction between *sequential* and *simultaneous* games. A sequential game occurs where each player knows the action of their competitor *before* they make their decision – in a simultaneous game this is not the case. Note: it is not the temporal order of events that is important. Rather, it is the flow of information that determines whether a game is classified as sequential or simultaneous.

[75] Notice we used the word "unique" to highlight that a Nash Equilibrium solution contrasts that of a "strictly dominant" solution in the sense that Nash Equilibrium refers to the highest possible risk/reward payoff over the entire life of the game and not just to the highest risk/reward tradeoff at a particular stage of the game.

the evolutionary decision path investors take in making their choices. Mathematically, such calculations have obvious Bayesian probability connotations in the sense that any payoff at a given node is contingent upon the array of probabilities determining the outcome of earlier possible choices. This is a concept often seized upon in projecting the potential array of solutions under many artificial market models and indeed our own microfoundations of an "artificial market" analysis presented in Chapters 5 and 6.

The difference between mixed and pure strategy games

Mixed and pure strategy games can be easily distinguished within the concise mathematical framework we presented earlier. For instance, suppose a player has to choose from *pure strategies* $st_1, \ldots, st_n$ then a *mixed strategy* can be represented by applying a probability distribution over $st_1, \ldots, st_n$ so that

$$\Pr = pr_1 st_1 + \cdots + pr_n st_n, \tag{3.6}$$

where $pr_1, \ldots, pr_n$ are all nonnegative and $\sum_1^n pr_{\eth} = 1$ where $\eth$ denotes the player's chosen strategy. If all the $pr_{\eth}$'s are zero except one (e.g., $pr_{\eth} = 1$) then Pr effectively represents a *pure strategy*.

Assuming players choices are made independently of each other, then the payoff function for a mixed strategy game can be defined as

$$\pi(\Pr) = \sum_{st_1 \in St_1} \cdots \sum_{st_n \in St_n} pr_{st_1} pr_{st_2}, \ldots, pr_{st_n} \pi(st_1, \ldots, st_n), \tag{3.7}$$

where $st_1 \in St_1, \ldots, st_n \in St_n$ represents the set of all strategies pursued.

Still, extensive form games (even taking into account the mixed versus pure strategy distinction) represent but the "tip of the iceberg" when it comes to intertemporal analysis within Game Theory. Of particular interest to most "new" view artificial market protagonists are not just "one-shot" games where each player's strategy does not extend beyond the terminal nodes of the particular game in play, but also the concept of "repeated games." Repeated games occur where the same situation is revisited by existing players thus allowing for the development of elaborate strategies where *learning* on a player's behalf can condition their "beliefs" of what a competitor may do in response to a particular information signal. It is here that the concept of limited (rather than perfect) information comes into its own and it is here that a virtual cornucopia of interesting "bounded rationality" manifests.

For example, in the case of the ubiquitous "Prisoners Dilemma" game, a repetitious scenario would engender a rational course of action as a "non-confession" response on behalf of both prisoners. In short, both the prisoners would *learn* that colluding is the best joint-strategy and (hopefully) develop an element of mutual trust.[76] This is totally at odds with the static game outcome that is so often cited in many an undergraduate textbook. In a sense, a repeated game allows for the acquisition of more information through the observance of a competitor's action as they respond to actions by oneself – and concurrent payoffs associated with each joint action. Such learning allows players to formulate "reputations" for their competitors that ultimately helps in future decision-making. As a consequence, a repeated game – via its allowance for repetition – allows limited information players to gradually "learn" much of the information that perfect foresight individuals are assumed to possess at each and every moment in time.

This convenient framework thus represented a perfect theoretical foundation for contrasting the instantaneous perfect foresight investment outcomes assumed under the traditionalist finance theoretical framework vis-à-vis the more realistic "bounded" informational world where there is an obvious dichotomy between public and private information. As a consequence, a diverse array of interesting strategic behavior was thus able to be explored as information itself (via one's actions) effectively became a bargaining chip in a repetitive game. We will use precisely such strategic reckoning in our modeling of the analyst community's choice of topic for information byte production as we develop the microfoundations of our own evolutionary model of the marketplace in Chapter 6.[77] For the moment however, let us provide a little more background on the concept of "Prisoner's Dilemma."

A ubiquitous game – Prisoner's dilemma

Suppose that the police have arrested two people whom they suspect have committed a crime. A detective makes the following offer to each prisoner: If you will confess, implicating your partner and providing police with further information about the crime, and *your partner does not confess*, then you'll go free and he'll get four years. If you both confess, you'll each get five years. If neither of you confess, then you'll each get two years for the crime.

Assuming the prisoners are kept apart and cannot collude, there is an overwhelming desire to "rat" on one's accomplice in the sense of being the first to confess and thereby getting off "Scot-free." The "rational" decision for both prisoners is therefore to confess (hoping their partner does not) but in doing so they will both

---

[76] For an example of such an "adaptive" version of Prisoner's Dilemma, refer to Kreps *et al.* (1982) or Holland (1995).

[77] That said, when exploring the microfoundations of analyst behavior in Chapter 6, we opt for what are known as "Evolutionary Games" where reputations are built not through a sequential revisiting of the "game" between rival analysts but rather reputations are built via the format of the information bytes that analysts produce. We will discuss the concept of Evolutionary Games a little later in this section.

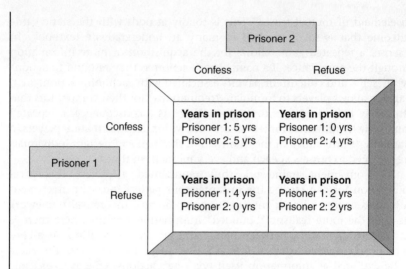

Figure 3.8   The prisoners dilemma conundrum

suffer the maximum penalty – represented in the upper left cell – of five years in prison. The "deal" works well for the police since in this way they get the maximum information about the crime (both prisoners offer additional information as they confess) and both the villains end up behind bars for the maximum duration. A "win/win" for the police, a "lose/lose" for the criminals.

But what of our earlier mentioned "Evolutionary Games." Evolutionary Games are basically "repeated games" with the added condition that the actual rules of the game are dynamically changing – or *evolving* – through time. Interest in this field first gained recognition with the pioneering work of John Maynard-Smith (1972). However, it was not until with the publication of his latter book – in 1982 – that the concept appeared to garner wider appeal. Maynard-Smith's work focused on using Game Theory as a means for explaining evolutionary biology – a nice fit with our own pioneering methodology. In short, he identified genes as "players" in a repetitive game – a serendipitous insight which has strong parallels to the micro-component informational modelling framework we will present in the following chapter.[78] For Maynard-Smith, a successful strategy under Game Theoretic conditions where the rules of the game are constantly

[78] See Maynard-Smith, J. (1972) "Game Theory and the Evolution of Fighting" *On Evolution (John Maynard-Smith)*, in pp. 8–28, Edinburgh University Press, Edinburgh; and Maynard Smith, J. (1982) *Evolution and the Theory of Games*, Cambridge University Press, Cambridge.

changing through time can be defined as that course of action that had the highest potential for replication – again, a seminal insight that draws upon Darwinian principles of natural selection and indeed will be used extensively in our Evolutionary Stable Strategy search presented in our distributional form analysis (Chapter 6) and our *applied* analysis (Chapter 7).

But what precisely do we mean by "highest potential for replication"? According to Maynard-Smith, a steady-state – or, using the correct terminology, an "Evolutionary Stable Strategy" – would occur when no individual gene (Player I) playing one strategy could improve its reproductive fitness by switching to an alternate strategy and no "mutant" gene (Player II) playing a different strategy could establish itself ("invade") the existing population.[79] Subsequently, this outcome represents the best over-arching strategy for the Player I gene simply because its reproductive fitness is greatest by following the prescribed Evolutionary Stable Strategy set of decision rules. Supplanting "genes" with analyst information bytes, memetic strings of information or even entire trading strategies allows us to extend Maynard-Smith's powerful equilibrium concept into an array of intertemporal (evolutionary) decision-making within our own evolutionary model of the market where the rules of the game are constantly changing. Indeed, it should be recognized that up until the formalization of Maynard-Smith's equilibrium mindset, what could be classified as "new" view challenge no. 3 could only be regarded as being in its infancy simply because try as they might, "new" view modelers of games of limited information (like ourselves) ran into intractable problems at the intertemporal/repetitious game level. This was because, as stated earlier, from a theoretical perspective "learning" by investors in response to new information implied the rules of the game were constantly changing across time. The traditionalists thus had the upper hand when it came to intertemporal modeling of strategic-like investment decision-making behavior along the lines extensively documented within Merton (1992). All this changed with Maynard-Smith's insights (especially those in regard to equilibrium) and "new" view challenge no. 3 was thus able to fulfill its destiny in becoming a veritable hotbed of academic endeavor – much to the dismay of some diehard traditionalists.

---

[79] *Non-kin altruism* – in the sense that an organism voluntarily lowers the expected fitness of a chosen strategy to promote the success of another organism – is an interesting contradiction to this rule of "success." At times this appears to occur in nature and is at odds with a pure Darwinian solution. *Reciprocal altruism* – where self-interested individuals adopt a cooperation stance in a repeated game to build a "reputation" that ultimately works in their favor – is better understood and occurs frequently in primates – see Trivers (1971), Byrne and Whiten (1988). The sacrifice of resources to promote the chances of survival for one's offspring is totally consistent with the concept of developing an *Evolutionary Stable Strategy* and occurs relatively routinely in nature – see Daly and Wilson (1983).

But were there any additional benefits? Indeed, there were. For starters, apart from the obvious Darwinian beauty of such a simple solution mechanism for the selection of an optimal evolutionary strategy, an added advantage of Maynard-Smith's concept of equilibrium was that it lowered the onus upon "rationality" as the dictum via which players make choices. In short, the pursuit of an Evolutionary Stable Strategy was a pure "survival of the fittest" type of solution mechanism – a beautiful result that guides much of our strategic modeling. Gone was the excess baggage of mapping utility functions in an endeavor to homogenize investor decision-making. Rather, a much simpler premise reigned supreme in Maynard-Smith's world – survival. It was the ability to survive to play another day that drove many of Maynard-Smith's outcomes and it was this, combined with the associated ability to allow for the presence of limited information, that ensured Evolutionary Game Theory a place at the theoretical high table in explaining some erstwhile perplexing investor behavior (in particular, behaviors of similar nature to those outlined in the previous Behavioral Finance section). It is for this very reason that we have seized upon this methodology in our strategic modeling presented in the ensuing chapters.

For example, a ubiquitous element in most market interchange is the presence of *asymmetric information* – that is, the information-set of buyer and seller is not identical.[80] Under such conditions strategies need to be developed that either:

1 leverage the information benefit one possesses; and/or
2 attempts to offset any informational deficiency one has by developing a particular strategy aimed at getting a better informed individual to "reveal their hand."

Many artificial models of markets are based upon attempting to explore the dynamics of such strategic informational exchange.[81] To do this, many artificial models of the market split the investment community into two "warring tribes" – sometimes labeled as "technicians" versus "fundamentalists." The model is then calibrated to observe the evolutionary phenomena of buy/sell interaction between the two contesting fields of thought as the

---

[80] As Malkiel (1996) argues, some elements within the field of research known as "technical analysis" are fundamentally based on the premise of asymmetric information. That said, it seems somewhat ironic that this more "realistic" interpretation of true market behavior – that of unevenly informed agents – should be labeled by the finance profession as the study of "market imperfection" when it is actually a more accurate reflection of the true state of private/public information dispersal at any given point of time.

[81] One of the most famous of which is the Santa Fe Artificial Stock Market Model so named after the august Santa Fe Institute situated in New Mexico in the USA – see Arthur *et al.* (1997).

comparative informational advantage of one group versus the other waxes and wanes in response to a series of (generally exogenous) informational shocks.[82] Indeed, the modeling of the "survivability" of particular trading rules under such conditions forms much of what can be categorized as "artificial market" studies *per se*. Even the term "Evolutionary Finance" has up until the publication of this book has been generally associated with a particular brand of such models and we indeed use such principles in our own microfoundation modeling presented in Chapter 6.[83] That said, as intimated earlier, we ultimately view this particular definition of "Evolutionary Finance" as limiting as it appears to be focused more on the practice of Genetic Algorithm/Genetic Programming techniques than our own "structured information" purview. In particular, we would argue that "Evolutionary Finance" as a thought paradigm should be motivated more by trying to better understand the biological-like molecular foundations of the way information is produced, agglomerated and interpreted within financial markets than on calibration studies alone. Calibration, while useful, is not an ends in itself and our own work attempts to "dig a little deeper" in the Evolutionary Finance genre by exploring the synergies between the way information is categorized and modeled in the biology/genetics sphere and the way the actual fabric of information can be better understood within the finance profession.

That said, it must be said that the work that has been done to date in the field of artificial markets – even that done under the "Evolutionary Finance" banner – has been extremely useful in dismantling some of the more outlandish belief structures perpetuated by the traditionalist finance school of thought. Thanks to this "new" view challenge no. 3 many microfoundation models of markets now have entirely new principles underlying their construction.[84]

- For starters, as consistent with our Game Theoretic discussion presented so far, many artificial models of the market presently purview the "market" as a repetitious game between a number of players (investors) with the market price being the reflection of the interaction between the participants of the game.[85]
- Further, such models typically allow players to learn from their competitors behavior and to build-up "reputations" of likely actions from competitors in response to certain price information or actions on behalf

---

[82] See Brock *et al.* (1992) or Darley and Kauffman (1997) for excellent examples of this.
[83] For examples of this earlier work under the "Evolutionary Finance" banner refer to Farmer and Lo (1999) and Hens, Schenk-Hoppe and Stalder (2002).
[84] For a comprehensive overview of the general principles underlying artificial market models see Brock (1997).
[85] See for instance Jefferies *et al.* (2002).

of the player's themselves. In fact, many trading strategies embedded within such models are built upon these preconditioned "beliefs."[86]

- Also, within this particular field of research it is generally regarded that the objective of the market "game" is replication as consistent with Maynard-Smith's musings. That is, once their position is set, market participants want as many people as possible to "buy" into their particular investing thematic thus perpetuating their idea "gene" and forcing the market price in a favorable direction – a tact we will build upon extensively in our own (evolutionary) artificial model of the marketplace.

- From a holistic standpoint, such artificial market models with trading rule-based microfoundations have an overwhelming tendency to be deterministic rather than stochastic in foundation and therefore provide a natural fit with the Complexity Theory principles espoused in "new" view challenge no. 1.

- Finally, consistent with Evolutionary Game Theoretics, many models of artificial markets allow for the rules of the game to be constantly evolving in the sense that institutional arrangements, response functions of the various players, the players themselves, or even the choice of market instrument – equity, convertible, option and so on – all undergo significant change through time as players adopt the most effective strategies to influence price. In other words, "optimal" trading strategies are in a constant state of adaptation and *evolution* in response to a constantly changing market environ.[87] Again, a tact we will build upon extensively in our own (evolutionary) artificial model of the marketplace.

As we said at the start of this section, it is the convenient marriage of both Behavioral Finance-based heuristic deterministic rule-based microfoundations, along with a holistic nonlinear perspective that is analogous to the work carried out under the Complexity Theory banner, which implies "new" view (challenge) $1 + 2 = 3$. Deterministic models of artificial markets – especially those that are "evolutionary" in their principles – come part way toward providing a framework consistent with our own Evolutionary Finance purview but still fall short by failing to explicitly take into consideration the actual texture and fabric of information itself. To truly understand "real world" financial markets, one needs a comprehensive framework for assessing how the microfoundations of market sensitive information are built "from the ground up."

---

[86] Our own "artificial model of the market" however, adopts a different approach in that reputations are built on the format of the information bytes that analysts produce and not through "learning" behavior *per se*.

[87] Indeed, even some traditionalists begrudgingly acknowledge that the market is in a constant state of evolution – although they use this as a reason why *not* to engage in active investing. See for instance, Malkiel (1996).

Which leads us quite nicely into the next section – just where does our own interpretation of Evolutionary Finance fit into the existing "new" view genre? What is our contribution relative to the studies that have preceded us? As will be outlined in the following section – and indeed, discussed in great detail for the remainder of this book – we intend to pursue the biological-like/evolutionary analogy toward the market in much greater depth than any of our forebears via our comprehensive molecular modeling of the actual micro building blocks of information itself. As stated earlier, our own purview of Evolutionary Finance is much more structured and deep relative to earlier usage of the terminology. Indeed, we feel the examination of "information" not as a texture-less generic concept but rather as a constantly evolving entity of which there are a colorful array of different hierarchies and classifications (from bytes, to memes, to themes, to overall market sentiment) represents a complete watershed for financial thinking – and in particular, for *applied* finance "best practice" as well.

3.4   So where does Evolutionary Finance fit-in to the "new" view genre?

Having spent years in "real world" financial markets observing the galling frequency with which many practitioners have given little accordance to the labors of earlier academic thinkers stretching the limits of the knowledge horizon, we are again at pains to point out that without such enduring academic efforts our own purview of Evolutionary Finance could not have been born. We reiterate once more that in true spirit with Sir Isaac Newton's maxim "If I have seen further it is by standing on the shoulders of giants" that we have benefited immensely from such academic insights. But likewise we feel our treatise represents a significant advance upon the analysis that has preceded us given that we go to the very heart of the evolutionary way in which the diverse array of information that exists in the marketplace at any one point in time somehow comes together in a meaningful manner to influence asset prices. We illustrate how this information (and the analyst community that produces it) has a natural tendency to bind into separate clusters so that the overall asset price impact is decidedly nonlinear. Indeed, the institutional structure that investors have created in financial markets not only condones such an outcome but actually enforces it as well.

Do we draw upon any particular "new" view paradigm more than the others in forming such a viewpoint? Certainly in Chapters 5 and 6 – where we present our own evolutionary (artificial) model of the marketplace and its subsequent distributional form implications – we leverage extensively from the academic literature categorized as "new" view challenge no. 3.[88]

---

[88] Indeed, as stressed earlier, up until the publication of this book the term "Evolutionary Finance" had generally been associated with artificial market models where a

But likewise, our model draws heavily upon the principles of Complexity Theory and nonlinear dynamics as well – which were outlined in "new" view challenge no. 1.[89] In addition, as we are by default making the distinction between public and private information in the evolutionary information production framework presented within this model, we are also forced down the path of "bounded rationality" as espoused in "new" view challenge no. 2. Ultimately, we use this concept to some effect in our later modeling of the strategic decision-making on behalf of the analyst community presented toward the end of Chapter 5 and over the early to middle stages of Chapter 6.

To conclude therefore, rather than confine ourselves by saying that our own purview of Evolutionary Finance is beholden to any one "new" view interpretation, we prefer to argue that what will be presented in the following chapters represents the fruit borne from an eclectic mix of the components within each of the "new" view challenges 1, 2 and 3. Information, market structure and investor behavior are all inexorably intertwined. Our insights and modeling presented in the following chapters reflect this. We would only add that too much time and effort has in the past been devoted to market structure and investor behavior and not enough on the actual fabric of information itself. It is here that our greatest contribution resides – simply through our modeling of the creation and formation of information into a structured array of bytes, memes, themes and even overall market sentiment as being the outcome of a biological-like molecular evolutionary process. The basics of this new paradigm for understanding the centrality of information in financial markets will be outlined in Chapter 4. We hope you find it interesting.

---

process of "natural selection" screened for the emergence of evolutionary dominant trading strategies. Our own purview of Evolutionary Finance drills much deeper into the evolutionary genre of financial markets – especially as it applies to information (as Chapter 4 will amply illustrate).

[89] In fact, the link between information and nonlinear dynamics was explored in the late 1970s and early 1980s by a group who Gleick (1998) labels as the "Dynamical Systems Collective" of whose principle members were Robert Shaw, Doyne Farmer, Norman Packard and James Crutchfield. Still, this group viewed "information" as a generic concept – an intangible input into their dynamic system of equations and not the diverse evolving entity as we define it.

4
The Mechanics of Modeling Information as an Evolutionary Process

As stated in our introduction, it is the ambition of this book to move away from treatment of "information" as a generic term and look at the actual microstructure of information itself. For example, what are the underlying building block components that make up what we as investors perceive to be information, how does this information come together in a meaningful way, what incentives are there for analysts to produce this information and what are the feedback mechanisms in the analysts choice of topic? All are worthy questions that will be answered in due course. Indeed, this chapter houses the "boiler room" of answering the first two conundrums. We do this via the establishment of a framework for understanding the actual building blocks of information – piece-by-piece. Then we go on to investigate the consequences of our approach. As we have hinted at upon numerous occasions in the preceding chapters, we believe there is an analogous link between the micro-foundations in the treatment of information in the evolutionary/biological sphere and the treatment of information within financial markets. It's just a matter of drawing this striking similarity to the academic community's (and investor's) attention. Once this is done, a whole host of interesting results manifest.

As we have said before, we believe the "idea" and the "gene" have a lot in common. In fact, we feel that some of the more exciting recent developments in genomics and bioinformatics have much to offer the fields of finance theory and finance practice.[1] It is the purpose of this chapter to lay

---

[1] In fact, we feel this cross-fertilization between disciplines is as potentially path-breaking as the (much older) hybridization between biology and physics – see von Bertalanffy (1950) and Kremyanskiy (1960) but also the early stages of Glaser (2001) who gives nice historical perspective on the fascinating field of biophysics. Further, our assertion should not surprise those familiar with much of the details of the "new" view finance literature – some of which has been documented in Chapter 3.

the foundations of this revolutionary evolutionary way of thinking via a thorough investigation of the actual texture and fabric of information itself. So where to start? Well, even diehard traditionalist finance theorists would agree that information and price are inexorably linked – after all, such a premise forms the basis of their much-loved "fair-game" model. So modeling the "production" and "consumption" of information must be the first port of call in any attempt to develop a wider understanding of the market process itself. But in doing so, one must at first appreciate how the individual components of information necessarily bind together. In short, if one is to walk away from the simplistic assumption that information arrival is IID, then one must ultimately have some mechanism for understanding how the various "bytes," "memes" and "themes" of our information categorization somehow interrelate until a cohesive whole of overall "market sentiment" is formed. In essence, what we are in effect looking for is a financial market equivalent of Watson and Crick's double-helix – a structure around which smaller components of information can assemble to form larger informational entities.

It is here that our earlier stated serendipity of the complementarities between the "idea" and the "gene" plays its all-encompassing role. Drawing heavily upon the seminal insights of Dawkins, Farmer, Gould and Kauffman we highlight how the deterministic building blocks of financial information come together in much the same way that the C, A, G, T nucleotides form the complex genetic coding of DNA and from this, even larger entities. Once these elementary building blocks of information and market pricing are well understood, then one has a much better appreciation of the driving forces behind market dynamics overall. Indeed, once this fundamental realization is made that information and pricing are ultimately the outcome of an evolutionary biological-like molecular process then one can better appreciate that the entire "market" as we know it is largely defined by its information servicing criteria – its production, its dispersal, its interpretation and consumption. That said, before we can even attempt to outline such an ambitious multifaceted philosophy toward the informational foundations of the marketplace we

For instance, Farmer and Lo (1999) highlighted that Darwin cited both Thomas Malthus and Adam Smith as sources of inspiration for the principle of natural selection. Ergo, the analogous relationship between finance and biology should be seen as something of a revisit (albeit from a different direction of causality) to an age old idea. Thankfully, recent developments in the bioinformatics area have allowed us to pursue this dream. Indeed, it is somewhat surprising that it has been less than 100 years (since 1909 to be precise) since the British biologist William Bateson first gave the science of inheritance a name – genetics. Without the rapid development in understanding in this area since then (and its concurrent formalization) we would not have been able to pursue the hybridization between biology and finance as pursued in this book.

must at first categorize the actual evolutionary building blocks of information itself.

4.1 Evolutionary information basics: Memetics and the contribution of Richard Dawkins

Let us start by stating right from the outset that we by no means want to leave readers with the impression that an evolutionary perspective toward information is necessarily an *entirely* new phenomenon. Rather, our task here is to formalize the concept within the finance literature – building upon some of the seminal work that has already been completed to date. In particular, there are plenty of academic giants who have previously likened meta-physical concepts to formal evolutionary processes – Sir Karl Popper and Herbert Spencer come immediately to mind. Indeed in the finance world, Doyne Farmer and Andrew Lo have perhaps came closest – but certainly not the full way – toward our own evolutionary approach and consequently both have played a decisive role in shaping our thoughts.[2] That said, the first "giant" upon whose shoulders we will rest for a short while is Sir Richard Dawkins from Oxford University. Why? In his book *The Selfish Gene* (1976) Professor Dawkins devotes an entire chapter to the analogous links between biological evolution and cultural evolution.[3] In particular, Professor Dawkins coined the concept of "memes" as the analogous metaphysical cousin to the ubiquitous gene.[4] A serendipitous insight from which we will leverage upon extensively in the ensuing pages.

What exactly is a "meme"? Well, according to Blackmore (1999) who has written extensively on the subject, a meme is "An element of culture that may be considered to be passed on by non-genetic means, esp. imitation."[5] It is the latter "non-genetic" part of this definition that would-be Evolutionary

---

[2] What differentiates our analysis from Farmer's and Lo's however, is our specific concentration on the centrality of information – in particular, information driven by evolutionary forces – within financial markets. This prime (informational) focus of our work has allowed us to drill much deeper in exploring the analogous link between evolution in the biological world, and evolution in the financial world. We do this by modeling the biological-like molecular componentary of financial information itself – thereby providing a formal evolutionary microstructure to explain an observed evolutionary macrostructure within financial markets.

[3] Indeed, the analogy also extends to human language. To quote Watson (2004) p. 254 "There are, of course, the obvious parallels between genes and words."

[4] As Professor Dawkins (1976) p. 192 put it "We need a name for the new replicator, a noun that conveys the idea of a unit of cultural transmission, or a unit of imitation. 'Mimeme' comes from a suitable Greek root, but I want a monosyllable that sounds a bit like 'gene'. I hope my classicist friends will forgive me if I abbreviate mimeme to meme."

[5] Even Darwin acknowledged the genetic nature of culture – "A language, like a species, once extinct, never reappears" (Darwin 1859, p. 422).

Finance theorists find potentially the most puzzling – particularly given our discussion to date. Does this represent an apparent contradiction to our cause of more thoroughly exploring the analogous genetic-like links between modeling physical evolution and modeling information evolution? To clarify Blackmore's view we decided to refer back to Professor Dawkins seminal text *The Selfish Gene*. Upon doing so, it is soon made only to clear what is the source of the apparent confusion. Professor Dawkins was continually at pains to divorce readers of the notion that somehow cultural evolution and genetic evolution are irreconcilably intertwined. That is, he did not wish to leave readers with the impression that a certain society's gene pool ultimately determines its cultural, scientific and economic success. Hence the "passed on by non-genetic means" inclusion in Blackmore's definition of a meme. In particular, one should be clear that while it is acceptable to argue that there are obvious analogous links between metaphysical and physical *models* of information transmission, it is anathema to attempt to extend the analogy much beyond this. Taken to the extreme and linking actual cultural evolution with genetic evolution is a potentially explosive issue. This is especially the case given the twentieth-century problems in the Western World with eugenics, Nazism, global warfare and so on. Far better to concentrate on the analogous links between the models themselves and not get drawn into polemical interpretations of the actual hard and fast policy implications of a broad-ranging analysis between genetic and metaphysical evolution. Given our particularly narrow frame of reference in only dealing with the treatment of information in financial pursuits, it is hardly likely that we could be accused of falling into such a trap. Indeed, we stress that our focus on the analogous nature in which information is produced, dispersed and interpreted in a biological-like molecular fashion is (and always will be) at the *meta-physical* – not physical – level.

That said, what is also clear in a re-reading of *The Selfish Gene* is precisely how accurate Blackmore (1999) was in stressing Professor Dawkins' interpretation of memes as being heavily dependent upon the concepts of an "element of culture" and "imitation." For example, Professor Dawkins highlights that national dress is a good example of memetics in action – a classically obvious imitative representation of an "element of culture." But while this may be a good starting point, we still feel this terminology is perhaps a little limiting for our purposes. For starters, we ask can we better define the term "element of culture"? After all, such a sweeping statement has the potential to include themes, beliefs, strategies, religion, processes, laws (either social or those of nature), rules of conduct, social mores, methods of practice – indeed "knowledge" in its most general form. Can we be more definitive – especially given the finance-orientated motivation for our study? Is it possible to systematize the actual classification of information into various forms so that the basic universal (read "mathematical") laws via which this information somehow comes together to influence price activity can be better understood? What sequence of information bytes typically result in a substantial alteration in price? How is this information agglomerated into memes and themes or even market sentiment? To what extent is market related information economically produced and what is the motivation for this behavior? Answering these questions would certainly help form the basis for a better understanding of the universal laws that govern the evolution of what investors come to perceive as "information" and through this, the evolution of market prices.[6] From an alternate tact, can we better define the process of "imitation" so that this too can be somehow formalized into our set of universal laws for evolutionary information dynamics within financial markets?[7] An ambitious charter we know, but one well worth pursuing given the earlier cited centrality of information within the financial market mechanism.

To his credit, Professor Dawkins gave us some seminal insights here too from which to build upon in our endeavors. For example, Professor Dawkins highlighted that some general principles of successful memes were:

- their "replication ability" – that is, how easily they are passed on[8];
- their "fidelity ability" – that is, how well they maintain their integrity as they are passed on; and
- their "externality ability" (our term) – that is, how they reflect a specific microstructure that enables the binding of various segments of information together so that the sum of a given information-set (and importantly, the *impact* of the sum of the components of this information-set) can at times be greater than its component parts.[9]

[6] Assuming one has a basic understanding of genetics, it is obvious where we are heading with such a hypothesis. What we are effectively searching for is the metaphysical cousin to the C, A, G, T sequencing that makes up organic DNA. After all, depending upon the sequencing, these four organic DNA nucleotides could just as easily determine eye color as they do skin tone. As Blackmore (1999) herself stated "Genes are instructions for making proteins, stored in the cells of the body and passed on in reproduction. Memes are instructions for carrying out behavior, stored in brains (or other objects) and passed on via imitation." What we are aiming to find out are the universal laws as to how these memetic instructions are loaded and indeed, imparted to other individuals.

[7] Blackmore (1999) highlights that imitation is a genetic trait that all humans (and indeed other higher species) carry. In fact, Dawkins (1976) highlighted the way memes replicate from mind to mind is "like parasites infecting a host" – an analogy we will draw upon in Chapter 5.

[8] It was via his focus on imitation as the basis for human learning that Professor Dawkins was able to illustrate that mutation is high in memes – all one need do is look at the outcome of a "postman" game among children to see this "mutation" in action in a highly entertaining form. Indeed, Professor Dawkins believed that wrong ideas in meme mutation eventually get "weeded out" – a potentially controversial hypothesis to which we will return to a little later.

[9] The first two attributes of successful memes are obvious in their comparative advantage for surviving the rigors of a natural selection type process. The last attribute is a

Combined with a little algebra and some calculus this certainly provides a nice foundation for attempting to mathematically formalize the universal laws of information in financial markets. It is the last point in particular, that potentially has merit for understanding the all-important information/price dynamic – especially those in nonlinear form – since it attends to the principle of information agglomeration. So it is toward the task of formalization that we now turn but before we do, let us leave you with this wonderful quote from Professor Dawkins as perhaps a fitting tribute to his seminal insights:

> When we die there are two things we can leave behind us: genes and memes. (Dawkins 1976, p. 199)

There is little doubt that Professor Dawkins memetic legacy will be long-lasting.

4.2 Moving past the elementary: Taking the evolutionary information concept further into the field of finance

By way of experiment, take a coin (any coin), two pieces of white sticky tape and a black biro. Stick the two pieces of tape (which should be roughly the same size) to either side of the coin and with the biro in hand, write a large "S" on one-side and a large "B" on the other. Now flip the coin 100 times (1000 times if you are feeling up to it) and write down the results of each "S" and "B" as they sequentially appear on the upward facing side. Next, examine the number of "S" (for "sell") and "B" (for "buy") signals you have written down and count the number of back-to-back sequences as they materialize – 1B, 1S, 2Bs, 2Ss, 3Bs ... and so on. Plot the distribution function of this series and what do you have? The ubiquitous Gaussian (Normal) distributional form.

This is IID at its simplest. The outcome of one coin toss has absolutely no effect on the outcome of another coin toss – assuming that both pieces of sticky tape weigh roughly the same amount. Add Fama's fair-game principles to establish a direct link between information and price and what you have here is the bedrock foundation underlying the entire traditionalist finance paradigm as to the way information and asset prices interrelate. Forget about employing thousands of research analysts whose job it is to attempt to predict the price movements in various assets – the market is way too informationally "noisy" (read "random") for these people to have any *consistent* success. In fact, all you need is a primate with an opposing thumb (for flipping the coin) and who possesses enough cognitive ability to mark a chalkboard and you

little more subtle in its "survivorship advantage" but when one thinks that a meme (or information byte for that matter) with exceptionally high externality potential has the ability to bind to a whole host of information, then it stands to reason that its survivorship potential will be significantly enhanced.

have the makings of a sound investment process for any would-be "research department" at an investment bank or fund manager.

Surely the traditionalist cause must be more complex than this? The short answer is no. While there have been many sophisticated appendages to traditionalist finance arguments, this should not obfuscate the core (somewhat simplistic) beliefs that these traditionalists hold true as to the way information and asset prices interrelate on a day-to-day, minute-to-minute basis. This is not to say that the traditionalist finance math is not elegant or may be useful for proving some esoteric point, but as any good mathematician will testify, when it comes to the "real world" your outcomes are only as good as your assumptions. Sure, if enough people *believe* in the traditionalist models then ergo, the market's movements will actually mimic such fundamentals – indeed, we witness large swathes of such logic at work in the derivatives market everyday.[10] But likewise, there are equally important major segments of the market where investor action is, at given points in time, belligerently and consistently diametrically opposed to the traditionalist finance (random movement) view. What we are essentially talking about here is the army of active investors who intrinsically believe their insights can *consistently* outperform the Strong EMH/IID random model. As we highlighted in Chapter 3, the fact that erstwhile "rational" investors willingly finance and encourage such alpha-orientated research intimates a widespread belief by *practitioners* within financial markets that the traditionalist notion of IID information arrival is erroneous in the extreme.

So what are the alternatives? It is insufficient to criticize something of such historical longevity as the traditionalist finance mindset without at least attempting to proffer an alternative interpretation. Let us do this in such a way that in effect challenges the much-loved traditionalist finance notion of IID by attacking each of its component parts – and at every interval, suggesting an alternate rationale.

Point of attack #1. Information is *interdependent* not independent. For example, a particular US GDP release may not have its full market impact upon asset prices until the arrival of associated information – such as the release of US Current Account numbers. That is, there is a degree of *latency* within each information byte that will only be unlocked with the arrival of more information. As memes and themes gradually form in response to the agglomeration of various information bytes, a definite dependence emerges between asset prices at time t and asset prices at time $t + 1, t + 2 \ldots$ and so on. This is in obvious contradiction to Strong EMH (and IID) principles. Why? Informational latency highlights the potential for technical trading rules premised on such intertemporal interdependencies to be *consistently*

---

[10] Although perhaps the ubiquitous presence of volatility smiles illustrates a persistent recalcitrance by derivative markets to fully conform with the traditionalist finance utopian view at all moments in time.

effective and as demonstrated in Chapter 2, the sustained profitability of such rules constitutes an emphatic refutation of the Strong EMH principle. Indeed, extending the informational interdependence concept even further and considering the potential for the imperfect distribution of information among investors, then it is possible to surmise that the realization of any information latency may at times only be achieved via the actions of the better informed investors influencing others. This is done primarily through price signals. In short, the better informed investors' actions will generate a price movement that spurs other investors to investigate the veracity of a particular theme. This, in turn, has the potential to unlock the full latency potential ingrained in a whole array of information bytes. The net result? Market prices moving substantively on days when there has been little in the way of formal data releases. These are not the actions of "noise traders" but rather the behavior of rational investors gradually coming to grips with the realization of an emergent theme.

Point of attack #2. Consistent with the interdependence notion expressed in point 1, the *sequence* of information arrival is important. A given US GDP figure may have an entirely different latent market impact depending upon the sequence of information that either follows it or precedes it. If for example there is a crescendo-like rise then abrupt fall in the market's belief in a particular emergent theme – thanks to a rapid sequence of strong ancillary support information but then an emphatic dismissal of this emergent theme – then it stands to reason that the pattern of pricing behavior would be significantly different than had the ancillary support information been more evenly spread over a broader breadth of time. Ergo, even though the net sum of supportive information is the same in both cases, an investor's interpretation of events would be less impassioned in its backing of the particular theme in question if the periodicity over which the supportive information has arrived is relatively wide. Consequently, an intermittent sequence of supporting information could have an entirely different influence in unlocking the latency potential across a given array of information bytes than had the same supportive information been condensed over a much shorter time-span. Indeed, consistent with the following third point in our attack upon IID, it will be established that the sequence of information has the potential to alter the distributive consequences of a given information byte upon the net sum of existing information simply because of the presence of Complex nonlinear relationships between each information byte within any given information-set.

Point of attack #3. Consistent with the sequencing notion highlighted in our point of attack #2, we would argue that information is *heterogeneously* (rather than identically) distributed depending upon its time of arrival and underlying characteristics. As will be displayed in the ensuing pages, it is the externality ability reflected in memes as given information bytes bind with each other that causes a nonlinear response function on behalf of investors. Just as certain molecules have greater ability to bind to other molecules, so too do certain information bytes have superior ability to agglomerate into memes and thus reinforce a range of possible themes. Going back to our earlier example of a strong US GDP result, this particular information byte could just as easily be a contributor to a "weak $US theme" (thanks to concern about a possible widening current account position) or a "strong $US theme" (thanks to expectations of rising interest rates and asset prices). The differing interpretation depends upon the associative information that a strong US GDP release has to bind with at a given point in time. This associative information is reflected in memes (which are short sequences of information bytes). As memes are formed, the latent information in existing information bytes is effectively unlocked – at times in a nonlinear fashion, so that the sum of the price impact of the information bytes represented within a particular memetic string can at times be considerably greater than its component parts. Subsequently, it is a *combination* of the arrival sequence of information and its underlying "externality" characteristics that memes effectively unlock which is ultimately the prime determinant of a given information byte's distributive characteristics both across time and across the entire information-set. Should a particular information byte have a high "externality potential" and what's more, this particular information byte arrives in an informational sequence that is especially amenable to this externality potential being unlocked – it forms a key component of a particularly strong meme – then not only will the distributive characteristics of this information byte be different from others within the information-set, but also across time the entire information-set's distributive form will itself alter thanks to the nonlinearity potential of new (and latent) information being progressively unlocked in a more efficacious manner thanks to the presence of this strong meme.[11] Needless to say, this too will be reflected in an altered pattern of price behavior and in particular, prices will have a tendency to exhibit a distributive pattern significantly different to what would have been the case had the information byte's format been more consistent with the traditionalist finance assumption of IID. In short, one should expect a greater recurrence of kurtotic/significantly skewed distributional forms for both information and price under an interdependent heterogeneously distributed information paradigm than would be expected had information itself conformed to the conditions espoused under the traditionalist IID norm. Interestingly, this is precisely what we observe in the real world – especially for price action over shorter, higher frequency data horizons. So while *heterogeneously* distributed information may be a far more difficult problem to wrestle with mathematically than its IID (Gaussian

---

[11] It is possible to look at this from "nature/nurture" type of perspective – "nature" is the externality potential exhibited by a particular information byte while "nurture" is the associative informational environment into which this information byte is thrust.

distribution) forbear, it is nevertheless a far more interesting premise just the same.

So how to formalize the earlier notions as to the characteristics of information/price dynamics under our Evolutionary Finance paradigm? Here is where we strike perhaps one our more enduring links with evolutionary principles. Following in the footsteps of Farmer (2000) Table 4.1 illustrates the analogous referencing one can use between existing biological/evolutionary terms and our own corresponding Evolutionary Finance terminology.[12] We like to think of it as a "synonyms table." As can be seen, themes are made up of component memes which in turn are made up of information bytes. This occurs in much the same fashion that DNA is made up of genes which in turn are made up of the chemical bases (C, A, G, T). Themes drive market sentiment which is then reflected in asset pricing and overall index performance. Likewise, a gene representation determines chromosome behavior which in turn determines cell structure and the organism's phenotype. In fact, we would go as far as to argue that just as there is a genetic information genotype underlying each and every organism's phenotype, there is an analogous financial information genotype underlying each and every market price phenotype. It is our objective to ascertain this sequenced genetic coding of financial information – as information bytes come to comprise subgroup memes which in turn, form larger theme strings. In short, what we are effectively searching for is the "informational genome" of both individual asset prices and for the market as a whole. We feel such an identification of the actual biological-like component structure of the sequenced information underlying each and every asset price phenotype is a powerful result that has many implications for finance – not only from a theoretical modeling perspective but also from an applied perspective as well. But to take our analogy further we need to formalize our "informational genome" framework in much greater detail – this is the charter of the following section.

4.3   The building blocks of our evolutionary approach toward information in finance

How to encompass the aspects of Table 4.1 into a comprehensive model of information evolution? At first we need to formalize the actual building blocks of financial information itself – how bytes form memes, how memes

---

[12] A key distinction in Table 4.1 is that between genotype and phenotype (see vertical axis). In biology, the genotype is a complex set of genetic information (genes) encoded in the DNA of an organism. The phenotype is the physical appearance of the organism itself. The development of the phenotype as directed by genes is called morphogenesis. In Evolutionary Finance, the phenotype of an asset is its price, its genotype is the information content embodied in that price.

Table 4.1   The definitional foundations of Evolutionary Finance vis-à-vis evolutionary biology

Evolutionary Biology	Evolutionary Finance

Genotype
Level 1

Bases (C, A, G, T)	Byte
Four bases make up the double-stranded molecule known as DNA (deoxyribonucleic acid) that looks a little like a spiral staircase. Each strand of DNA is comprised of a sugar-phosphate backbone and numerous base chemicals attached in pairs. These "bases" as they are known, effectively make up the stairs of the spiraling staircase and come in the form of cytosine (C), adenine (A), guanine (G) and thymine (T). The bases act as letters of the genetic alphabet and it is through their sequencing that genes are formed. For example, around 3 billion bases arranged in sequence form the approximately 35,000 genes that comprise the human DNA molecule. Within this molecule, all the information required to "build" each human being is stored.	The fundamental building blocks of all information. Bytes are encoded into sequences in accordance with the information-set within which they operate. For financial information, bytes are encoded in the form of specific action responses – Buy (B), Neutral (N) and Sell (S). There is both a contemporaneous (1st letter) and latent (2nd letter) aspect to byte information. So according to Evolutionary Finance principles, a byte can take either one of nine forms: BB, BN, BS, NB, NN, NS, SB, SN, SS. For financial information a byte can refer to an analyst's research report, a financial commentators story or even an official release from a government statistical bureau. Some bytes have more "externality potential" than others. That is, they have a superior ability to bind information together.

Level 2

Gene	Meme
Comprised of base sequences, genes are the functional units of heredity. Genes store information which is then converted into signals for building a specific protein and thus cells – governing everything from vital organ tissue to hair color	Comprised of byte sequences, memes are the functional units for the way information is stored and relayed. Memes unlock the latent information in bytes. Just as particular letters of the alphabet make up certain words, so too do particular bytes form certain memes. Some memes are stronger in their impact than others but without every necessary component of a memetic sequence in place, the latent information within constituent information bytes will go untapped.

Level 3

DNA	Theme
First described by James Watson and Francis Crick in 1953, DNA represents the complete aggregation of an	An agglomeration of memes, a theme is a broad reference given to a group of memes (sometimes from a variety of

Continued

Table 4.1 Continued

Evolutionary Biology	Evolutionary Finance
organism's genetic information. It is comprised of the entire gene sequence which, in turn, represents the entire base sequence.	assets). This classification of alternate memetic information into associative thematic groupings helps humans encode information bytes (and their larger form of agglomeration – memes) for storage and retrieval within the human brain. A new theme is announced with the publication of a particularly seminal information byte – one with exceptionally strong externality potential.

Level 4

Chromosome	Sentiment (Bull/Bear)
Part of the cell that contains genetic information. It is comprised of a tightly packed coil of DNA – humans having 46 such tightly packed DNA coils in each cell nucleus.	An aggregation of themes that comprise an overall interpretation of information toward a given subject (in our case, the market) at a given point in time.

Phenotype
Level 5

Cell	Asset prices
The basic structural and functional unit of all organisms. The nucleus of a cell contains all the genetic information. Collections of cells form an organism's tissues, blood and organs.	The object via which one's interpretation of all financial information at a given point in time is embodied. The most readily observable response to changes in an investor's perception of the informational state is alterations in asset prices.

Level 6

Organism	Index/market
What we see as the "final product" of genetic information. As the highest level of aggregation of the encoded information, it is also the most obvious physical manifestation of the genetic blueprint.	An agglomeration of assets into specific (meaningful) groups. When most people refer to the "market" they are usually referring to an index (e.g., S&P500) or a group of indices (e.g., global bonds).

form themes and then in turn, how themes are reflected in overall market sentiment. This conscious gradation toward ever increasing agglomerations of sequenced information requires some form of mathematical framework if it is to be organized into a meaningful schematic. It is the objective of this section to develop such a mathematical structure. In doing so, we will present our argument sequentially by moving through each successive information level (1–6) represented in Table 4.1.[13] Readers will no doubt quickly surmise that as we progress through each of these levels we are in fact sequentially increasing the level of aggregation of sequenced information within the overall market informational set. We do this until we finally reach a position where only a binary choice of classification is available – a bullish or bearish perspective – which we classify as the prevailing "market sentiment." We then go on to identify how this information actually interacts with price – levels 5–6 in Table 4.1. This is done by examining how our derived informational genome for the market is reflected in both individual asset price phenotypes and for the overall price phenotype of the entire market itself. So now over to the building blocks themselves.

Building block level 1: Information bytes

Following on from Table 4.1, let us assume information bytes (φ) are encoded (for financial interpretation purposes) into one of three forms – "B" for buy, "N" for neutral and "S" for sell. Further, let us assume that there is both a *contemporaneous* and *latent* aspect to each information byte so that, in effect, each information byte can take one of nine forms

$$(\varphi) = \{BB, BN, BS, NB, NN, NS, SB, SN, SS\}, \tag{4.1}$$

where the first letter constitutes the contemporaneous information signal and the second letter the latent signal. A typical financial information genome ($\sum \varphi_a$) is thus comprised of a finite string of information bytes as reflective on a particular asset (a) at time t

$$\sum \varphi_a = \begin{pmatrix} SS,NS,SN,BN,BB,BS,SN,SN,BB,BS,SB,SB,BN,BS,NN,NB,NN, \\ BB,SS,SB,NS,SB,SN,SS,SB,SN,NN,NB,BN,SB,SB,BB,BN,NB, \\ NN,SN,BB,BB,BB,SB,SN,BN,BN,SB,SS,NS,SN,NN,NN,NN, \\ BS,BB,BN,BS,BB,BN,NB,SB,BB,BN,NN,NB,BN,BS,BB,SB, \\ NB,BN,SB,NN,NB,BN,BS,SB,BB,\dots\dots\dots\dots \end{pmatrix}$$
$$\tag{4.2}$$

where a subscript "Ï" rather than "a" denotes the informational genome for the entire market – more on this later. Subsequently, the change in any asset's price reflects both the release of contemporaneous actionable information – "buy" or "sell" signals – and the net sum of actionable latent information enforceable at time $t(\Delta)$.

However, before moving on, some additional features of Equation (4.2) are worth noting. For starters, a particular information byte's form – the "buy,"

---

[13] This is not to say that there need only to be six levels within our analogous mapping of information in finance to an evolutionary biology referencing structure. Indeed, if one wanted to drill down further, the actual "words" within analyst information "byte" publications could effectively represent "atoms" and even the vowels that make up such words could be described as "quarks."

"neutral" or "sell" signal – is directly dependent upon what asset it is related to. It is entirely possible that a "BB" information byte signal for a given asset in response to a certain "US GDP" release could just as easily be an "SS" signal form for an alternate asset. For example, a bond asset's information byte form for a given economic release will typically differ from an equity asset's information byte form – given the underlying pricing characteristics of these financial instruments. In other words, φ_a is asset specific – hence, the subscript "a" in the notation.

Building block level 2: Memes

A meme is a subset of the information genome ($\sum \varphi_a$) and represents a specific finite sequence of information bytes. Any given information genome for a given asset (a) is thus comprised of both memetic (M) and singular (Ξ) information

$$\sum \varphi_a = \{M, \Xi\}. \tag{4.3}$$

What differentiates memetic from singular information is that the informational impact (II) of memetic information is greater as it binds to the other information already contained within the information genome. Indeed, memetic information bytes have the potential to bind to both contemporaneous and latent information alike and even have the potential to bind to future information as it arrives as well. They do this via their externality potential (x) – provided they arrive in the "correct" sequencing environment for this externality potential to be utilized. Indeed, it is the *combination* of these two forces which in effect brings previously dormant latent information signals into the investor's frame of reference.[14] In short, correct memetic sequencing and externality potential (x) enables "sleeper" latent information to awaken and thus have a (potentially nonlinear) bearing upon asset prices at any given time t.[15] As briefly mentioned earlier, it is the merger of these two "nature" and "nurture" forces which act as the binding glue to help form memes. "Nature" represents the externality potential (x) of certain information bytes contained within the information genome. "Nurture"

[14] In Chapter 5 we will embellish this concept considerably further as we measure the II of the net memetic externality weight for a given information set in terms of its influence upon the price of a representative asset (i) and then distinguish between its impact in contemporaneous (Ω) and latent (Λ) form.
[15] Indeed, it is possible to regard the externality potential (x) of certain information bytes as having an epigenetic-like effect in the sense that they may control meme formation not necessarily immediately around them within the informational genome sequence, but rather further along in the code. A certain information byte may result in a "new way of thinking" about a given information sequence. This sequence may come much later in the informational genome. This control from a remote part of the information genome can be likened to epigenetic effects witnessed in biological evolution. We will return to this analogy in the following section.

is the sequenced informational environment into which these externality potential bearing information bytes are thrust.[16] Without the correct combination of both events, memes will not form and all information within the informational genome could be regarded as being singular – and thus consistent with IID principles. Thankfully this is not the case and the information genome can be regarded as being comprised of both singular and memetic information.

As we mentioned earlier, since memetic information binds latent information (and potentially future information as well) its II is larger than singular (independent) information. Further, given the potential for memetic strings to bind with each other via the externality potential (x) conduit – in essence, small memes merge to form larger memes thanks to the arrival of a key piece (a key information byte) of the memetic sequence – then there is the definite potential for nonlinear IIs within our system.[17] Consequently, assuming an equal number (k) of memetic and singular information bytes within a given information genome, the following is true

$$\mathrm{II}_M\left[\sum_k \varphi_a x\right] > \mathrm{II}_\Xi\left[\sum_k \varphi_a x\right], \tag{4.4}$$

where order matters so that $\varphi_{a1} x, \varphi_{a2} x \neq \varphi_{a2} x, \varphi_{a1} x$.

Building block level 3: Themes

A theme (j) is a subset of all memetic information (M)

$$j \subset M, \tag{4.5}$$

where the number of themes in existence (Γ) for a given market informational genome ($\sum \varphi_{\ddot{i}}$) is always less than the total number of memes.[18] Themes

---

[16] Such sequencing alone has the potential to generate an impressive diversity of memetic sequence permutations into which the "nature" element of an externality potential bearing information byte can be thrust. For example, Mainzer (1997) illustrated in referring to the human genome sequence that even though genes in nature are seldom more than 1000 sequential positions, if there are 4 symbols to each position there are 4^{1000} alternative gene permutations – or in scientific notation, 10^{600} possibilities. He states: "In order to get an impression how huge this number is, we should recall that the content matter in the whole universe corresponds to 10^{94} and the age of the universe is less than 10^{18} seconds."

[17] Such results are consistent with the earlier discussed Complex nonlinear dynamic systems detailed in Chapter 3. Indeed, the arrival of a key information byte allowing an intermittent group of small memes to suddenly become a large meme can be regarded as a classic case of the "tipping point" in action. For a very readable exposition of this phenomenon refer to Gladwell (2000).

[18] Notice we use the subscript "$\ddot{i}$" to denote the market information genome rather than the subscript "a" which applies to the information genome for individual assets.

are formed on a classification basis as investors encode memetic information bearing certain characteristics into various thematic groupings.[19]
Subsequently, both the externality potential (x) and the correct memetic
sequencing of information will have a significant role to play in the formation of themes as well. This is simply because it is the combination of these
two factors that affect not just the formation of a meme but also (by-default)
the meme's characteristics too.

A key attribute of thematic information is that it is generally comprised
of the memetic sequences from a *variety* of individual asset information gen-
omes. The appearance of a particular meme for a given asset may be generally
associative with the appearance of an alternate meme for a different asset. In
the investor's mind both memetic strings are thus categorized into associative
information classes which we know as "themes." As will be illustrated later
in this chapter, this is a fundamental attribute of the way the human brain
stores and retrieves information.

Given their information byte microfoundations, the arrival of new themes
$(\dot{\Gamma})$ is coincident with the publication of a particularly path breaking
information byte that has exceptionally strong externality potential (x).
This information byte forms the kernel of an emergent theme as it binds
large sequences of memetic information together across a variety of different assets.[20] The appearance of these significant memetic strings then
have knock-on effects (via the investor's categorization process) to theme
formation.

Building block level 4: Market sentiment

Market sentiment (H) is a subset of the number of themes in existence (Γ)

$$H \subset \Gamma, \tag{4.6}$$

where market sentiment can take on either the binary "bullish" or "bearish"
form

$$H = \{\text{Bullish, Bearish}\}. \tag{4.7}$$

[19] Such memetic groupings are immediately analogous to what we observe in nature.
Watson (2004) describes the process where the presence of certain genes "switch on"
other genes until a certain "switching hierarchy" manifests where, left to its own
devices, the biological equivalent of a "brick wall" is produced from a "pile of bricks."
The argument also works in reverse where a pre-existing knowledge of the association between memes can assist in the search for the early stages of a particular theme
being built. We will put this knowledge to good work in our "early stage signaling" of
theme building in our applied analysis of Evolutionary Finance principles presented
in Chapter 7.
[20] The implications that this "binding formation" of memes into larger entities (emergent themes) has for asset prices is immediately analogous in the biological world to
the polygenic (multiple gene) effects that certain combinations of genes have for the
appearance of certain observed characteristics in human behavior.

This binary outcome reflects the highest level of aggregation for financial information. Again, it represents the outcome of the investor's neurological need to classify information, this time in relation to themes. The net balance of themes as having a predisposition to either a "buy" or "sell" orientation will determine the final "bullish" or "bearish" market sentiment decision.

Building block level 5: Asset prices

Finally, let us discuss how information relates to price – in effect, the phenotype of the financial information genotype. At any time t, and assuming (consistent with Fama's "fair-game" assertion) that the market price of a given asset reflects the sum of all available information,[21] then the price of any given asset (a) can be expressed as the sum of all information bytes available in the market as reflective on that particular asset (a) and scaled by a net memetic externality potential weight (ϕ)

$$p_a(t) = \int_0^t \phi \varphi_a x(t)\, dt, \tag{4.8}$$

where

φ_a is the information byte as applicable to asset (a);
x the externality potential contained in each information byte; and
ϕ is the net memetic externality potential weight.

This net memetic externality weight takes into account the sequencing of information and the way it unlocks the externality potential (x) within each information byte. It thus represents a convenient mechanism to reflect the net sum of memetic II across a particular asset's information genome ($\sum \varphi_a$).

Building block level 6: The market index

Summating the entire information-set across *all* assets (A) within a given market at time t and weighting each asset by the appropriate market capitalization weight forms the market index ($\ddot{I}$)

$$p_{\ddot{I}}(t) = \int_0^A \int_0^t \phi \varphi_a x(t)\, dt, \tag{4.9}$$

[21] Note that in contrast to Fama's "fair-game" model we have omitted the terminology "expected" from market prices as we are not concerned just yet with the aspects of the public versus private information dichotomy. Further, we also omit the terminology "market relevant" from our informational assessment as we allow for the embodiment of *all* information – even null (NN) information – into the information-set that determines asset prices.

where

Ï represents the market index of all asset prices in a given market;

φ is information byte signal as applicable to each asset within the index;

ϕ the net memetic externality potential weight;

x the externality potential contained within each information byte; and

A is the market capitalization weighted sum of all assets that constitute the given market index Ï.[22]

This in essence, represents the price manifestation of the market informational genome we mentioned earlier. Which, in turn, concludes our discussion of the building blocks of an evolutionary approach toward information in finance. In the next chapter we put these building blocks to work by developing a comprehensive model of information production/consumption in the financial arena and examining the asset price consequences of perpetual information byte output by the analyst community. However, before embarking upon this mission we should at first embellish our analysis by examining some of the implications of adopting an evolutionary approach toward information in financial markets. For starters, in the above we have made with subtlety certain assumptions as to the way information necessarily bolts together. Given our evolutionary context, these assumptions are nontrivial and need to be elaborated upon further.

4.4   Some consequences of our evolutionary approach toward modeling information

The gradualist versus punctuated equilibrium debate

As any student of evolutionary biology will only too readily testify, at present there is a raging debate within this particular field of academic endeavor. The "gradualist" versus "punctuated equilibrium" controversy has been a godsend for many a popular science writer. We too have inadvertently stumbled into this polemical area through our modeling of the actual building blocks of an evolutionary approach toward information in finance. How precisely did we manage this? Well, basically we have displayed a definite bias toward one of these two competing viewpoints in the modeling we have presented in the previous section.

By way of background, we provide a brief summary of both viewpoints below.

• First, there is the "gradualist" perspective that evolution occurs in a relatively smooth fashion with mostly (but not always) minor mutations. These mutations at times confer some advantage within a given

---

[22] Market capitalization is not the sole asset weighting procedure that could be used here. GDP or even an equal weighted system could be used just as effectively in the construction of the market index.

environment and thus favorable mutations will be positively "screened for" via the process of natural selection.

• Second, there is the "punctuated equilibrium" perspective that evolution proceeds more spasmodically with large-scale mutations both in form and across the breadth of species occurring over relatively short epochs followed by periods of an elongated quiescence in evolutionary activity over what are labeled as longer term "equilibrium" periods of time (hence, the concise nomenclature "punctuated equilibrium").[23]

The gradualist camp has been most notably championed by none other than the aforementioned Professor Richard Dawkins of Oxford University, while the punctuated equilibrium school was up until recently championed by the late Stephen Jay Gould of Harvard University.[24] This debate – sometimes acrimonious – not only had the makings of an Ivy-league transatlantic rift but also constituted a fundamental juxtaposition as to how life on earth (as we presently know it) actually evolved. Indeed, the battle lines between the "Dawkins versus Gould"/"gradualist versus punctuated equilibrium" perspectives could even be drawn in terms of "reductionist versus holistic thinking" – as briefly described in Chapter 3. It seems no stone has been left unturned by this – as yet unresolved – debate.

For example, Morris (2001) argued that the gradualist school of thought is fundamentally reductionist in their purview since they contend that *individual* genes (as screened for by natural selection) will determine the traits that make up an organism. Consequently, all one need do to "build" a given organism is to identify the right set of genes that match a given set of traits. To contrast, Morris (2001) argued that Gould and the "punctuated equilibrium" protagonists were more holistic in their perspective. This is primarily because of their arguing that the presence of "spandrels" (a type of externality in the genetic code as a result of the evolutionary process) results in the need to also take into consideration the way that genes *interact* with each other in the building of any organism – a type of "sum is greater than the part's" purview.[25]

[23] According to Gould, such short bursts of intense evolutionary activity have a duration that can be measured in tens of thousands of years – perhaps 10,000 to 50,000 years – rather than millions of years. The rest of the time, natural selection actually acts to keep speciation relatively *stable*. As will be argued a little later, this is a classic Complex nonlinear mathematics result. Darwinists however, would favor a smooth continuous form for a mathematical representation of the process of evolution.

[24] Gould first published his theory of punctuated equilibrium in 1972 "Punctuated Equilibrium: an Alternative to Phyletic Gradualism" under joint authorship with Niles Eldredge. An elegant portrayal of this theory is available in Eldredge's book *Reinventing Darwin* (1995). For Dawkin's perspective, see our earlier referenced *The Selfish Gene* (1976).

[25] For example, Winston (2003) raises the interesting evolutionary conundrum as to why deleterious neurological disorders such as depression seem to have survived the rigors of evolutionary screening. At a reductionist level, one would expect such

From a mathematical standpoint, Gould's approach should be seen to be consistent with the presence of Complex nonlinearities in the order of magnitude of mutations. Why? Simply because the earlier mentioned inter-dependencies between genes (in particular, the functional form of these interdependencies) provides a nice mathematical explanation for the perceived intermittency in the "jumps" in the evolutionary process between various equilibrium points. In short, the concept of "punctuated equilibrium" is entirely consistent with the Complex dynamics we presented in Chapter 3. This is a point that has been expanded upon in considerable detail by Kauffman (1993, 1995). Indeed, this Complex nonlinear outcome is in fact augmented by the externality effect represented by the presence of spandrels under Gould's framework which, in effect, assists in mathematically formalizing any nonlinearity.[26]

So just where do we fit into this debate? Given our sequence driven/externality orientated building block structure for an evolutionary approach toward information in finance (which, as we will display, has definite nonlinear potential) we can be definitely categorized as being sympathetic to Gould's "punctuated equilibrium" ideas.[27] In particular, it is our formalization of the "externality potential" (x) which we argue is displayed by certain information bytes, which forces us toward this outcome. As explained earlier, it is

---

"imperfections" to be removed. However, if one adopts a holistic perspective of the human mind then it may be that such negative emotions are a necessary by-product of the overwhelming complexity of human thought. In other words, it is possible that we have to make sacrifices in the form of some minor imperfections at the individual component level for the "greater good" of the way that we as humans think. This interdependency argument is consistent with the punctuated equilibrium purview of a holistic approach toward the process of evolution. More will be presented on the links between evolution and the human mind – and in particular, the human mind's information processing ability – later in this chapter.

[26] Indeed, this externality driven Complexity analogy can be applied not just to the evolution within a single organism but across a range of organisms as well. As Kauffman (1995) nicely summarized "Speciation and extinction seem very likely to reflect the spontaneous dynamics of a community of species. The very struggle to survive, to adapt to small and large changes ... may ultimately drive some species to extinction while creating novel niches for others. Life, then, unrolls in an unending procession of change, with small and large bursts of speciations, small and large bursts of extinctions, ringing out the old, ringing in the new ... , these patterns ... are somehow set organized, somehow collective emergent phenomena, somehow natural expressions of the laws of complexity."

[27] But is there a hole in our analogous argument for likening information in finance with information in the biological realm? After all, nonlinear jumps between two evolutionary price equilibria sometimes entail a "revisiting" to a previous steady-state price level. At first thought, one would not think there is an analogous "revisiting" in evolutionary speciation in the biological world but interestingly there are just such effects in some exceptional cases – see Whiting *et al.* (2003) "Loss and Recovery of Wings in Stick Insects" *Nature*, 241, pp. 264–67.

this crucial factor that enables information bytes to form memes when thrust into a sequenced information environment conducive to meme formation. The most obvious nonlinear observance of this information aggregation phenomenon is the unlocking of previously dormant latent information so that the II of memetic information is considerably greater than that of singular information – on a byte for byte basis. Further, as also mentioned earlier, it is important to recognize that it is not necessary for memes to form in direct proximity along the information genome to the particular information byte possessing high externality potential (x). The "externality control" of a given information byte can be exercised from relatively remote areas of the informational genome if for instance the particular information byte in question results in a new way of thinking about future information sequences. This epigenetic-like effect is entirely consistent with Gould's premise of spandrels (which are argued to typically reside in non-genetic areas of the genome sequence).[28] It is for this reason – in particular, the format of our assumptions – that we have inadvertently journeyed into an academic area well and truly signposted *"Here there be Dragons."*

The "rights" and "wrongs" of information evolution

The consequences of our evolutionary approach toward information in finance do not stop with the gradualist versus punctuated equilibrium debate. Loaded into the underlying assumptions of the analysis presented in Section 4.3 are some fundamental precepts about the way information actually *evolves* through time. Take for example the implications of our approach for the often-stated premise that investors always benefit from the provision of more information. Is this actually the case? Are there evolutionary "dead-ends" to the spread of biological-like formed memetic information that should be considered before making such assertions? Indeed, the existence of speculative bubbles seem to intimate as much. Does our framework have anything to say about the "rights" and "wrongs" of information evolution so that the dispersion of erroneous information can possibly be circumvented?

Well for starters, as Blackmore (1999) acknowledged, one cannot categorize the evolution of memes as having a predetermined conscious sense of "right or wrong" – in fact, it is just as easy for bad ideas to spread as good ideas.[29]

---

[28] Indeed, to be sure, Watson (2004) has described such regions as evolution's "wild frontier" with mutations aplenty. It may be that similar epigenetic-like effects occur in finance with seemingly innocuous information presented in the past suddenly having a dramatic "meme enhancing" effect when coupled with the presence of recent information.

[29] This is true of the biological world as well. Winston (2002) provides a nice example of suboptimal evolution – the human retina lies behind a layer of blood vessels and nerves which generates a blind spot. This is an obvious design flaw of human evolution that doesn't appear in large cephalopods such as squid and octopus.

Subsequently, there is little doubt that the system of equations presented in Section 4.3 can be just as efficacious in explaining the contagion of information that will ultimately prove deleterious to long-term investment performance as it does to explaining the spread of more worthwhile investment philosophies. Sure, as Professor Dawkins argued, it is highly likely that wrong memes will eventually get "weeded out" as part of the process of natural selection choosing only advantageous mutations – but all this involves an element of time. Mistakes often get made in the short term and furthermore as our discussion on Behavioral Finance in the previous chapter demonstrated, it is important to remember that what actually constitutes a "mistake" often not only involves a degree of judgment in terms of time but also who ultimately are the winners and losers.

As we have witnessed all too regularly in financial markets, sometimes certain investors are willing to "go along" with the belief structure of the crowd knowing very well that such expectations are likely to prove to be erroneous in the long term, but at the same time recognizing the potential for short-term gain in trend following behavior. Such is the "greater fool you" theory of investing. This strategy is perfectly "rational" for those skillful enough to pursue it successfully on a recurrent basis. Indeed, perceptions of "rationality" and "irrationality" under such circumstances ultimately reside on whether investors recognize the ephemeral nature of these momentum strategies (i.e., there is a judgment of time involved). Also, while such strategies may be "rational" and "right" for those adroit enough to get in and get out of the trade at the correct points in time, they are equally "irrational" and "wrong" for those who fail in this undertaking. That is, as identified earlier, there is also a judgment of who are the winners and who are the losers when it comes to classifying the "rights" and "wrongs" of information evolution. It's a simple case of the "winners are the grinners" under such conditions and to these individuals such information is undoubtedly "right" even though to the greater portion of the investment community (the losers) such information will always be – with the benefit of hindsight – classified as "wrong."

How to ensure one is a winner rather than loser under such a market environ? It is just as likely that a memetic sequence exists for signaling the optimal time to pursue such a strategy as it does for the embarkation upon what fundamentalists would argue to be more "sound" investment approaches.[30] In fact, trying not to be harsh, one could label momentum strategies as being definitely "parasitic" in nature – at least from an evolutionary perspective. These parasitic outcomes of rampant opportunism at another's expense are an endemic feature of evolution both in real and

---

[30] The term "fundamentalist" in this context refers to those who advocate the use of long-term valuation methodologies as the sole motive for investing.

artificial ecosystems alike.[31] Rather than being evolutionary "dead-ends" such strategies should be seen to be one of the natural outcomes of the evolutionary process. There is no "right" and "wrong" in such results. They are simply a reflection of the optimal strategy for certain investors to pursue given the present evolutionary stage of the information genome for the market. Indeed, while the memetic genesis of each speculative bubble may eventually get "weeded out" through time as the majority of investors learn that the premise of their beliefs has proven to be ill-founded, the actual parasitic response of the few who correctly diagnose the presence of a speculative bubble and then (successfully) use momentum trading to exploit this phenomenon is likely to be handed down from generation to generation.[32] In other words, rather than trying to regulate against such activity, one should see such bouts of "irrational exuberance" as a natural outcome of the evolutionary market process.

But this still hasn't answered our question – how best to ensure one is on the right side of this regularly reoccurring lucrative trade? The trick in developing such winning parasitic strategies is ascertaining the memetic string that signals the optimal time to pursue momentum investing – but in turn, that doesn't rely upon the actual fundamental cause of the speculative bubble itself as a trigger (as this will be constantly changing). Rather, for such a strategy to be pursued successfully on a recurrent basis any devised memetic string signaling device must rely upon the generic impact that the presence of a speculative bubble has upon asset prices at large – a difficult but not insurmountable task. We will elaborate upon how to ascertain these all-important memetic strings in much greater detail in our *applied* Evolutionary Finance analysis presented in Chapter 7 of this book. Suffice to say now, we feel that the ecology of the market in terms of speciation of possible investment strategies will prove to be just as startling in its diversity under our Evolutionary Finance banner as does the biological world constantly amaze in its diversity under the principles of Darwinism.

So by way of summary, to be regularly successful in choosing the correct strategy for a given asset over a chosen duration of time from the diverse ecology of possibilities, one must have a firm grasp as to how information evolves over time. There is no "right" or "wrong" in this process only winners and losers. Winning evolutionary strategies are ultimately "right" while those strategies that fail from an evolutionary perspective will become the

---

[31] For example, Thomas Ray has been a pioneer in creating biological-like replication within computer programs. An interesting outcome of his research was the emergence of "parasitic" like code structures. Ray hypothesized that parasites were a natural by-product of the evolutionary process. Indeed, it appears highly likely that parasitic investment strategies are a natural outcome of an evolutionary approach toward information within the field of finance.

[32] In this instance, a "generation" reflects the average periodicity between speculative bubbles.

aforementioned "dead-ends." It's simply a case of the better adapted (at least in terms of *understanding* the evolutionary drivers of market information) outperforming the ill-adapted – as should be expected under the auspices of natural selection. Unfortunately, up until now, a framework for describing this process within the field of finance has been lacking. Subsequently, this is one area where we feel our Evolutionary Finance approach can make a huge contribution.

The stock and flow of knowledge

But what does the framework presented in Section 4.3 have to say about the process of information production itself? Are there any implied assumptions as to the "manufacture" of information buried within our analysis? Interestingly there are evolutionary interpretations to be drawn here as well. For starters, let us concentrate on the strategic behavior of information providers. As will be elaborated upon in considerable detail in Chapter 5, strategic evolutionary behavior is rampant within the analyst community. Nowhere is this more evident than in the pursuit of thematic based research. The choice of topic for a particular information byte, its final recommendation, how this is contributory or contrarian to the prevailing market consensus, are all crucial strategic decisions undertaken by analysts on a day-to-day basis. Modeling this within the context of the framework presented in Section 4.3 provides the opportunity for the study of various evolutionary dynamic principles such as predator/prey responses in topic selection, bull/bear positioning for the differentiation of research product and responding to the opportunity for contrarian swings in market sentiment.[33]

But all this is grist for the mill for our modeling to be presented in Chapter 5. What is important to recognize here and now is that a natural by-product of our assumptions in Section 4.3 as to the continuous production of information bytes is a tacit recognition that the information genome itself will be continuously expanding across time. In other words, the stock of information (and possibly knowledge itself) is ever rising. A rising stock of information has the potential to result in an expanding knowledge base if one assumes theme diversity expands "hand-in-glove" with the pace of information. Why is theme diversity an integral part of a rising stock of knowledge? Simply because if all information bytes were on the same topic, then overall "knowledge" *per se* would not be increasing even though "information" is.

---

[33] The possibility for predator/prey strategic responses on behalf of analysts within our evolutionary framework should really come as no surprise. Lewin (1999) – in the spirit of Thomas Ray's earlier work – illustrated that predator/prey relationships not only apply to the physical world but also in the artificial world created within computer simulations of artificial life systems. Subsequently within the elaborate framework for the evolutionary modeling of financial information presented in the previous section, there are a plethora of avenues for pursuing the formalization of strategic information provision by analysts – with predator/prey dynamics representing just one of these outcomes.

Thus it is important to form a link between information overall and the subject matter of that information if we are to say something meaningful in regard to the relationship between information and knowledge. Forging a direct connection between theme diversity and "knowledge" in its most general form is thus a convenient way to do this. Subsequently, as the following shaded section illustrates, should analysts decide to embark upon a career of information byte production – which we assume they do, using strategic evolutionary behavior to guide them in their choice of topic for each information byte – then the stock of information (and possibly "knowledge" in its most general form) will be on a perpetually upward trajectory – that is, provided theme diversity too is rising commensurately.[34]

The ever-expanding stock of information (knowledge)

For simplicity, let us assume analysts produce *all* information and there is a one-to-one link between information byte production and theme diversity (an assumption we will later drop in Chapter 5).

Ergo, we assume analysts produce both information and knowledge (K) in a manner consistent with strategic evolutionary principles at a rate η. The amount of information (knowledge) in the world is thus governed by the equation

$$\dot{K} = \eta K. \tag{4.10}$$

Separating variables gives

$$\frac{dK}{K} = \eta \, dt. \tag{4.11}$$

Integrating both sides gives

$$\ln K = \eta t + c, \tag{4.12}$$

where "c" is a constant of integration. Assuming (rightly or wrongly) that we are not alone in the universe and some omnipotent "being" had at least some information (knowledge) before the genesis of the analyst community, then the amount of information (knowledge) at time $t = 0$ is a non-zero integer represented by K_0. Thus the final solution to the growth of information (knowledge) under this rather simplistic interpretation of our Evolutionary Finance principles is

$$K = K_0 e^{\eta t}. \tag{4.13}$$

This can be represented in graphical form as shown in Figure 4.1.

---

[34] For the stock of knowledge to actually fall, we would need to introduce some form of entropy within our modeling process. While this is not explicitly undertaken, it is cited as a worthwhile avenue for future research. For example, Shannon and Weaver (1963) demonstrated that the more new information entering a system, the faster the rate of decay (entropy) of the existing information-set.

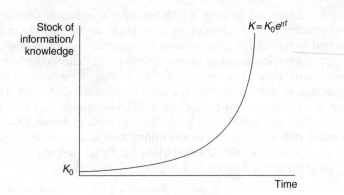

Figure 4.1   The ever-expanding stock of knowledge

In Chapter 5 we will expand this concept considerably by removing the assumption of a one-to-one link between information byte production and expanding theme diversity. In particular, we introduce a model where analysts are perpetually economically rewarded for introducing new ideas and concepts through the pursuit of thematic based research (i.e., the production of information bytes in a thematically purposeful way) where there are direct incentives to expand theme diversity. We then allow for the potential for variations in the rate of information production and the rate of theme diversity production (η) by formalizing various scenarios for strategic evolutionary behavior on behalf of the analyst community in terms of the choice of topic for their information bytes – which in some cases results in the emergence of a new theme. Subsequently, we are able to formalize alterations to the rate of knowledge accumulation (K) simply because knowledge itself is related to this time-variant pattern in theme diversity. In particular, switching between new and old themes in their (continuous) information byte production results in wave-like patterns emerging in the upward trajectory of (K). Subsequently, we are able to formalize into one interconnected framework a whole host of interesting behavioral outcomes ranging from strategic analyst behavior, to market (and therefore price) responses to theme variation, to alterations in both the structure and fabric of knowledge itself. This is but the tip of the proverbial iceberg when it comes to reaping some of the benefits of our evolutionary approach toward information in the financial sphere.

But can we state more about the flow of information without recourse to the sophisticated model we present in Chapter 5? Sure, as we have stated earlier, in Chapter 5 we will present various scenarios for strategic evolutionary behavior on behalf of analysts altering the flow in the provision of thematic information (and therefore "knowledge" itself) – but so far this is just an analysis (albeit in-depth) of the supply response. Can we state anything about the demand response? In particular, are there any "demand-pull" principles regarding the investor take-up of the ever-rising tide of analyst provided information that we can elaborate upon at this introductory stage of our analysis? The good news is that there is a considerable body of literature dealing with this very premise. What we are referring to here is the ubiquitous "S-shaped" diffusion pattern for information dispersal – a model that is often cited as the "base-case" scenario for illustrating the take-up of new information.

Diffusion is an often-encountered principle in metaphysical and physical phenomena that appears across a spectrum of intellectual thought – ranging from physics to chemistry to engineering to economics and finance. The concept of information diffusion (as embodied in the form of a new technology) was first introduced by Ryan and Gross (1943) but then popularized within the field of economics and finance by Zvi Grilliches (1957, 1960). Each of these studies considered the take-up of new hybrid corn varieties across the US mid-West. That said, the principle of diffusion itself has now been extended to a plethora of information dispersal examples including (among other things) consumer acceptance of new generation mobile phones, to internet usage, even to the adoption of rap music by white Americans.[35]

Depending upon specification, the pace of diffusion can be either fast or slow but the generalized pattern remains the same – an elongated S-shape for the cumulative take-up of the observed "diffusive" variable across time. Why is this the case? In short, two characteristics typify a diffusion model – a finite population and a (constrained) exponential growth function. The "diffusion" variable typically represents the relative proportion of the finite population who has taken onboard the new information at a rate dictated by the (constrained) exponential growth function. By observing the order of magnitude of the second derivative of the particular diffusive variable with respect to time it is possible to see the drivers of the characteristic S-shape pattern that emerges.

But this still doesn't shed any light on why an S-shaped pattern should appear in preference to other well-documented graphical representations of dynamic forms? Well, at the early stages of diffusion, the finite population does not provide a major constraint upon the exponential-like take-up of the diffusive variable – whether this represents the use of new hybrid corn varieties within a geographic locale, the appeal of rap music to a particular audience, or the conversion of a proportion of the analyst community believing in a particular investment theme. So for a short period of time, the initial stages of the diffusion pattern resemble more the ever-increasing function presented in Figure 4.1. Gradually however, the finite population begins to reach saturation point and the pace of diffusion begins to slow. As stated earlier, in mathematical jargon this implies that the second derivative

---

[35] See Rodgers (1995) for an extensive discussion of the diffusion model and a colorful array of examples of the take-up of new products (which, in effect represent a classic embodiment of new "information").

flips from positive to negative as the finite population constraint is neared. It is this combination of events that gives the evolutionary time path of the diffusive variable its characteristic "S-shaped" pattern as in the simplest of terms, the upward spike represented in Figure 4.1 is repelled as it approaches the finite population ceiling and thus the function gradually "bends" back down toward flat.

Adding to our conviction for this particular format for new theme dispersal is an analogous body of research that is generally supportive of the S-shaped diffusion pattern as representative for the "demand-pull" by the analyst community for new investment themes. What we are referring to here is the considerable amount of literature that has been put forward within the field of finance on the presence of "information cascades."[36] Here, the "cascade" is effectively started by one "new theme believer" analyst talking to another (and thereby converting) two more analysts (total converts = 3), who then each impart this information to two more analysts (total converts = 7), then these four "newly converted" analysts each talk to another two analysts (total converts = 15) ... and so on.[37] The path of the stylized information cascade is represented in Figure 4.2. In short, it represents a classic example of a geometric progression, but what can be readily seen from Figure 4.2 is that

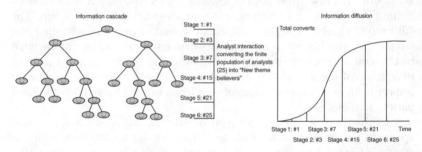

Figure 4.2 An information cascade for new theme contagion between analysts

[36] Sometimes such information cascades result in the presence of our earlier mentioned Complex nonlinearities. The link between Complexity Theory and information cascades (sometimes known as "information avalanches") is nicely summarized by Bak (1997); "Complex behavior in nature reflects the tendency to evolve in a poised critical state, way out of balance, where minor disturbances my lead to events, called avalanches, of all sizes.... The evolution to this delicate state occurs without design from any outside agent. The state is established solely because of the dynamical interactions among individual elements of the system: the critical state is *self-organized*." For an exposition of a relatively simple model of an information cascade, see Banerjee (1992, 1993).

[37] The geometric scalar chosen here is by way of example only. Obviously the higher the scalar, the faster the pace of diffusion (and vice versa).

pretty soon it becomes difficult to find an analyst who actually isn't "in" on the story. This is where the finite population of analysts begins to limit the pace of contagion of the new theme and therefore the information diffusion pattern acquires its characteristic S-shape.

In the box below, we illustrate the formalization of such an information cascade/information diffusion pattern – but interestingly, it is not instigated by analyst "agents" who do the diffusing. Rather, we use the attributes of information itself – in particular, the strength of externality potential (x) that a particular information byte possesses – to act as the diffusive control variable. Why? In short, it is the strength of externality potential (x) that acts as the coagulating force in meme formation that is related to a particular theme (j). It is therefore this innate characteristic of information itself that in essence can act as a driving force in the contagion of any given theme (j). Basically, externality potential (x) is a formal embodiment of the more a given theme gets "talked about" and "related" to other events (in terms of formation of associative memes) it therefore governs the rate of a theme's uptake within the investment community.[38] However, as a particular theme's dominancy begins to grow, it starts to approach the saturation point for the collective investment community's mindset. When this occurs, the pace of diffusion begins to slow.[39] It is here that our externality potential driven (memetic-based) diffusion process acquires its characteristic "S-shape" – as will be illustrated by the following.

The diffusion of memetic information[40]

Let us assume that a new theme (j) is "announced" via the publication of a seminal information byte that has exceptionally strong externality potential (x). Further, let us assume there is a fixed population of investors (Z) with the proportion of investors believing in this new theme being represented by (Θ).

The stronger the externality potential (x) the greater the pace of meme formation – especially when this particular information byte is thrust into a sequenced information environment conducive to meme formation. Assuming a simple linear relationship between meme formation related to theme (j) and the take-up of this theme by the fixed population of investors (Z), then the following

---

[38] Notice here that we made the switch between "investment community" and "analyst community." As will be displayed in the Chapter 5, this switch is inconsequential for our purposes as we assume that the total number of analysts equates to the total number of investors. Needless to say, this does not always have to be the case but for modeling purposes it does help reduce the complexity of our system of equations.

[39] We will build upon the concept of theme dominancy (and the consequences this has for strategic evolutionary behavior on behalf of analysts) considerably in the next chapter.

[40] For a comprehensive overview of the mathematics underlying diffusion-based methods, see Crank (1975).

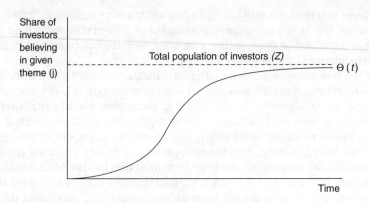

Figure 4.3 The memetic-based diffusion of a given theme

logistic equation represents the time dependent evolution of the share of investors believing in this particular theme at a given point in time[41]

$$\dot{\Theta} = x\Theta \left(1 - \frac{\Theta}{Z}\right). \tag{4.14}$$

Separation of variables and integration gives the solution to the above equation as

$$\Theta = \frac{Z\Theta(0)e^{xt}}{Z + \Theta(0)\left(e^{xt} - 1\right)}. \tag{4.15}$$

The evolution of (Θ) can thus be represented graphically as shown in Figure 4.3.

Subsequently, we have formalized is the ubiquitous "S-shaped" diffusion pattern for information dispersal into a simple model where the characteristics of information itself – in particular, the externality potential (x) embodied within each information byte – governs the pace of the memetic-driven uptake of a given theme (j). This is a powerful result that we will revisit upon in many occasions in the ensuing pages as we further embellish our Evolutionary Finance principles.

As stated earlier, we will elaborate upon this particular pattern of theme dispersal (along with its implications for asset prices) in much greater detail in the calibrations for our comprehensive evolutionary model of the

---

[41] We assume here for convenience that each meme is effectively the same size and meme formation – as governed by externality potential (x) – is a smooth continuous function. Should we assume a more "lumpy" pace of meme formation then Figure 4.3 would still basically resemble the characteristic "S-shaped" diffusion pattern in terms of its trend path but the pace of investor take-up would be decidedly less smooth so that the diagrammatical representation would be less smooth also.

marketplace to be presented in Chapter 5. That said, we hope we have provided at least the preliminaries of some of the forces governing both the stock and flow of information (knowledge) within our comprehensive Evolutionary Finance approach toward the study of the actual microstructure of information in the financial world. Further, we hope we have provided some insight as to the implications of the assumptions we have made in our formalization of this informational microstructure in Section 4.3. As stated at the outset, these assumptions were nontrivial and we hope we have displayed as much in the preceding pages. But before we further elaborate upon our comprehensive evolutionary model of the marketplace itself, we must at first understand how investors actually *interpret* information that is the result of our analogous evolutionary/biological-like building block process. Do investors actually "think" in a manner that is supportive of our revolutionary viewpoint? As will be seen, the framework we have presented so far is surprisingly sympathetic to neurological studies as to the way humans actually process information. This is good news for us but definitely bad news for the Strong EMH/traditionalist finance paradigm that's for sure.

4.5   How investors interpret Evolutionary Information

One of the serendipitous beauties of our evolutionary approach toward information in finance is that not only is it analogous to the way researchers have described evolution in the biological world, but it is also analogous to the way neuroscientists have described the human thought process. We as humans approach information in a systematic way. We break information down into its component parts, we sift for what information is relevant and what information can be discarded, we look for associations between various stratum of information both for interpretation and for forming inter-linkages so that the retrieval of this information is easier at a later date. As will be displayed, these key attributes are inexorably consistent with the structured approach toward information under our Evolutionary Finance framework.

Investors look to build themes from memes, and memes from bytes. They can't help but think this way because their entire information interpretation experience is founded on the premises of categorization and association. "Bytes"/"memes"/"themes" are merely a convenient nomenclature for describing this human thought process within a context that is readily understandable within the financial arena. Indeed, the consistency between our approach and the neuroscience approach is so stark that it seems almost unbelievable that a detailed study of the actual microfoundation building block structure of information within finance has not been undertaken until the publication of this book. Certainly we as humans do not treat information as though it is the outcome of some random number generator, so why should we overlook the wealth of evidence from the neuroscience area pointing to a systematized piece-by-piece structured approach toward information processing by the human mind in favor of the random-based IID principles that are assumed to apply under the Strong EMH/traditionalist framework?

To illustrate our point, we will at first provide a cursory examination of investor perception – detailing the strong consistency between the neuroscience perspective of the human mind's information processing regimen and our own Evolutionary Finance perspective. We then go on to extend our analysis by exploring in greater detail the nature versus nurture debate as to the way humans systematically approach information. In particular, we highlight that while great strides have been made in the field of Behavioral Finance to better understand the "humanity" of the investor's thought process, this approach can still be criticized for being too "nurture" orientated in its foundations. Finally, we examine the ultimate human limits of information absorption itself and how this can at times act as a constraint on the optimal outcome for meme and theme formation.

The fundamentals of investor perception – similarities between our Evolutionary Finance purview and the neuroscientist's perspective on information processing

Let us start by stating right from the outset that there is no way we can even attempt to cover the complex area of human perception in the short space allotted here. Rather, what we will attempt to do is to outline a series of points that we feel are crucial in detailing how our structured approach toward information under our Evolutionary Finance banner is strikingly consistent with neuroscientists view as to how human's "make up their minds."

Point of consistency #1. Information encoding – the first thing that happens when we "perceive" information is that we encode it. Is what we have just experienced a sight, a sound, a sense of smell, a taste or a touch? In much the same fashion as we have encoded information into a series of actionable signals – "buy" (B), "neutral" (N) or "sell" (S) – under our Evolutionary Finance framework, so too does the human brain find it necessary to encode raw information into meaningful entities.[42] Indeed, assisting us greatly in this process is the obvious structure provided by language. In precisely the same way that bytes make up memes, and memes make up themes, phonemes (various combinations of sounds) make up words which in turn forms syntax. In effect, words give our experiences structure and meaning and thus help us to encode information.[43] Writing too plays an important role as it forms a record of such encoding and thus assists in the process of learning. To quote Winston (2003) "Language, like making and using a tool,

---

[42] For a detailed discussion of the mechanics of this process, refer to McClelland and Rogers (2003).
[43] For a discussion of the evolutionary linkage between human development and speech, see Aitchison (2000).

is a sequence."[44] It is thus this sequencing of encoded information that provides a key foundation stone for both the processing of information in the human mind and the formalization of the informational microstructure under our Evolutionary Finance approach. Ergo, it is little wonder that we should put this point of consistency at the top of our list for the symbiotic nature between the neuroscience perspective and our own structured approach toward information interpretation in the finance world.

Point of consistency #2. Information categorization – immediately after encoding information we categorize it. In short, the human brain breaks new information down into manageable interpretable "bits" before categorizing these various "bits" of information by their associations. This act of association is important for the storage and retrieval of information at a latter date.[45] For example, looking at a particular person may provide information as to gender, age, hair color, apparel, height, weight, even clues to where they have been recently (have they been swimming for instance) and where they have been in the past (e.g., they could be wearing a school ring). We use a variety of perceptions to make these judgments – sight, smell, sound and so on. The piece-by-piece subcomponent categorization and interpretation of information is entirely consistent with the "byte to meme" and "meme to theme" building block structure we have presented in Section 4.3. This is simply because the structure we have developed is based upon the process of association too. In the earlier example, the person in question may be associated with other individuals in our minds simply because they have red hair. The same process of association occurs when we categorize various memes into different themes – a cyclical recovery story, a long-term demographic story, an exchange rate volatility story and the like. Indeed, a well-known example of such categorization at work is the so-called "P300 response". This occurs when volunteers are given a list of words with strong associations but also one exception. Invariably, when the readers strike the exception there is a lagged 300 millisecond surge in EEG measurement as the reader processes the new information and the brain must work harder to "look for" where to categorize this associative information "outlier." Sorting a group

---

[44] Winston (2003) *The Human Mind: And How to Make the Most of It*, p. 191.
[45] The categorization of information into separate bundles is handled by separate areas of the brain. For example, Winston (2003) highlights that as information travels through the brain it is split into various streams – some information may go to the visual cortex, some to the limbic system for tagging in terms of memory and emotion. Within each of these regions information is split down further still before being married to corresponding associations. Within this process, the anterior cingualate gyrus (ACG) plays a key role by acting as a filter for incoming information – prioritizing some information while discarding other information that is not needed. This is a necessary evil to prevent information overload, otherwise we would not be able to drive down a busy street without running the risk of a serious accident simply because we are overwhelmed by the information onslaught.

of memes into themes requires precisely the same neural structure. In fact, we are neurologically "hard wired" to categorize. During the first stages of life we produce new neurons at the staggering rate of 250,000 per minute. Initially these neurons connect with each other in a seemingly random fashion but during our early infant years apoptosis takes hold and neurons are selectively culled.[46] It is only the neurons that have built up the strongest associations that survive – this is done by being repetitively called upon to categorize information "bits" in accordance with their associative similarities. Very seldom do we associate cheese with a pen so it is unlikely that such a neuron connection would survive. This process of apoptosis is sometimes referred to as "Neural Darwinism." Without doubt, we call upon this "associative hard wiring" in our interpretations of financial information in precisely the same fashion that we use this associative neuron network to interpret other forms of information. What we have done in the microstructure building block framework that we have presented in Section 4.3 is to give this associative behavior a formal mathematical reasoning using a terminology familiar to most of us in the finance profession so that we can better investigate the implications of the human mind's categorization and association process across a variety of observed financial phenomena.

Point of consistency #3.   Information latency, memory and the externality potential – hand-in-glove with the process of information categorization and association goes the unconscious act of information storage. Interestingly, there are parallels to be drawn here as well between the neuroscience perspective of information processing and our own Evolutionary Finance purview. In particular, we refer to the presence of information latency. Quite often a major crime case is not solved until one crucial piece of the puzzle falls into place and the full associative network of information becomes apparent. Our mind uses precisely the same information latency effect in drawing its own interpretations – indeed, a major constituent of creative thought relies on this very premise (the "eureka" effect). Creative genius often relies upon a sequence of subconscious associations until some trigger – quite often a mistake – awakens this network of sleeper latent information and the full beauty of the interconnected information-set can be appreciated in its entirety. This fundamental aspect of the way we as humans "think" is

---

[46] By way of background, each neuron in the brain is connected to up to 10,000 other neurons by tendril like roots. These trendrils come in two forms – dendrites and axons. Dendrites receive information, axons send information. The information is "sent" via the release of chemicals from the neuron. Mainzer (1997) highlights that there are $10^{10^{11}}$ associative neuron interconnections within our neural subsystem. This represents a huge amount of processing power for encoding, categorizing and processing information. Neural network algorithms represent but a limited approximation of this information processing power. These computer algorithms will be discussed in greater detail in Chapter 7 in our *applied* perspective on Evolutionary Finance.

exactly the reason why we have explicitly formalized a latency structure into our microfoundation building block framework for information outlined in Section 4.3. But the parallels do not stop here. Generally the stronger the emotion associated with a memory (in effect, a latent "bit" of information) the better its recall.[47] Most people can tell you precisely where they were when they learned of the shocking events of 9/11. In memory, this emotive linkage is called long term potentiation (LTP). This "emotional scalar" attachment to our memories is entirely synonymous with the concept of externality potential (x) being an integral feature of each and every information byte. Obviously, it will be a combination of latency and strong externality potential (x) that helps form strong memes but this occurs in precisely the same fashion that a combination of latency and strong emotional attachment helps form strong memories. The similarities are striking. In short, in both cases it is a combination of the two factors – externality potential (x) for Evolutionary Finance, emotion for neuroscience – with a form of informational latency that is crucial in forming the all-important associative jig-saw puzzle that helps us to interpret our informational environment.[48]

Point of consistency #4. Learning and meme selection – synonymous with the above, strong memories and strong memes must at times withstand the process of selection. In neuroscience this is labeled as "learning." Getting to match the right associations between new and old information is effectively how we "learn." This selection process is entirely consistent with the way we select memes – we gradually learn the correct associations. As stated earlier, there are no philosophical "rights" or "wrongs" in this process, just correct and incorrect interpretations of the given state of the informational environment. That said, sometimes this process of deducing the "correct" memes for a given point in time over the inevitable evolution of the market's informational genome can be expedited. As we stated earlier, Professor Dawkins (1976) made considerable headway in stressing the importance of imitation as a mechanism for meme selection. Neuroscience too highlights the importance of imitation as a mechanism for expediting learning.[49] Indeed, Winston (2003) illustrates that just as humans are neurologically "hard wired" to categorize, they are also "hard wired" to imitate each other – some neurons within the brain only fire in response to the actions of other humans. This is especially important in learning difficult concepts such as language and the appropriate responses to universal facial gestures such as smiling

[47] In terms of neural structure, the amygdala plays a key role here as it "tags" incoming information with an emotional value.

[48] Consistent with the above, memory can be enhanced by "training" the mind for association. For example, the "Method of Loci" uses a stylized journey to assist individuals in recalling a string of facts.

[49] Attenborough (1980) highlights that complex imitation as means of learning distinguishes higher order – human-like – apes from their lower order brethren.

and frowning. But imitation is not the only way we learn. The earlier cited "eureka" effect intimates how creativity (via association) at times relies upon surpassing (and even circumventing) the rote process of imitation.[50] It is interesting that both imitation and creativity form an integral part of the strategic analyst behavior cited in our model in Chapter 5. Just as these processes are essential to the way we learn to interpret each other's behavior, they are also crucial to the way we select and develop superior investment strategies. What we have done in the evolutionary building block information framework we have presented in Section 4.3 is to put forward a formal mathematical representation which allows us to better understand this fundamental human (and therefore investor) action. In fact, in our *applied* analysis presented in Chapter 7 of this book, we take this process even further by developing a framework that predicts how other market participants will actually "learn" a new investment strategy. This effectively keeps our readers one step ahead of the pack in developing evolutionary strategies that take into account the learning/imitative behavior of others.

Point of consistency #5. Shaping our informational world – finally we come to the institutional setting that we create to assist us in our information processing role. Just as there are countless examples of classification nomenclature that assist us in categorizing general information – the chapters of a book for instance, the demarcations of isles in a supermarket, the sorting of magazines at a newsstand – so too we have created a classification nomenclature in finance. Micro/macro, growth/value, technical/fundamental are all labels which assist us in classifying the research output by analysts.[51] But the classification does not stop here. Interestingly, such categorizations can also be reflected in floor layout within fund managers or investment banks. Many of us have walked onto a large trading floor and have been initially bewildered by the seemingly overwhelming amount of information presented to our senses. At closer inspection this "informational noise" begins to form a structure as we witness various clues to assist us in our categorization efforts. Small national flags on trading desks illustrate the geographic proximity of where the desk "trades," what news services are up on the traders screens, even the size of their wristwatches (fixed income traders generally favor big chunky sports chronographs) all provide us with insights to help

---

[50] For example, Peters (1996) highlighted that early Artificial Intelligence algorithms were consistently beaten by Grand Masters as they failed to possess the human characteristic to associate. Rather, they sought to go through every possible permutation of response by referencing identical games in history rather than their human counterparts who were able to refer to similar (but not exact) historical precedents and devise new, superior strategies.

[51] As stated earlier, typical artificial models of the market break the "market" into different trading groups. For example, Rodriguez *et al.* (2002) break the market into two warring factions of "fundamentalists" and "chartists" in their artificial market model of deterministic Chaos. This in effect, represents an informational (and subsequent institutional) classification typically undertaken within the market.

sort and classify what is initially an amorphous mass of information.[52] Once this is done the information becomes much more manageable and indeed we start to build up a schematic representation of a fully categorized trading floor within our heads. "FX guys to the left, derivative guys straight ahead, CBOT traders to the right within this group" and so on. We cannot help but think in this way just as we cannot avoid building an elaborate financial institutional structure that by its very labeling is assistive for us in our efforts to categorize information. It is simply a function of the way our brains are structured. We will dwell upon the relationship between institutional structure and its information servicing role in much greater detail in Chapter 8. Suffice to say here, the metaphysical building block structure that we have provided in Section 4.3 for an evolutionary approach toward information in finance is entirely consistent with the evidence we observe in the physical world – both within the finance industry and beyond – simply because it too is nomenclature based. The labeling of various stratum of information into component bytes, memes, themes and even overall market sentiment has been quite purposeful. All these terms are already in common usage. All have been used at various times to describe that amorphous mass of what we perceive as financial "information." What we have done in Section 4.3 is to give this information a structure so that we can better understand it – in precisely the same fashion that the trading floor problem was effectively "sorted" in our minds in the earlier example. In this respect the nomenclature we have chosen, and indeed the act of choosing nomenclature itself, is consistent with neuroscience observations as to the way we as humans think. Yet again our Evolutionary Finance perspective and the neuroscientists perspective provides a nice marriage of convenience in better describing (and understanding) our informational world.

The nature versus nurture debate on information processing – moving beyond the Behavioral Finance purview

Without doubt, a number of the earlier points illustrate that it is ultimately a *combination* of both nature and nurture that determines how we as investors "think", for example, it is both neurological "hard wiring" and our actual experiences that help us to form the all-important associations for categorizing and retrieving information.[53] With respect to this "humaness" embodied

---

[52] As stated in Chapter 3, it is interesting to observe the Complex nonlinearities that emerge when modeling the interactions between such groups. At times even relatively unsophisticated creatures can display surprisingly intelligent behavior when drawing upon the processing power of the collective group – see for instance, *Swarm Intelligence: From Natural to Artificial Systems* by Bonabeau *et al.* (1999).

[53] Interestingly, it was Francis Galton who first coined the terms "nature" and "nurture" to describe the different forces in shaping human behavior in his treatise *Hereditary Genius: An Inquiry into Its Laws and Consequences* in 1869. Unfortunately, it was Galton too who was to give the "eugenics" movement its name.

within all investor decision-making, Behavioral Finance has made great strides at countering some of the more outlandishly stringent belief structures purveyed by the Strong EMH/traditionalist finance school of thought. That said, as touched upon in Chapter 3, Behavioral Finance can still be criticized for treating information as though it were some intangible commodity. In particular, Behavioral Finance has displayed little understanding of the microfoundation building blocks for what we as investors come to perceive to be "information" in the first place. These concerns aside for the moment however, there is one additional criticism that can be leveled at Behavioral Finance in its erstwhile noble endeavors. In particular, we ask the polemical question – is Behavioral Finance necessarily too "behaviorist" in its interpretation of investor decision-making?

To begin with, let us take an evolutionary psychology perspective on investor decision-making. Evolutionary psychology is a rapidly emerging field of research in the social sciences arena.[54] Basically its philosophy is to argue that the human brain has evolved on a functional basis in much the same fashion as any other human tissue – in short, the human brain is a by-product of natural selection. Consequently, it can be argued that certain elements within the human brain are triggered into action primarily to respond to various primal needs – fight/flight/fright, sex, communication and so on – simply because such responses are generally associated with basic survival/reproduction type behaviors. These responses (and the neurological hard wiring that accompanies them) are thus positively "screened for" by the natural selection process. Indeed, it is an increasingly accepted fact that these instinctual responses are already ingrained within one's neurological processes at birth.[55] But do such "low brow" instinctual responses also have a role to play in "high brow" reasoning – such as investment decision-making – as well? The short answer is – yes they do.

According to Robert Winston (2002) "Instinct, then, is essentially that part of our behavior that is not learned."[56] Indeed, Pinker (1994) identified that commensurate with natural selection forces governing the development of the human brain, both reasoning and rational thought are themselves (by default) the by-products of natural selection and therefore not completely "learned." In other words, one can argue that the basics of even notoriously "high brow" neural actions are to some extent hereditary (read "instinctual") – at least in the sense that they reflect the neurological "hard

---

[54] See Cartwright (2000) for a comprehensive introduction to evolutionary psychology principles.
[55] This purview contrasts diametrically with Behaviorists – a group of psychologists who believe the human mind is infinitely malleable from birth and is thus primarily molded by learning and our ongoing life experiences. In other words, Behaviorists believe that humans are born a *tabula rasa* – or "blank slate" so to speak.
[56] See Winston (2002) *Human Instinct: How our Primeval Impulses Shape our Modern Lives*, p. 6.

wiring" of the evolution of the human mind.[57] True, learning and experience will always play a predominant role in the final outcome of any higher level thoughts, but it should also be noted that humans evolved to the top of the phylogenetic totem pole primarily because of a *combination* of their nature and nurture talents. In other words, we would argue that Behavioral Finance may be lacking in its interpretation of the "humaness" of investor behavior simply because of its failure to take into consideration the notion of "financial instinct."[58]

Surely, as we will readily admit "instinct" does not govern a full 100 percent of investor action – but neither does it have a zero influence either. Indeed, we would go as far as to argue that try as investors might, they cannot avoid at least some degree of neurologically ingrained instinctual undercurrents within their actions simply because (like every living organism) they are the sons and daughters of their primitive ancestors. As demonstrated at numerous stages in the preceding subsection, we have evolved to think in a certain way. In short, a certain algorithmic sequence of "encoding/associating/memorizing/learning" has been "hard wired" into the human brain and this constitutes the universal foundation for the way we approach problem solving.[59] Indeed, just as this "hard wiring" affects our purposeful customization of the institutional structure that surrounds us so that we can better assimilate information, so too have we been "hard wired" to think about our investment decisions in a particular way. This all-too-human instinctual neural hard wiring undercurrent to our behavior has so far been largely overlooked by the Behavioral Finance theorists. Indeed, we would go as far as to say that a natural "evolution" in the interpretations of investor behavior would intimate the incorporation of instinctual type studies into various analyses of investor decision-making in the not too distant future – as Figure 4.4 demonstrates.[60]

---

[57] Indeed, we are in good company in making such assertions. To quote Watson (2004) p. 405 "Behavior, too, has been critical to human survival, and therefore sternly governed by natural selection."

[58] As an aside, it is important at this stage not to confuse this nature/nurture "demand" response of information interpretation with the earlier mentioned nature/nurture "supply" response. This "supply" response occurs when an information byte with extremely strong externality potential (x) – nature – is thrust into a sequenced informational environment conducive to this externality potential being utilized – nurture. We will outline a little later how ultimately this "supply" response and the investor interpretation "demand" response do indeed come together.

[59] Indeed, "algorithmic" is the correct terminology here. Researchers have illustrated particular instinctual responses that appear to have a generalized algorithmic neurological solution sequence. See for instance, Barkow *et al.* (1992); and Hirschfeld and Gelman (1994).

[60] Indeed, it is interesting that a relatively new branch of "neurobiology meets finance" study has recently emerged – aptly named "neuroeconomics." Initial studies

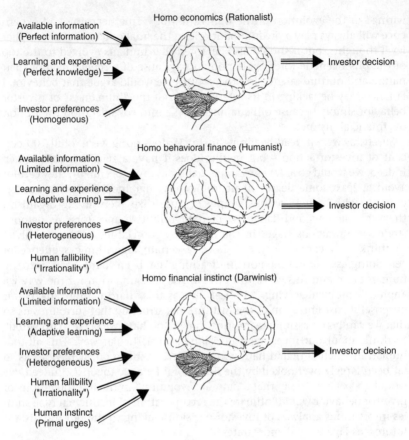

Figure 4.4   The evolution of interpretations of investor behavior

An interesting concept one might say, but how to prove our nature *plus* nurture hypothesis for investor decision-making? This is certainly not an easy conundrum to wrestle with. To start with, it takes roughly 1000–10,000 generations (20,000–200,000 years) in evolutionary time for efficacious mutations to arise and spread within the human race. Subsequently, there is little value in looking for a neurological evolutionary response to the complex problem of better derivative pricing.[61] Rather, in sympathy with the previous

as to the neurological attributes of investor decision-making have proven extremely interesting – see Lo and Repin (2002) for an example.

[61] This would also assume that derivative traders are not only academically sharp but also reproductively successful. While this may seem fanciful, there is evidence of a link between wealth and the sheer volume of one's offspring (and therefore genes) throughout history.

subsection, it is memories with the strongest emotion that have the greatest LTP (recall potential) that provide us with the strongest lead. The archetype instinctual responses fight/flight/fright are generally extremely strong emotions so searching for memes containing the investor equivalent of these is likely to provide us with some of the more reliable memetic strings to associate with particular phenotype pricing behavior (and hopefully, pricing prediction ability). In other words, associating a fight/flight/fright response with one of our encoded nine possible actionable information byte forms – BB, BN, BS and others – and then identifying this key information byte within a given (particularly strong) memetic sequence is likely to yield us reliable predictive capability. This is because appropriately encoded instinctual strong emotions are likely to invoke an investor response that implies we have found an information byte with extremely strong externality potential (x).[62]

We will discuss these issues in far greater detail in our *applied* analysis presented in Chapter 7. Suffice to say now, we feel the incorporation of instinctual responses in addition to learned behavioral responses is an essential element for providing a more thorough representation of the true complexity of investor decision-making. Understanding the primal forces behind the constant battle between nature versus nurture, frontal lobe (responsible for higher order thought) versus basal ganglia and amygdala (responsible for emotion) in driving investor decision-making is crucial in forming a better understanding of those all-important "fear versus greed" market dynamics. The evolutionary microfoundation building block structure toward information within finance that we have presented in Section 4.3 allows us to formalize such behavior in a manner that better enables us to quantify and classify the stronger emotion instinctual responses and by so doing, better understand this crucial aspect of observed market volatility. We do this via the identification of information bytes with extremely strong externality potential (x). It is these "emotion charged" information bytes that have a high potential to form pivotal memes, themes and even swings in overall market sentiment. Take for example the "fear" factor – the typical investor response to financial crises. Anyone who has experienced such a fear

---

[62] It is here that both the "demand-side" and the "supply-side" of the nature versus nurture debate comes together under our Evolutionary Finance framework. In essence, the investor's demand-side nature/nurture condition is split between instinctual and behavioral responses (respectively). Here a "nature" response helps create an information byte with exceptionally strong externality potential (x). Likewise, from the supply-side perspective, the nature/nurture debate refers to an information byte with exceptionally strong externality potential (x) – nature – being thrust into an informational environment that is conducive to this externality potential being utilized – nurture. Subsequently, the demand-side nature/nurture condition has a direct bearing on the supply-side nature/nurture condition via the externality potential (x) link.

gripped market "event" and (with the benefit of hindsight) examined the surprisingly innocuous genesis information byte that acted as a harbinger to this exceptionally strong investor emotional response will realize the importance of undertaking analysis such as ours so that we can better develop forecasting methodologies to predict such outcomes. Thankfully, we believe we have now created precisely such a framework (as represented in Section 4.3) that at least goes part of the way (if not the full way) toward allowing this important task to be achieved.

Setting the bounds – the limits on investor information absorption capacity

This brings us to the final element in our analysis of how investors interpret evolutionary information – the ultimate limits on our ability to absorb information. According to Winston (2003) the human brain takes in 100 billion billion "bits" of information in over a lifetime but can only store 100 thousand billion bits. We are therefore forced into discarding some of the information we perceive. As stated earlier, the anterior cingulatae gyrus (ACG) plays a key role here by acting as the initial filter for incoming information and discarding what we don't need. In terms of our nomenclature specified for informational genome, one would think such information would be encoded with an "NN" form – but this is not the case. Null "NN" information does have the potential to be useful. In building a meme it is the *totality* of information that is important. So while an "NN" information byte signal may not be actionable on its own, when it is associated with additional information bytes it could, in turn, form a key constituent of a particularly powerful meme.[63] For example, while a given economic release may be a benign influence for certain assets, when this information is married to a preceding "BB" and following "SS" signal, it could form a key constituent of a memetic string that ultimately provides investors with insight as to how to be positioned (overweight, at benchmark, then underweight) over the duration of the economic cycle.[64] Subsequently, an "NN" encoding for information does not mean that this information is deemed unsuitable by the ACG, rather some information in our various interpretations of the world that surrounds us must actually be discarded *before* it even enters the encoding process.

[63] This contrasts with the presence of what are known as "introns" (or "junk DNA") within biological DNA sequences. These base sequences are not regarded as "actionable" genetic code. To contrast, some "NN" byte signals – when combined with the presence of other information bytes – can still be regarded as useful information and thus comprise an integral component of an "actionable" meme.

[64] It is probably useful at this juncture to make the distinction between "information" and "data." "Data" refers to information from which inferences can be drawn. Therefore, it not only refers to information which is encoded, but actionable components of this encoded information. In other words, "data" refers to the informational byte components that make up memes.

This presents us with a problem – how to ensure the proverbial "baby is not thrown out with the bathwater." In other words, investors may be selectively discarding what on the surface is initially perceived to be "useless" information without realizing that when combined with ancillary information this seemingly "useless" information may have the potential to become extremely powerful. In fact, this is why a portion of the law enforcement community regularly undertakes training to learn to absorb information that on first principles one would think should be discarded – simply because at times it is the innocuous information that represents the key breakthrough needed to solve a major crime. As far as investors are concerned, this oversight is also important simply because – as we have outlined in Chapter 3 – investors typically use simple (deterministic) heuristic rules of thumb to help guide their investment decision-making. Such simplistic rules of thumb (by definition) all too frequently fail to take into consideration the necessary information for plotting the emergent success and subsequent failure of such rules. This leaves avenues open for opportunists such as ourselves who have a far more comprehensive "structured information" approach toward the prediction of the emergence, maturation and decline of heuristic based trading rules to effectively "prey" upon other investors information processing fallibilities. Such fallibilities are an all-too-human shortcoming in response to the inevitable "information absorption threshold" that each investor necessarily pushes against each and every day in the market as they attempt to synthesize that vast quantity of information presented to them. To further illustrate our point, we detail the selective screening of information that all investors undertake by mathematically formalizing a stylized "information absorption threshold" for the human brain box below.

Imposing limitations – Thresholds on information absorption

Assuming U represents the upper threshold on how much information a given human being can physiologically absorb and assuming knowledge (K) can be acquired by an individual at rate l, the simplest way to represent the uptake of information by any one individual is

$$\dot{K} = l \left(1 - \frac{K}{U} \right) K \quad U, l > 0. \tag{4.16}$$

Separating variables gives

$$\frac{dK}{K(U - K)} = \frac{l}{U} dt. \tag{4.17}$$

Integrating both sides gives

$$\int \frac{dK}{K(U-K)} = \frac{l}{U}t + c, \tag{4.18}$$

where "c" is a constant of integration. Using the method of partial fractions (and taking the antilogarithms of both sides) gives

$$\frac{K}{U-K} = be^{lt}, \tag{4.19}$$

where $b = e^{cU}$ (yet another constant of integration). Assuming the individual has at least some knowledge when they are born (an assumption entirely consistent with our financial instinct approach) then the amount of knowledge at time $t = 0$ is a non-zero integer represented by K_0 (and therefore $b = K_0/(U - K_0)$). The final solution to the growth in the absorption of knowledge by any human being is therefore

$$K = \frac{U}{Ce^{-lt} + 1}, \tag{4.20}$$

where

$$C = (U - K_0)/K_0. \tag{4.21}$$

In graphical form this can be represented as shown in Figure 4.5.

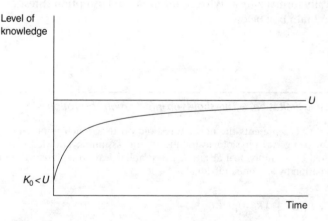

Figure 4.5 The pace of information absorption by the human mind

Compare this outcome with the ever-expanding stock of knowledge represented in Figure 4.1 and it becomes all too apparent that we as humans are forced to automatically discard some of the information we perceive.

So how to overcome this problem of not being able to see the proverbial "forest for the trees" – and even unconsciously culling these "trees" as well? As we stated earlier, the answer resides within this book. In short, overcoming this problem is what our entire "structured information" Evolutionary Finance framework is primarily about. We do this by making investors aware of what is important and what is not from a near-infinite sequence of the market (or even for an individual asset's) information genome. This is done by providing investors with a comprehensive framework for better understanding the actual *structure* of how information interrelates.[65] Investors are thus less likely to discard what on the surface appear to be innocuous pieces of information (but ultimately prove to be "informational gems") if they better appreciate how each byte of information interacts with other information bytes. Hence our preference for a holistic approach toward modeling the II of information as opposed to a reductionist approach. Unfortunately, up until now, a comprehensive framework for understanding this process – piece-by-piece – has not been available. Thankfully this shortfall has now been addressed by our modelling efforts in Section 4.3 and indeed the true power of our Evolutionary Finance approach may ultimately reside in bringing to light these less obvious previously ignored information bytes that form a key piece of the memetic sequence jig-saw puzzle. Everybody knows the blindingly obvious information byte signals and their appropriate asset responses – we are generally taught such things at business school – but it is in the understanding of the less obvious that the greatest potential for alpha-generation resides.

To display this, Chapter 5 we will take our evolutionary building block formalization of information in financial markets one step further by outlining a complete evolutionary model of the marketplace. The results are as interesting as they are powerful. They certainly put to rest the concept that "information" can be treated as some colorless textureless generic form. Indeed, we believe the formation of a better understanding of the actual microfoundation building block structure of information itself could potentially become a key component of any course in investment management. So now, over to our comprehensive evolutionary model of the marketplace.

---

[65] The fact that we are not alone in developing better methods for individuals to cope with the "information overload" of the modern world may be in part an explanation as to why the "Flynn effect" (so named after the New Zealand psychologist who discovered it) of upwardly trending global IQ is now a widely recognized phenomenon.

5
Putting it Altogether – An Evolutionary Model of the Marketplace

Building an evolutionary model of the marketplace is not an easy task. Those familiar with intertemporal modeling and high level calculus will immediately recognize the potentially intractable problem presented by our foundation assumption that there is a less than perfect distribution of information (including knowledge) across investors. If we assume investors are imperfectly informed – as is necessary for analyst research output to have any real value – then we run the risk of violating the all-important transversality conditions that are necessary for a general equilibrium "closure" to a continuous form infinite horizon model where there is a finite number of economic agents.[1] We therefore could be forced down the path of using a much simpler (discrete) overlapping generations (OLG) type of framework which while appealing, is certainly not as elegant as its more sophisticated continuous form cousin. That said, an OLG model does have its uses too – especially when one wishes to look at the analyst community's contribution to (and responses from) the presence of speculative bubbles. Decisions, decisions – which framework to choose? We opted for the easy way out and decided to present both types of modeling framework within the body of this book.

To start off with, in this chapter we will present an evolutionary model of the marketplace under continuous form/finite agent/infinite horizon auspices. We do this because of our liking of its mathematical splendor. To contrast, in Appendix 2 we present a decidedly simpler – yet nevertheless appealing – OLG framework that investigates some of the results that will be

---

[1] Why? Tirole (1982) demonstrated that such conditions – even though realistic – have the potential to force a departure of the actual price of an asset from its "fundamental" value. It is this condition which causes a violation of transversality conditions under an infinite horizon, finite population general equilibrium model.

presented in this chapter but in a more user-friendly environ (which is one of the principle advantages of this class of model).[2] Still, we do recognize that our infatuation with the mathematical elegance of the infinite horizon model does come at a considerable cost. In particular, to get around the potential transversality condition problem that invariably arises when we assume an imperfect distribution of information, we need to "trick" the model by confining the intertemporal choice to the analyst research sector only. In short, the model presented in this chapter does not cover the plethora of consumption/investment decisions that a representative investor will no doubt face on a day-to-day basis, it only covers their research output wants and desires. Further, when it actually comes to this research sector, we assume the representative investor who effectively forms the lynchpin of our model has the ability to perfectly project forward the net worth (at least in terms for them) of analyst research output. Everything else in their world however is uncertain – hence their need for analyst research output to assist this "representative agent" in their decision-making. This is especially the case when it comes to interpreting the (exogenous) volatile financial world toward which analyst research output is focused. Yet again, this demonstrates the voracity of the assumption of a representative investor's insatiable appetite for a never-ending stream of analyst research. So basically, we are putting forward an infinite horizon representative agent model where the only thing this poor soul is sure about is their lack of knowledge and therefore they know they will always need analyst research to assist them in their investment decisions. We do not feel that these conditions are major shortfalls of the model – especially when it comes to its outcomes – but nevertheless, they should be stated up-front.

In terms of the model itself, basically it is divided into three parts. The first component deals with ensuring analysts are adequately rewarded for their research activity over our infinite horizon timeframe. As stated earlier, this would not be the case had we assumed a "blanket" assumption of perfect knowledge across all sectors of the economy. After all, since our fundamental hypothesis is that analysts actively seek opportunities to "manufacture" information that is interpretive of the (exogenous) volatile financial market, then a framework needs to be put in place that illustrates the economic rationale for analysts undertaking this behavior. Ergo, in sympathy with our comments earlier, in Section 5.1 we construct an intertemporal "information consumption" optimization problem as consistent with the intertemporal framework provided by Ramsey (1928). We then derive the solution to this intertemporal information consumption problem which has the particularly appealing attribute of taking explicitly into account everexpanding theme variety. Indeed, the actual model itself is an adapted

---

[2] Both models have an "economic growth" foundation as ultimately we view information production as synonymous with the process of economic growth in general.

version of Grossman and Helpman's (1991) endogenous growth model for the economy at large as interestingly we find their methodology particularly amenable to exploring both the investor's insatiable desire for more information and the ever-expanding theme variety that goes hand-in-glove with the growth of knowledge overall. It is these information demand conditions which are satisfied at each point in time by the construction of thematic-orientated information bytes by the analyst community.

The second part of our model deals with what analysts are actively attempting to explain in their research endeavors via their manufacture of these information bytes. Having established that analysts will be adequately rewarded for their research efforts, the next step is to attempt to explain precisely what form their thematic-orientated research takes and indeed, how this relates to asset prices overall. This is the charter of Section 5.2. In this section, a number of possible strategic behaviors on behalf of analysts are discussed – in terms of their choice of topic for their information bytes – and are then linked back into the intertemporal optimization model presented in Section 5.1. Furthermore, the asset price implications of such behavior are thoroughly explored. As will be seen, this is where the true beauty of the continuous form infinite horizon framework comes to the fore as an interesting range of strategic behaviors (and subsequent asset price responses) are identified.

Finally, in Section 5.3 we discuss in detail our preferred marriage of the strategic analyst behavior/asset price dynamics expressed in Section 5.2 and the investor's intertemporal information consumption equilibrium conditions expressed in Section 5.1. The result is a unified evolutionary model of the marketplace where analysts are perpetually rewarded for developing thematic-orientated insights and ideas in a purposefully strategic manner which, in turn, form larger informational entities via a molecular-like informational building block process. It is this "information building" process that subsequently has a deterministic impact upon the evolution of asset prices. The calibrations of our preferred model which links this evolutionary-like information creation and dispersal process to actual asset price dynamics have a number of notable features that are typically mirrored in "real world" asset price movements. In particular, we:

1 highlight the tendency for increasing analyst uncertainty as to the prevailing form of market sentiment to result in an increasingly volatile pattern for asset prices;
2 we illustrate the ubiquity of jump diffusions to new steady-state equilibrium positions as analysts 'herd' from one perspective on market sentiment to another; and
3 we demonstrate the potential for substantive shocks to the market system when feedback between asset prices and analyst strategic behavior results in price movements that can only be described as calamitous.

All will be discussed within a framework that takes into consideration the way information bytes bind together to form memes which, in turn, bind together to form themes and overall market sentiment. So now over to our evolutionary model of the marketplace itself.

5.1 Stage I: Developing an intertemporal optimization model of information production/consumption and solving for general equilibrium conditions

The investor's information consumption function – measuring information appetite

Let us assume that investors display a voracious appetite for new information. This appears quite a plausible assumption to make given the plethora of news services, magazines, newspapers and research organizations all servicing the market's apparently insatiable demand for information. How to incorporate this tendency for information insatiability within a sophisticated intertemporal framework? To start with let us turn to the ubiquitous "infinitely lived representative agent model" – in this case, as developed by Grossman and Helpman (1991) – where it is assumed there is a fixed population (Z) of investors who each share identical preferences. Furthermore, let us assume that each individual investor maximizes their own utility from receiving a given stream of information over an infinite horizon. Assuming for simplicity that we will only be dealing with the information aspect of investors wants and desires, the representative investor's utility maximand can therefore be formalized as follows

$$U_t = \int_t^\infty e^{-\rho(\tau-t)} \log D(\tau)\, d\tau, \tag{5.1}$$

where $D(\tau)$ represents an index of information consumption at time τ, and ρ is the subjective discount rate applied to future information flow.[3] The natural logarithm of this "information consumption index" effectively measures the instantaneous utility of the representative investor at each and every moment in time from receiving an infinite stream of information.

Given our earlier cited assumption of the representative investor's voracious appetite for information, let us attempt to formalize the information demand condition a little further by assuming the representative investor's utility is directly related to "theme variety" (as stated in Chapter 4) with their preferences extending over a hypothetical infinite range of themes indexed by $j \in [0, \infty)$. This ensures there is always an incentive for analysts to undertake information byte manufacture which may or may not result in new themes emerging – more on this later.

---

[3] That is, investors prefer more information today rather than tomorrow.

While on the topic of the investor's information demand function, note that at any point in time the total number of themes available for the representative investor to select from is static. Assuming this given theme variety at a given point in time can be categorized by the interval $[0, \Gamma(t)]$, then the representative investor's information utility index across this given array of themes can be represented as follows

$$D = \left[\int_0^\Gamma \varphi(j)^\alpha dj \right]^{1/\alpha} \quad 0 < \alpha < 1, \tag{5.2}$$

where $\varphi(j)$ denotes consumption of information byte φ that contributes to theme j and α is the elasticity of demand for this particular segment of information.[4] An important property of Equation (5.2) is that the elasticity of substitution between any two competing themes, ε, is constant and greater than one

$$\varepsilon = \frac{1}{(1 - \alpha)} > 1 \tag{5.3}$$

thus indicating an expanding range of market opportunities for those analysts who are entrepreneurial enough to be willing to embark upon an avenue of research that could promote the emergence of a new theme – the proverbial "big call" as it is known in the industry.[5]

Since we are only dealing with the information aspect of the investor's long-run optimization problem, we can effectively solve the instantaneous utility maximization condition represented in Equation (5.2) by constructing a budget constraint of total income at each and every particular point in time being equivalent to

$$E = \int_0^\Gamma p_\varphi(j)\varphi(j)\, dj, \tag{5.4}$$

where $p_\varphi(j)$ is the component of asset prices that constitutes a reward to the analyst who manufactures information byte φ that contributes to theme j.[6]

---

[4] Dixit and Stiglitz (1977) have highlighted a number of useful properties of this form of utility function. In particular, cross-preference stability between competing themes in the investor's mind is effectively ensured by a diminishing marginal utility from consumption of more information bytes related to a particular theme ($0 < \alpha < 1$).

[5] It is important here to recognize that an "analyst" *per se* could just as easily be a news-presenter, a desk salesperson, an academic, even a financial tipsheet publisher. The term does not refer only to the generic labeling applied to individuals employed in the research departments of sell-side and buy-side institutions. Rather, the term applies to all financial information providers who draw at least some portion of their income from their information creation role.

[6] The implication here is that, as stated at the outset, this is purely an information generating and producing society – no other wants and desires are taken into consideration. This is a necessary simplification assumption for our model.

Lagrangean optimization yields the following optimal demand condition for information byte $\varphi(j)$ where the representative investor's income and maximal utility conditions are equilibrated

$$\varphi(j) = Ep_\varphi(j)^{-\varepsilon} \Big/ \int_0^\Gamma p_\varphi(j')^{1-\varepsilon} dj'. \tag{5.5}$$

Since Equation (5.4) is effectively measured in nominal price terms and a necessary condition for intertemporal equilibrium is for the *real* supply of information to equate to *real* demand, then an additional requirement for intertemporal equilibrium is

$$D = \frac{E}{p_D}, \tag{5.6}$$

where p_D represents a hypothetical aggregate price index across all indices reflecting the equilibrium asset prices at which the per unit information supply (of each information byte) equates to per unit information demand. Remember, as consistent with Figure 2.3 in Chapter 2, under imperfect information conditions (Weak/Semi-Strong EMH), a portion of each asset's price reflects the reward to information providers. This provides a convenient bridge between the "well-informed" world of the investor's decision in terms of their recognition of the need for analyst research and their less "well-informed" world of the remaining share of asset prices about which they remain highly uncertain – hence the need for analyst research. So while the p_D share of the aggregate asset price index that persists for the economy may be relatively "well-behaved" across time and thus moves in a reasonably predictable manner, the greater percentage of the remaining share that makes up each asset's price within the overall index is highly volatile and therefore needs to be explained by the collective analyst community's research output.

For an instantaneous equilibrium at each point in time p_D is approximated by

$$p_D = \left[\int_0^\Gamma p_\varphi(j)^{1-\varepsilon} dj \right]^{1/(1-\varepsilon)}, \tag{5.7}$$

where, as consistent with our comments earlier, a price basket across all indices is set to a level so that the rewards to analysts from information manufacture equates to the rewards (read "utility") to investors from consuming this information. The information supply/demand optimization problem is thus easily broken down into an instantaneous optimization condition – represented in Equation (5.5) – and an intertemporal condition. Substituting Equation (5.6) into our original intertemporal optimization maximand in

Equation (5.1) yields

$$U_t = \int_t^\infty e^{-\rho(\tau-t)} \left[\log E(\tau) - \log p_D(\tau)\right] d\tau. \tag{5.8}$$

The above highlights that the intertemporal utility of the investor from the provision of information by the analyst community evolves in accordance with the difference between the growth of the aggregate income analysts receive from information byte manufacture (E) and the growth in the diversity of themes – as determined (in real terms) by the equilibrium price index at each point in time. Assuming that investors can borrow or lend freely at an instantaneous interest rate to finance such entrepreneurial behavior on behalf of analysts, then a necessary condition for dynamic equilibrium is

$$\frac{\dot{E}}{E} = r - \rho. \tag{5.9}$$

Choosing the convenient numeraire of $E(t) = 1$ so that nominal spending by investors on the information reward component embodied within asset prices remains constant at each and every moment in time allows us to identify the intertemporal equilibrium condition for investors in terms of their information consumption as

$$r(t) = \rho \quad \text{for all } t. \tag{5.10}$$

Importantly however, this is only one side of the coin – the demand-side – of our model for the intertemporal equilibrium conditions in the provision of information within financial markets. What about the supply-side? To ascertain the necessary intertemporal equilibrium conditions for information (and theme) manufacture on behalf of the analyst community we must now turn our attention toward the nuts and bolts of information production itself. What are the optimal conditions under which analysts should operate so that they are effectively generating ideas at a sufficient enough level at each and every point in time so that an equilibrium is formed with the insatiable appetite of investors for new strategies? Further, what is the optimal allocation of resources for this endeavor? It is toward answering these questions that we now turn.

The mechanics of information production

Within our highly simplified model of information provision within financial markets it is assumed analysts face a choice of either:

- devoting their time toward producing information bytes to embellish an existing theme; or, alternately

- opting to try their hand at compiling the necessary facts and figures to enable them to produce an information byte that will represent the vanguard of a new theme for market participants to focus upon – in short, an information byte with exceptional externality potential (x).[7]

Let us deal with the former choice – embellishing an existing theme – first. For starters, it is assumed for simplicity that the production of information requires no capital input, only labor input, (L), is needed. Under conditions of general equilibrium, α represents both the marginal utility investors derive and the marginal product analysts achieve from an information byte within the given theme range $[0, \Gamma(t)]$. Across the analyst community we assume there are constant returns to scale for research so the information production function for the manufacture of information bytes contributing to existing themes takes on the relatively simple form

$$Y = L\alpha, \tag{5.11}$$

where Y is the total information byte output (aimed toward existing themes) at a given point in time.

Using this production function, analysts manufacture a diverse array of information bytes aimed at embellishing the existing theme structure. It is assumed that at any one point in time there is at least one analyst producing information relating to each preexisting theme. This ensures investors can select from a veritable smorgasbord of themes an information byte necessarily suited to satisfy their thematic information wants and desires – as consistent with the notion expressed in Equation (5.2).

Analysts who concentrate on producing information bytes that contribute to these existing themes attempt to maximize profits consistent with the following

$$\pi(j) = p_\varphi(j)\varphi(j) - w\varphi(j), \tag{5.12}$$

where the first term on the RHS of Equation (5.12) represents the total revenue from producing information bytes for theme (j) of quantity $\varphi(j)$ and the second term represents the total cost of such theme production.[8] Subsequently, analysts maximize profits by charging for their research a price which takes into consideration both wage costs (w) and the marginal product

---

[7] In Chapter 6 we will outline the Game Theoretic underpinnings motivating analyst choice toward both endeavors.

[8] In Chapter 6 we will build upon the profit incentive for analysts to produce information bytes by examining the Game Theoretic implications for analysts profits (payoffs) from producing differing formats of information bytes under a highly competitive environ.

of their labors (α). This yields a "per byte" pricing equation of

$$p_\varphi(j) = \frac{w}{\alpha}, \tag{5.13}$$

where, as consistent with our discussion in Chapter 2, p_φ constitutes a share of the aggregate price of any asset – we will dwell more on this in Section 5.2. This profit maximizing pricing behavior on behalf of analysts thus implies as per "thematic" operating profit of

$$\pi = \frac{1-\alpha}{\Gamma}, \quad \text{where } E = 1. \tag{5.14}$$

Generating new themes: How it is done and the implications of this behavior

What happens if an analyst decides to devote some portion of their time toward attempting to produce a new theme rather than slavishly producing information bytes under the old theme regime? If an analyst decides to try their hand at generating a new theme by producing an information byte with exceptionally strong externality potential (x), then it is assumed they can generate new thematics at the rate

$$\dot{\Gamma} = L_\Gamma \eta, \tag{5.15}$$

where

L_Γ is the amount of labor time devoted to new thematic research; and η is the productivity in the new thematic research area.

Consistent with our comments in Chapter 4, and assuming a direct link between the stock of knowledge (K_Γ) and the number of new themes that have previously been developed, then

$$K_\Gamma = \Gamma, \tag{5.16}$$

subsequently it is possible to model the growth of η as some function of K_Γ

$$\dot{\eta} = f(K_\Gamma). \tag{5.17}$$

Assuming the most simple form – an assumption we will relax in Section 5.2 – of direct proportionality between the stock of knowledge, the number of themes in existence and the analyst's productivity in producing new themes so that

$$K_\Gamma = \Gamma = \eta, \tag{5.18}$$

and noting the cost of new theme development is effectively $wL_\Gamma dt$ while the value of this activity is $v(L_\Gamma \eta)dt$, then value maximization by the analyst community requires L_Γ to be set as large as possible whenever $v\eta > w$ or zero when $v\eta < w$. Equilibrium in this sector of analyst research endeavor thus necessitates

$$\frac{w}{\eta} \geq v \quad \text{with equality at } \dot{\Gamma} > 0. \tag{5.19}$$

Assuming constant returns to scale and free entry into new theme search, then this equality condition is satisfied. What does this imply? At any point in time, there will always be a sufficient incentive for a share of the analyst community's time to be spent searching for new themes. However, this percentage representation of the collective analyst labor market is effectively bounded by the value investors place upon new theme arrival. Crucial to all this is the variable η – the productivity of analysts in the new thematic research area – which gradually improves as the number of themes (Γ) expands.

How should one interpret this relationship? Well, in short, analysts learn from the past mistakes of others which helps them to better exploit possible themes going forward. Just as Moore's Law is effectively a productivity relationship highlighting a rise in a computer's information processing power, so too there is a productivity dividend to reap from more efficient information processing on behalf of analysts. For us, this link is formalized via a proportional relationship between the stock of knowledge (K_Γ) at any given point in time and the productivity of analysts in the new thematic research area (η). That said, as mentioned in the introduction to this chapter, in Section 5.2 we will investigate in greater detail the potential array of functional forms that could be considered as possible alternatives for this link between the stock of knowledge (K_Γ) and the productivity of analysts in new theme search (η). We do this via categorizing new theme research into various types and then examining how this relates to asset prices overall.

The financing of analyst information manufacture

All the above would effectively come to naught if there weren't some form of financing arrangement to ensure analysts engaged in new theme research are somehow rewarded for their efforts – at least in the form of some dividend payment system. Further, our model requires a sophisticated enough capital market structure to be in place to allow investors to smooth their intertemporal information consumption via the purchase of both equities and bonds. In short, to solve for general equilibrium we require a financial sector to be loaded within our framework and the purpose of this subsection is to do just that.

First, to intertemporal consumption smoothing. Since, as stated earlier, we are only dealing with the information aspect of the representative investor's wants and desires, we assume that this individual uses the per theme profit return represented in Equation (5.14) in their calculations of the distribution of their savings between bonds and equities to facilitate intertemporal consumption smoothing for the stream of information they expect over their infinite life span. An "equity" under these conditions represents a share in a "research firm" which in essence, constitutes the portion of an analyst's time devoted toward the construction of a new theme.

It is assumed that each analyst is effectively "branded" by investors with the theme they create and thus have intellectual monopolistic rights over its dispersal within the market – primarily via the publication of information bytes. These intellectual monopolistic rights are non-transferable. Further, it is important to realize that for the moment our "equity market" refers only to research sector equity prices – because as yet we are not considering wider asset prices which will be explicitly introduced into our modeling in Section 5.2.[9]

Subsequently, we are faced with a reasonably simple intertemporal optimization problem of attempting to match the information demands of the representative investor with the information supply conditions of the analyst. Under general equilibrium conditions it will be displayed that analysts and investors are one and the same – they just happen to be treated as opposite sides of the production/consumption problem until the very end of our model. But for the moment, we are still working under assumption that these "two sides of the same whole" are being treated as though they were separate entities.

Ergo, for the representative investor, the return on our "research sector only" equity is the average per theme profit return (π), assuming all profits are paid in the form of dividends, plus any expected capital gains or losses, ($\dot{v}$). Why do such capital gains and losses arise? A quick examination of Equation (5.14) illustrates that as the number of new themes (Γ) increase, the operating profit per thematic brand declines. Consequently, we assume the representative investor has perfect foresight as to the evolution of ($\dot{v}$) to take such capital losses into account in their financial decision-making.[10]

This makes the derivation of our equilibrium financing condition relatively simple. Since the return on bond investment is simply the nominal interest rate (r) multiplied by the value of bonds held (v) and for simplicity it is assumed that stocks and bonds are perfect financial substitutes within the representative investor's portfolio, equilibrium in our simplified financial sector requires that the return from holding research firm equity is the same

---

[9] Indeed, it is important here to make the distinction between the "endogenous" form of equity – in the embodiment of a share in a research producing firm – with the "exogenous" variety which represents the external equity market that the analyst community is attempting to explain.

[10] In fact, this is the only perfect foresight that our representative investor possesses.

as the return from holding bonds

$$\pi + \dot{v} = rv. \tag{5.20}$$

But what of the incentive structure for analysts engaged in new theme research? As we mentioned earlier, under general equilibrium conditions these individuals are one and the same as the representative investor. That said, while they are still being treated as separate entities it is perhaps useful to think of these analysts as attempting to maximize the stock market value of their "research firm." This entails setting information byte prices at a level dictated by Equation (5.13) so that the discounted present value stream of future profit can be represented as

$$v(t) = \int_t^\infty e^{-[R(\tau)-R(t)]}\pi(\tau)d\tau, \tag{5.21}$$

where $R(t) = $ cumulative discount factor applicable to profits earned at time t. Note that by differentiating Equation (5.21) with respect to t we ascertain Equation (5.20).

Equilibrium in the labor market: Matching analyst supply to demand

Next, we turn our attentions to the labor market. Using (L) to denote the aggregate analyst labor time available at each and every moment that is under consideration within our intertemporal framework, it is a simple case of matching this aggregate analyst labor supply to aggregate labor analyst demand to achieve our the final sectoral equilibrium condition within our model. We do this before moving onto general equilibrium estimation – and thereby effectively solving the model overall.

Essentially there are two parts to the demands upon analyst time – demand for analysts in manufacturing information bytes under the existing theme regimen and demand for analysts to devote time toward new theme research. We will deal with each separate demand requirement in turn next.

First, let us focus on the demand for analyst time to produce information bytes relating to existing themes. Since we have earlier chosen the convenient numeraire of setting total income at each and every point in time being equivalent to $E(t) = 1$ and we know that the profit maximizing price per byte in the provision of information relating to existing themes (p_φ), it is straightforward to ascertain the total demands upon analyst time in this sector of the research market as $(1/p_\varphi)$. Why? Simply because $(1/p_\varphi)$ effectively represents the number of information bytes sold. As for the demands upon analyst time devoted to new theme research (L_Γ), rearrangement of Equation (5.15) quickly yields $(\dot{\Gamma}/\eta)$ as the demand condition in this sector of the research market. Subsequently, combining our two labor demand conditions with our labor supply condition gives us an overall equilibrium condition for the analyst labor market of

$$L = \frac{\dot{\Gamma}}{\eta} + \frac{1}{p_\varphi}. \tag{5.22}$$

Finally, note that since employment must be non-negative, a necessary side condition to the above is that the equilibrium price per information byte must satisfy

$$p_\varphi \geq \frac{1}{L}. \tag{5.23}$$

This is derived by setting $(\dot{\Gamma}/\eta = 0)$ in Equation (5.22).

General equilibrium: Putting all the pieces together to solve the model

Having methodically laid the foundations of each relevant sector within our model, we are now ready to provide a general equilibrium solution for information production and consumption within financial markets. Given $\Gamma(t)$ and $v(t)$ we can solve for equilibrium prices and resource allocation and by doing so satisfy our intertemporal equilibrium requirements. Therefore, our results are conveniently able to be expressed in terms of $(\dot{\Gamma})$ and $(\dot{v})$.

Combining Equations (5.13), (5.19) and (5.23) gives the boundary condition

$$\bar{v} = \frac{\alpha}{L\eta}, \tag{5.24}$$

which states that the development of new themes will only take place when the reward for such behavior is sufficiently high. Thus $\dot{\Gamma} > 0$ only occurs when $\bar{v} > 0$. To solve for $\dot{\Gamma}$ we use the labor market equilibrium condition represented in Equation (5.22), the equilibrium price for research condition specified in Equation (5.13) and the resource allocation maximization requirement in the new theme sector represented by Equation (5.19) to give the emergence of new themes under general equilibrium as

$$\dot{\Gamma} = \begin{cases} L\eta - \dfrac{\alpha}{v} & v > \dfrac{\alpha}{L\eta} \\ 0 & v < \dfrac{\alpha}{L\eta} \end{cases}. \tag{5.25}$$

Next we use the long-run capital market equilibrium condition identified in Equation (5.20) along with the intertemporal consumption maximization condition given in Equation (5.10) and finally the per thematic operating profit condition represented in Equation (5.14) to derive the evolving valuation that investors place upon analyst research output under conditions of general equilibrium as

$$\dot{v} = \rho v - \frac{1 - \alpha}{\Gamma}. \tag{5.26}$$

Our system of equations for the intertemporal optimization of information production and consumption within financial markets is now formalized into the dual differential Equations (5.25) and (5.26) along with the boundary condition represented in Equation (5.24). This completes our model of the optimal conditions for the intertemporal production/consumption of information within financial markets but then begs the wider question – how does this information output (whether of the "existing" or "new" theme variety) influence the wider economy asset prices that are presently external to our framework? It is to this question that we now turn in the next section of this chapter.

5.2   Stage II: Linking analyst research output to asset price dynamics

Having established a framework that effectively ensures analysts are adequately rewarded for research activity across time, the next issue to face is the implications of this research activity for asset markets overall. In choosing a link between analyst research output and asset prices, a number of potential functional forms spring to mind. We will consider just a few here but hopefully we will leave the reader with the indelible impression that the generalized model for the provision of market information that has been presented in the preceding section is amenable to a whole variety of interesting interpretations as to how information and asset prices ultimately interact.[11] That said, given our evolutionary focus on the way information bytes interrelate to form memes, themes and then overall market sentiment, most of the explanations provided in this section will have at their root foundations some form of biological/evolutionary dynamics. Indeed, we felt that such biological-based system dynamics provided the most fertile ground for developing a comprehensive model of the evolutionary principles underlying the formation of information in financial markets. Thankfully our intuition proved correct – as will be displayed by our preferred model described in detail in Section 5.3. But for now, let us more closely examine the building blocks of our evolutionary framework as to the way analyst information provision and asset prices actually interrelate.

Market sentiment oscillation

Let us now assume a particular functional form for the growth in productivity in the new theme research sector. This chosen functional form makes the key distinction of being *nonlinear* as opposed to the simple (linear) direct proportionality link between the stock of knowledge, the number of themes

---

[11] In fact, we will expand upon this array of models in even greater detail in Chapter 6 where we explore the microfoundations of analyst interaction using Game Theoretic analysis as a basis to build a deterministic driven model of Evolutionary Distributional Form.

in existence and the productivity of analysts in producing new themes – as expressed in Equation (5.18).

How is our newly hypothesized nonlinearity introduced? To do so, we formalize the novel concept of "market sentiment oscillation" in differentiating between the types of new themes introduced by the analyst community. In essence we argue there is some type of "bull/bear" new theme saturation threshold for investors in the sense that too many new themes being introduced at any one point in time under either the "bull" or "bear" banner generates an inevitable investor backlash where investors start to prefer a contrarian viewpoint to be expressed in new theme generation. This desire for a contrarian viewpoint eventually grows to such an extent that it becomes the dominant form of market sentiment prevailing in analyst new theme generation – so it too eventually suffers the same fate as its predecessor and the pendulum of "bull/bear" sentiment preference effectively swings back the other way.

How to formalize these cyclical swings within investor preferences for every new bull/bear thematics? The answer resides with a simple Sine curve, so what we now do is – taking into consideration Equation (5.18) – we replace Equation (5.15) in the model we developed in the preceding section with the following

$$\dot{\Gamma} = L_\Gamma \left[\left(\frac{1}{\alpha} \right) - \psi \sin(\eta) \right] \qquad \left(\frac{1}{\alpha} \right) > 0, \quad 0 < \alpha < 1 \qquad (5.27)$$

where

$(1/\alpha)$ is investor's underlying desire for new themes; and
ψ is investor's bull/bear saturation threshold.

How does the above alter the dynamics of our model? Well, for starters note that we now have extended what was previously a purely *quantitative* preference on behalf of investors for new themes into a *qualitative* preference as well. As $\dot{\Gamma} \to L_\Gamma [(1/\alpha) + \psi]$ the representative investor's appetite for a change in market sentiment begins to wane – they are happy with the prevailing orthodoxy of analyst opinion in existing theme production – and new theme production ($\dot{\Gamma}$) actually starts to slow down to the point where it eventually declines. This is a classic case of the market becoming dangerously too complacent with a dominant viewpoint. Thankfully however, $\dot{\Gamma}$ finally slows down to a point where the lower bound of the market's bull/bear saturation threshold begins to be reached as $\dot{\Gamma} \to L_\Gamma[(1/\alpha) - \psi]$ and investors begin to cry out for a contrarian viewpoint. Subsequently, the pace of new theme creation once more begins to rise. At each point in time Γ itself is monotonically increasing, as it is only the pace of new theme creation that alters under our bull/bear scenario. One way to display these results is with the aid of a simple calibration of Equation (5.27) itself.

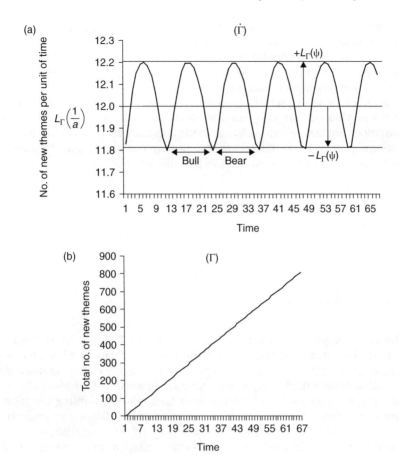

Figure 5.1 Calibration from our market sentiment oscillator – $(1/\alpha) = 6$, $\psi = 0.1$, $L_\Gamma = 2, \eta = 1$

If one looks closely, you can see the minor ripples in Figure 5.1(b). These ripples represent the swings in the pace of $(\dot{\Gamma})$. A cursory examination of various alternate calibrations of our market sentiment oscillator illustrated just how dynamically unstable such a feedback induced system is. Increasing the investor's bull/bear saturation threshold (ψ) incrementally illustrated increasing degrees of instability until a truly chaotic sequence manifested for $\psi \geq 2.0$. This stands to reason as the higher one raises ψ, the more draconian the representative investor's attitudes are toward their degree of comfort or discomfort with the ruling collective mindset on behalf of the analyst community. Incrementally raising ψ quickly approaches a point where negative feedback results in an almost schizophrenic investor response.

We will examine such chaotic sequences in greater detail in our preferred model presented in Section 5.3 but for the moment let us focus on deriving the new general equilibrium conditions for this particular extension to our model – by substituting Equation (5.27) directly into the generalized information provision model presented in the previous section.

As can be seen later, substituting Equation (5.27) into the model presented in the previous section does not alter dramatically our general equilibrium conditions but it does – via a change in the boundary condition and η itself – alter the pace of $\dot{\Gamma}$. So our general equilibrium conditions under a market sentiment oscillation framework now become

$$
\dot{\Gamma} = \begin{cases} L\left[\left(\dfrac{1}{\alpha}\right) - \psi\sin(\eta)\right] - \dfrac{\alpha}{\nu}, & \nu > \dfrac{\alpha}{L\left[(1/\alpha) - \psi\sin(\eta)\right]}, \\ 0, & \nu < \dfrac{\alpha}{L\left[(1/\alpha) - \psi\sin(\eta)\right]}. \end{cases} \tag{5.28}
$$

and

$$
\dot{\nu} = \rho\nu - \frac{1-\alpha}{\Gamma}. \tag{5.29}
$$

The inclusion of the (α/ν) term into the pace of $\dot{\Gamma}$ under general equilibrium does not dramatically alter the type of dynamics illustrated in Figure 5.1 – with the obvious exception that the order of magnitude in the calibrations will change. More important however is how these alternations in the pace of $\dot{\Gamma}$ feed directly into asset prices. Assuming the most simple of forms – a one-to-one link – between the publication of information bytes under the collective "bullish"/"bearish" sentiment banners and actual movements in asset prices, then the oscillations apparent in $\dot{\Gamma}$ will be reflected in asset prices too.[12] In other words, if we assume a bullish market sentiment information byte carries with it a "BB" information byte encoding while a bearish market sentiment byte carries a "SS" encoding, and assuming these binary "buy"/"sell" signals are reflected in asset prices at large, then the actual aggregate market index for assets presently exogenous to our enclosed information production/consumption framework will actually mimic the pattern of behavior exhibited by $\dot{\Gamma}$. In the following representation of our model we pursue this thought in considerably more detail by explicitly modeling the formation of encoded information bytes by the analyst community.

---

[12] Such a linkage has firm foundations – as exemplified by our exposition of Fama's "Fair-Game" in Chapter 2. Also, Maheu and Mccurdy (2004) have demonstrated such a linkage at the empirical level. Later in this chapter (and then again in Chapter 6) we will extend this assumption to allow for the presence of complex nonlinearities that typically arise in the presence of memetic information (M).

The bull/bear analyst cycle

What happens if the swings in market sentiment are not so much driven by investors desires for a change in consensus analyst opinion but rather by the actions of the analysts themselves? How best to formalize this "supply driven" response to the forces governing the cyclicality in market sentiment? In this particular subsection we aim to address this very question and by so doing presenting an endogenized version of the encoding of information bytes (φ_a) that we presented in Chapter 4.

Here we measure the analyst community's output of information bytes (which are then categorized into various themes) in terms of their *contemporaneous* and *latent* signaling effects – BB, BS, SS, SB – for a representative asset (i) that resides external to our model of intertemporal information consumption/production. For simplicity, we assume information bytes are categorized into separate themes on the basis of both this recommendation status for the representative asset (i) and the timing of each theme's creation. As for the representative asset (i) itself, it reflects the *average* overall (exogenous) asset market response to each individual information byte's recommended "buy" (B) or "sell" (S) signal. For the moment, we will ignore non-actionable "neutral" analyst recommendations so that the categorization of each information byte into various themes (j) can be detailed with the aid of the following dual differential equation system

$$\dot{B}_0(ij) = \xi B_1(ij) \quad \text{and} \quad \xi, \zeta > 0 \tag{5.30}$$

$$\dot{B}_1(ij) = -\zeta B_0(ij), \tag{5.31}$$

where

$B_0(ij)$ is the contemporaneous "buy" signal encoding for an information byte published under theme j and as reflective upon representative asset i; and $B_1(ij)$ is the latent "buy" signal encoding for an information byte published under theme j and as reflective upon representative asset i.

Within this framework a "minus buy" equates to a "sell" so there is no need for any formal representation of the sell signal "S" in either Equation (5.30) or (5.31). As each new information byte is produced, we assess its impact (encode it) in terms of its signaling for the representative asset (i). This broad buy/sell recommendation is then interpolated as either a generic "buy" or "sell" recommendation for the entire market for a particular information byte. It is the aggregate net balance of published information byte recommendations that will tilt the weight of existing themes toward an overall "bullishness" or "bearishness" assessment at any given point in time and will thus determine swings in market sentiment. Once more it is considerably easier to comprehend the overall dynamics of this system via a simple calibration of the differential equations themselves.

Assuming for simplicity a one-to-one linkage between the encoding of information bytes and actual movements in asset prices, it immediately becomes obvious that analyst "herding" toward either a net bullish or net bearish position generates cycles in the growth of prices for our representative asset *i* and through this, the growth in the overall market index. This predominantly cyclical behavior on behalf of analysts in the format of their asset price recommendations arises thanks to the presence of the negative feedback expressed in Equation (5.31). This negative feedback ultimately influences both the contemporaneous and latent aspects of analyst recommendations – hence the regular clockwise cycle between BB, BS, SS and SB. What rationale is there for such obviously pro-cyclical behavior on behalf of the analyst community? Consistent with the previous subsection, the presence of negative feedback toward the degree of net bullishness or bearishness expressed *en masse* by the analyst community could represent some form of saturation in terms of investor preferences for research falling under the umbrella of a particular form of market sentiment. Alternately, analysts themselves may induce such behavior as they attempt to "brand" themselves via a "big call" when they publish an information byte with exceptionally strong externality potential (x) – which, in turn, causes the emergence of a new (contrarian) theme. Under these conditions, the incentive for producing product differentiating contrarian views obviously grows the more the market becomes beholden to a particular dominant viewpoint. Still, irrespective of the forces that govern it, such clustering on behalf of analysts in terms of their opinions is a regular feature of what we observe in "real world" markets and proves to be a nice result for this form of appendage to our model.[13]

Finally, note that the paths expressed in both diagrams in Figure 5.2 illustrate the presence of increasing returns. This is because the fundamental model expressed in Section 5.1 has not been altered in any way by our suggested appendage of a "buy/sell" encoding of information byte output. Subsequently, the number of themes published (Γ) remains a monotonically increasing function – all we have done here is classify some of the dynamics that can assist in forming a better understanding of the actual format of new themes as they emerge. Likewise, by actually encoding information bytes and relating this to both theme formation and actual swings in overall market sentiment, we are able to form an effective bridge between the endogenous asset price sector of the model (the equity held in analyst research firms) and the exogenous (imperfectly understood) asset price markets of the wider economy. This is done in accordance with each information byte's recommendation for the representative asset (i) and it is through this that we form an interpretation for the growth in asset prices overall. Since we

---

[13] Indeed, this relative theme preference "clustering" exhibited by analysts will be developed far more extensively in Chapter 6. Here we describe the Game Theoretic underpinnings of analyst decisions in determining the "theme format" of a particular information byte and how this decision-making process can ultimately come to influence Evolutionary Distributional Form.

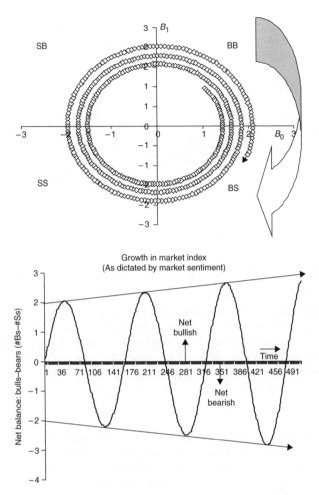

Figure 5.2 Calibration from the bull/bear analyst cycle – $B_0 = 1$, $B_1 = 1$, $\xi = 0.04$, $\zeta = 0.04$

have ensured in Section 5.1 that the intertemporal aspects of the information production/consumption decision remains effectively exogenous to any such encoding process, we have not altered the general equilibrium conditions represented in Equations (5.25) and (5.26) in any way with our suggested appendage to our model. All up, a nice outcome.

Research pulses

Inspiration rarely comes in a smooth uniform manner, but more in the form of short-sharp bursts. Further, inspiration is often followed by failure and idea degradation as what were once thought to be worthwhile avenues of research endeavor become discredited. How best to incorporate such "research pulses" into our framework of information production/consumption presented in Section 5.1? Again, we turn toward a dual differential equation system

$$\dot{\vartheta} = \delta_\vartheta + \beta_\vartheta \sin(\kappa - \vartheta) \tag{5.32}$$

and

$$\dot{\kappa} = \delta_\kappa + \beta_\kappa \sin(\vartheta - \kappa), \tag{5.33}$$

where

ϑ is the "inspiration" which assists analyst productivity in generating new themes;
κ the "degradation" which retards analyst productivity in generating new themes; and
δ, β is the constants affecting the rate of inspiration as opposed to degradation.

So far so good, but to complete this appendage to our model we need to incorporate such "research pulses" into the intertemporal framework presented in our information production/consumption model outlined earlier. The way we do this is to directly relate idea inspiration/degradation to the pace of new theme construction – as represented by the equation

$$\dot{\Gamma} = L_\Gamma(\vartheta - \kappa)\eta, \tag{5.34}$$

which replaces Equation (5.15) in our model presented in Section 5.1.

Now that we have the basics of our suggested adaptation to the model, we will once again call upon the process of calibration in attempting to explain the dynamics of the system of equations presented earlier in a fashion that is both user-friendly and concise.

As was the case with our earlier presented market sentiment oscillator, if one looks closely at Figure 5.3(b), you can see the minor ripples. These ripples represent the swings in the pace of $(\dot{\Gamma})$ thanks to the research pulses formalized in Equations (5.32) and (5.33) – and as dictated by the relative strength of inspiration versus degradation in these respective equations. Yet again, a cursory examination of various alternate calibrations of this dynamical system yielded chaotic results for $\beta_\vartheta, \beta_\kappa \geq 2.0$ due to the presence of negative feedback expressed in both Equations (5.32) and (5.33). As for general equilibrium, once we insert Equations (5.32–5.34) back into our original model presented in Section 5.1, and assuming Equation (5.18) – $K_\Gamma = \Gamma = \eta$ – still holds, then general equilibrium conditions for our intertemporal model of information production/consumption under an appendage of research pulses become

$$\dot{\Gamma} = \begin{cases} L(\vartheta - \kappa)\eta - \dfrac{\alpha}{\nu}, & \nu > \dfrac{\alpha}{L(\vartheta - \kappa)\eta}, \\ 0, & \nu < \dfrac{\alpha}{L(\vartheta - \kappa)\eta}. \end{cases} \tag{5.35}$$

and

$$\dot{\nu} = \rho\nu - \frac{1-\alpha}{\Gamma}. \tag{5.36}$$

Again, as was the case with our market sentiment oscillator, the inclusion of (α/ν) into the pace of $\dot{\Gamma}$ under general equilibrium does not dramatically alter the type of dynamics illustrated by our suggested appendage to the model (and illustrated in Figure 5.3) – in particular, "research pulses" still occur, only the order of magnitude of the calibrations change.

However, the implications for asset prices in the wider economy (presently exogenous to our model) under our "research pulses" mechanism is a little more subtle. Assuming the simplest of all possible interpretations with each "new theme" being contrarian to its predecessor, then such pulsating activity in new theme creation will also be reflective in the pattern of asset pricing (assuming there is a direct link between overall theme recommendation and asset prices at large within the wider economy). Obviously, if one were to assume a more complex relationship between each new theme's eventual appearance and asset prices in our (presently exogenous) wider asset market, then one could explain more elaborate behaviors on behalf of these asset prices in response to our hypothesized "research pulse" mechanism. Still, we feel the suggested appendage to our model in its present form nicely explains the overall concept and we flag possible extensions to this basic concept as a potentially worthwhile avenue of future research.

Contrarian thought contagion

Of all the suggested appendages to our model we have presented thus far, it is "contrarian thought contagion" that has the most obvious linkage to the dynamic systems research readily identifiable within the existing biological/evolutionary literature. Advocates within this specialist field of academic endeavor will soon recognize that the equations presented later are largely commensurate with the work of Lotka (1925) and Volterra (1931, 1937) in their attempts to explain predator/prey fish dynamics in the Upper Adriatic post World War I. Indeed, the rationale we use in presenting such a model here is that just as there is a constant battle between predator and prey in the natural world, so too there is a constant battle between contrarian and dominant thoughts (read, "ruling orthodoxy" thoughts) within the collective analyst community's mindset.

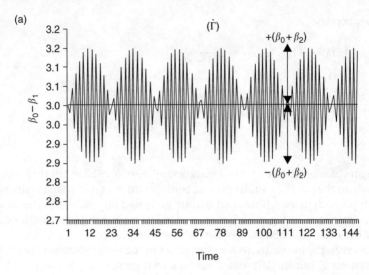

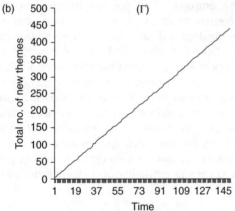

Figure 5.3 Calibration of "research pulses" – $\delta_\vartheta = 6$, $\delta_\kappa = 3$, $\beta_\vartheta = 0.05$, $\beta_\kappa = 0.1$, ϑ, $\kappa = 1$, $L_\Gamma = 1$

How to model this phenomenon? Let us start by hypothesizing that contrarian thoughts have a tendency to spread like an epidemic as they "infect" the recommendations espoused by analysts in the publication of their information bytes. Next, let us assume that there is in place a dominant form of market sentiment (υ) that can be either bullish or bearish at initiation and which is growing in its dominancy (as measured by the net share of information bytes consistent with this viewpoint) at a rate ι. In short, this dominant form of market sentiment effectively represents the ruling share of collective analyst opinion as to their bullishness or bearishness – thus forming the "ruling orthodoxy" viewpoint.

Given these conditions, it is relatively easy to model the "contagion" of a contrarian mindset into this ruling orthodoxy of collective analyst opinion by specifying a rate of converts (χ) as they come into contact with the (minority) contrarian grouping (ϖ) who subsequently waste no time in converting the "dominant sentiment" believers to their contrarian cause.[14] However, without a worthy foil, contrarians are assumed to decline at the specified rate (λ) simply because if one wants to be a contrarian, then it is helpful to have in place a dominant viewpoint to be contrarian against. This dominant/contrarian parasitic poaching of each other's members within the collective analyst community can thus be illustrated using the following dual differential equation system

$$\dot{\upsilon} = \iota - \chi \upsilon \varpi \tag{5.37}$$

and

$$\dot{\varpi} = -\lambda \varpi + \chi \upsilon \varpi, \tag{5.38}$$

where

υ is the dominant share of market sentiment as represented in published information bytes;
ϖ the contrarian share of market sentiment as represented in published information bytes;
ι the rate of "natural increase" in dominant sentiment share of published information bytes;
λ rate of "natural decrease" in contrarian sentiment share of published information bytes; and
χ is the convert rate to contrarian viewpoint as dominant sentiment analysts interact with contrarian sentiment analysts.

Again, as was the case with the suggested appendage to our model presented in the previous subsection, if we assume information bytes are categorized into separate themes on the basis of both their recommendation and the timing of their creation, we now have a "byte to theme," "theme to market sentiment" framework. Furthermore, the actual format of the analyst recommendations in this escalating structural hierarchy will be oscillating between various phases of "bullishness" and "bearishness" (dominant/contrarian) as the following calibration of our system in Figure 5.4 demonstrates.

However, there is one factor that is worth noting. While we once more observe an obviously cyclical behavior in the collective analyst allegiance toward any particular form of market sentiment, one important feature which differentiates the results presented here from either the "market

---

[14] Obviously, instrumental in such contagion is the degree of conversation between analysts. Indeed, Shiller (1995) indicated that conversation is critical in the contagion of popular ideas about financial markets.

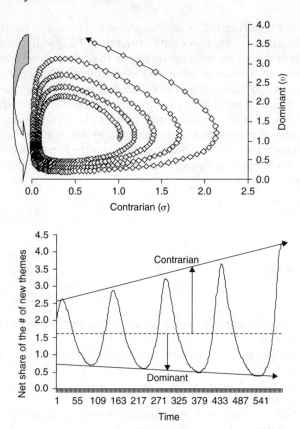

Figure 5.4 Calibration of contrarian thought contagion – υ, $\varpi = 1$, $\iota = 0.09$, $\lambda = 0.03$, $\chi = 0.08$

sentiment oscillation" model or the "bull/bear analyst cycle" model that were outlined earlier is that we now express our results in terms of the "net share of information bytes in existence." This chosen format for our modeling generates some particularly interesting interpretations for *negative* values of either υ or ϖ as either "negative share contrarian" or "negative share dominant" implies a switch in the original directionality of contagion expressed in Equations (5.37) and (5.38). Such a switch infers a "role-reversal" circumstance of dominants infecting contrarians rather than vice versa – obviously, the "poaching" of members from opposing sides of the thought spectrum cuts both ways. This only goes to show that ruling orthodoxies have a tendency to change through time simply because what starts out as a surreptitious poaching of analysts from the ruling orthodoxy (which, in turn, creates oscillations) eventually becomes so serious that what is perceived to be the "ruling orthodoxy" actually changes.

Extending time within our chosen calibration soon revealed that when such negative "share dominant" or "share contrarian" values appeared they tended to act as a harbinger for increasing instability within the system overall. Indeed, the further one extended time the more unstable our system became for our assumed calibrations. Once more, assuming a relatively simple relationship between the movement in asset prices and the net weight of analyst "bullish" or "bearish" opinion (in other words, market sentiment) then it is possible to argue that this oscillatory (and potentially chaotic) behavior exhibited in analyst views will also be reflected in asset prices at large.

Finally, let us close our analysis of this particular appendage to our model by confirming that since we have not altered any of the fundamental equations in the framework for information production/consumption presented in Section 5.1 – remember, we are talking about the *net share* of $(\dot{\Gamma})$ in this particular appendage to the model – we have not altered the general equilibrium conditions expressed in both Equations (5.25) and (5.26). Subsequently, we have yet another relatively straightforward adaptation to our model for linking the perpetual stream of analyst research output to actual asset price dynamics in the wider economy. Now we turn our attention toward detailing our preferred appendage to the intertemporal information production/consumption framework we presented in Section 5.1.

5.3 Stage III: Highlighting our preferred evolutionary model of the market – constructing the informational genome of asset prices

So what is our preferred model? Certainly all of the models presented in Section 5.2 have their relative merits – but many have drawbacks too. What we are aiming for is a model that links the infinite stream of analyst output established in Section 5.1 with asset prices – but sufficiently generalized so that we can explore a number of features we observe in "real world" asset markets. Further, a necessary condition for such a model is that it must not disrupt (at least too much) the nice general equilibrium conditions identified in our intertemporal model of information production/consumption. Does such a model exist? Taking an eclectic view of some of the key features of the suggested appendages to our model that have been presented in the previous section, we feel we have managed to formulate a worthy compromise. The basic foundations of our preferred evolutionary model in linking the analyst's proclivity for continual research output with the evolution of asset prices is represented with the following differential equation system

$$\dot{P}_i = \Omega B_0 + \Lambda \Delta B_1 + \omega P_a, \tag{5.39}$$

$$\dot{B}_0 = \varsigma_0 B_1 + \theta_0 P_i B_0 + \gamma_0 P_i B_1, \tag{5.40}$$

and

$$\dot{B}_1 = -\varsigma_1 B_0 - \theta_1 P_i B_0 - \gamma_1 P_i B_1, \qquad (5.41)$$

where

Ω = net memetic externality weight placed upon *contemporaneous* information;

Λ = net memetic externality weight placed upon *latent* information;

Δ = net sum of latent information enforceable at time t;

ω = degree of association between asset price (a) with representative asset price i;

ς_0 = tendency for complementary "research herding" by analysts in their recommendations – contemporaneous "buys" follow latent "buys" and contemporaneous "sells" follow latent "sells";

ς_1 = tendency for contrarian "research herding" by analysts in their recommendations – latent "buys" follow contemporaneous "sells" and latent "sells" follow contemporaneous "buys";

θ_0 = tendency for *reinforcing* contemporaneous price herding behavior by analysts in their recommendations – contemporaneous "buys" follow contemporaneous "buys" and are scaled by price movements (with the opposite applying for "sells");

θ_1 = tendency for *offsetting* contemporaneous price herding behavior by analysts in their recommendations – contemporaneous "sells" follow contemporaneous "buys" and are scaled by price movements (with the opposite applying for "buys");

γ_0 = tendency for *reinforcing* latent price herding behavior by analysts in their recommendations – latent "buys" follow latent "buys" scaled by price movements; and

γ_1 = tendency for *offsetting* latent price herding behavior by analysts in their recommendations – latent "sells" follow contemporaneous "buys" scaled by price movements.

So what does this rather complicated system of equations necessarily mean? For starters ς effectively controls the intertemporal relationship between analyst research output – or in other words "research herding." That is, by altering the value of ς_0 as opposed to ς_1 we are able to determine the extent to which analysts are either producing information bytes that have either the same final recommendation as those that have preceded it, or alternately producing information bytes with a different final recommendation from those that have preceded it. Conversely, both θ and γ control the degree of "price herding" behavior on behalf of analysts where the feedback from price activity influences the "buy/sell" recommendation decision made by analysts – in a positive manner for *reinforcing* price herding, in a negative manner for *offsetting* price herding. Further, by altering the values of Ω and

Λ we can determine the extent to which externality effects in contemporaneous as opposed to latent information will affect an asset's price – in short, these variables determine how important the net impact of memetic *sequencing* is in generating nonlinear price responses from both contemporaneous and latent signals. Finally, by altering the value of ω we can extend our model of representative asset prices to any asset price across an entire portfolio spectrum simply by altering the degree of assumed association between the representative asset (i) and price of the portfolio asset (a).

As can be seen, the beauty of our approach is that it is able to be generalized to a wide range of interesting observed market phenomena. In short, simply by altering the aforementioned variables we can model under a single framework: (1) the tendency for analyst herding in the their recommendations; (2) the potential for speculative bubbles and market crashes to reinforce this behavior – along with feedbacks between recommendations and price activity in both instances; (3) the desire for analysts to be contrarian and how this is helped or hindered by price movements; (4) the implications of varying degrees of memetic sequencing in generating nonlinear price responses to the information flow of both a contemporaneous and latent nature; and (5) the implication of feedback between asset prices within a specified portfolio of assets that are each following separate analyst behavioral regimes. And this list of alternatives is by no means conclusive. Further, since we are only *encoding* research output with our system of Equations (5.39–5.41) we have not altered in any way the general equilibrium conditions espoused in Section 5.1 in our model of intertemporal information production/consumption – a convenient outcome overall we feel.

So in short, we now have a model where analysts continually produce and disseminate information bytes and (in the process) sometimes create new themes – and importantly, are rewarded for doing so. Further, this perpetual research output flows into a system of equations that affects wider asset prices via its genome-like sequencing effect as the string of recommendations interrelate (via a memetic scalar) with the price of a representative asset. It is this all-encompassing byte to meme, meme to theme, theme to overall market sentiment explanatory power that makes this particular appendage to our model far superior to those provided in Section 5.2. What are some of the more interesting observations stemming from this comprehensive "information building" appendage to our model? To give the reader some insight as to the model's capabilities we have provided a brief representation of a number of calibrations of Equations (5.39–5.41) next.

Calibration 1: Increasing analyst uncertainty

Here we calibrate the model in such a fashion that – as consistent with the bull/bear analyst cycle presented earlier – there is a never-ending cyclicality in the rotation of analyst recommendations between BB, BS, SS, SB. However, since we have now endogenized a pricing equation we can directly observe

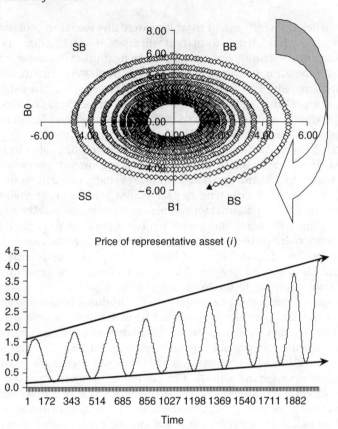

Figure 5.5 Calibration of increasing analyst uncertainty – $\Omega = 0.01$, $\Lambda = 0.007$, $\Delta = 1$, $\omega = 0$, $\varsigma_0 = 0.01$, $\theta_0 = 0$, $\gamma_0 = 0.01$, $\varsigma_1 = 0.02$, $\theta_1 = 0.01$, $\gamma_1 = 0$

what effect this boom/bust rotation in analyst preferences has upon the price of the representative asset i. In doing so, it is important to note that we have allowed for a greater degree of contrarian research herding behavior on behalf of analysts (in other words $\varsigma_1 > \varsigma_0$) so there is an increasing tendency for volatility within the system. As can be seen, this increasing degree of analyst "uncertainty" as contrarians respond with greater and greater disproportionality to the ruling orthodoxy recommendations makes for considerably more volatile representative price action the further one extends time (see Figure 5.5).

Calibration 2: Equilibrium price convergence

Here we take a different tact. Rather than analyst behavior being destabilizing, we take the approach that analyst recommendations actually help our system

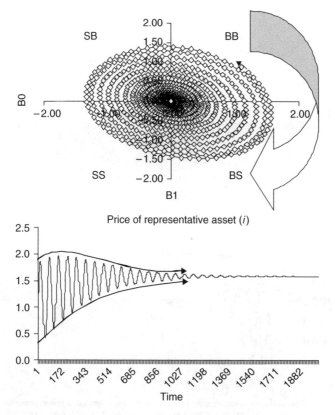

Figure 5.6 Calibration of equilibrium price convergence – $\Omega = 0.04, \Lambda = 0.04, \Delta = 1,$ $\omega = 0, \varsigma_0 = 0.1, \theta_0 = 0.01, \gamma_0 = 0, \varsigma_1 = 0.1, \theta_1 = 0, \gamma_1 = 0.02$

equilibrate. Both ς_0 and ς_1 are evenly matched. Rather, it is price action – and in particular, the strength of *offsetting* latent price herding behavior by analysts – that forces the system toward an equilibrium price as shown in Figure 5.6.

Calibration 3: Jump diffusion

Jump diffusion is a frequently encountered phenomenon in financial markets – can our preferred model appendage replicate this form of pricing behavior? The short answer is yes. This is done by ascertaining the correct balance between complementary (ς_0) and contrarian (ς_1) research herding by analysts to instigate an initial upswing phase of the diffusion process. Then, as the price of the representative asset (*i*) rises, this effect is counteracted by a marginally greater weighting for *offsetting* price herding behavior as opposed to *reinforcing* price herding behavior $(\theta_1 + \gamma_1) > (\theta_0 + \gamma_0)$ thus generating the

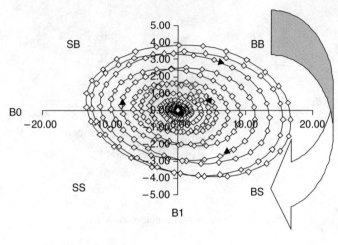

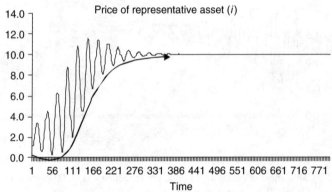

Figure 5.7 Calibration of jump diffusion – $\Omega = 0.06$, $\Lambda = 0.02$, $\Delta = 1$, $\omega = 0$, $\varsigma_0 = 0.9$, $\theta_0 = 0.001$, $\gamma_0 = 0$, $\varsigma_1 = 0.06$, $\theta_1 = 0$, $\gamma_1 = 0.01$

downswing (flattening) phase of the typical jump diffusion pattern as in Figure 5.7.

Calibration 4: Overwhelming sellers

What happens if things go horribly wrong? Just how stable is our system? Calibrating for excessive contrarian research herding on behalf of analysts generates the "doomsday scenario" result shown in Figure 5.8.

Calibration 5: Associative asset prices

Finally we come to one of the most interesting – and yet the most difficult – calibrations of the model. Up until now you may have noticed that in each calibration we set ω to zero as we wished to concentrate upon the forces

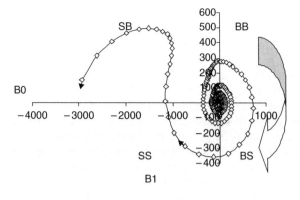

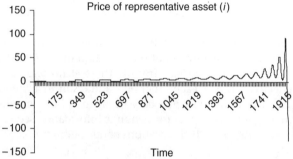

Figure 5.8 Calibration of overwhelming sellers – $\Omega = 0.01$, $\Lambda = 0.01$, $\Delta = 1$, $\omega = 0$, $\varsigma_0 = 0.003$, $\theta_0 = 0$, $\gamma_0 = 0.01$, $\varsigma_1 = 0.039$, $\theta_1 = 0.001$, $\gamma_1 = 0$

driving a single asset's price – the dynamics of the representative asset (*i*). Now we alter that assumption by allowing for *association* between asset prices operating under different analyst regimens.[15] This generates some of the most compelling (but also unpredictable) asset price dynamics of the model. Such results stand to reason as it was in response to the cacophony of informational "noise" generated by such interrelationships and the way that this, in turn interrelates with asset prices, that resulted in earlier researchers becoming overwhelmed as they attempted to describe the driving forces governing information/price dynamics. As explained in Chapter 2, their response was

[15] A far more simple associative regimen would be to assume that all assets within the market system are operating under the same analyst regimen. In such a case, all one needs to do to form a market portfolio of asset prices is to generate the asset price dynamics for the representative asset (*i*) and then interpolate these results to the wider market via the assumed association between assets ω – which in this instance would reflect the degree of correlation between each asset in the market and the representative asset (*i*). This particular calibration will not be undertaken here due to its inherent simplicity.

essentially to capitulate by arguing that information arrival and its resulting influence upon asset prices was primarily a "random" phenomenon. We illustrate here that far from being random, such a system is truly deterministic – it simply gives the appearance of being random primarily because (under some instances) the resulting price action is Chaotic.

To illustrate our point we outline a case of increasingly Chaotic behavior on behalf of movements in an overall market index. Our market index was constructed as the weighted sum of all asset prices contained within our system. To keep things as simple as possible our "market" is comprised of just two assets – with a 50 percent equal market capitalization weighting for both constituents. These assets are assumed to be operating under different analyst regimens in terms of the calibrated dynamics driving their price action. That said, both asset prices are argued to be associative in the sense there is feedback from asset (a) to representative asset (i). The results are shown in Figure 5.9.

So what does the overall informational genome sequence of our market index look like under the auspices of our set calibration? Allowing for simultaneous "buy" and "sell" signals from both asset (a) and representative asset (i) to effectively cancel each other out (and result in a "neutral" rating for the overall market index) when they occur in either contemporaneous or latent signaling form generates the genomic information sequence for our market index over the last 500 observations of our calibration exercise shown in Figure 5.10.

Are there any crucial sequences within this informational genome that appear to act as a harbinger to the increasingly chaotic nature of our model's results? In attempting to ascertain the key memetic sequences driving the market index price dynamics observed within our calibration, we sequenced each information byte for both asset (a) and representative asset (i) into pairs. We then compared the various permutations of these sequences vis-à-vis the market index price response. The results of the average response are represented in Figure 5.11. As can be seen, different memetic sequences have demonstrably different impacts on the performance of the overall market index. Further, some of the most dramatic responses were not the obvious candidates. Simultaneous "BBBB" and "BBBB" memes for asset (a) and representative asset (i) did not always generate the strongest overall positive price response. Nor did the simultaneous "SSSS" and "SSSS" outcome always generate the most dramatic negative price growth outcome.

Indeed, this type of result is precisely what our Evolutionary Finance framework is designed to expose, the way certain sequences of information bind together in what some would label as a "biological"/"genetic" molecular fashion to generate nonlinear informational impact (II) responses in asset prices – that are not always obvious from the outset. Without doubt, this represents but the "tip of the iceberg" and space limitations prevent us from exploring every possible permutation of the various calibrations

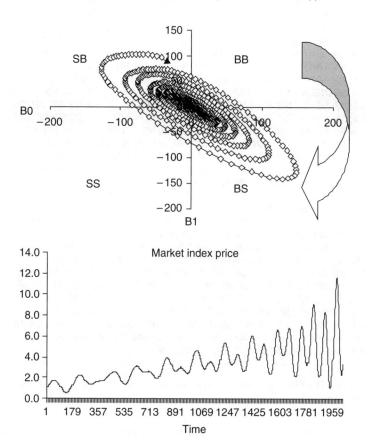

Analyst recommendations for representative asset (*i*)

Figure 5.9 Calibration of associative asset prices – representative asset (*i*) $\Omega = 0.01$, $\Lambda = 0.01, \Delta = 1, \omega = 0.01, \varsigma_0 = 0.045, \theta_0 = 0.01, \gamma_0 = 0.01, \varsigma_1 = 0.01, \theta_1 = 0.01$, $\gamma_1 = 0.01$; associative asset (*a*) $\Omega = 0.01, \Lambda = 0.01, \Delta = 1, \omega = 0, \varsigma_0 = 0.01, \theta_0 = 0.01$, $\gamma_0 = 0.01, \varsigma_1 = 0.063, \theta_1 = 0.01, \gamma_1 = 0.01$

of our preferred extension to the intertemporal model of information production/consumption outlined in Section 5.1. However, before closing, let us return to the issue of the identification of key memetic sequences – in particular the visual representation of such, as this will form an integral part of our *applied* analysis presented in Chapter 7 of this book.

As stated at the outset, what we are looking for is the financial market equivalent to Watson and Crick's double-helix DNA structure. A good place to start is via an examination of the inner core of the information genome itself. How is this done? We can do this by using a simple 3D trace plot of the

......, SB; SB; SB; SB; SB; SB; SB; SB; SB; SB; SB; SB; SB; NB; NB; NB; NB; NB;
NB; NB; NB; NB; NB; NB; NN; NN; NN; BN; BN; BN; BN; BN; BN; BN; BN; BN;
BN; BN; BN; BN; BN; BN; BN; BN; BN; BN; BN; BN; BN; BN; BN; BN; BN; BN;
BN; BN; NN; NN; NN; NN; NS; NS; NS; NS; NN; NN; NN; NN; NN; NN; NN; NN;
NN; NN; NN; NN; NN; NN; NN; NN; NN; NN; NN; NN; NN; NN; NN; NN; NN;
NN; NN; NN; NN; NN; NN; NN; NN; NN; NN; NN; BN; BN; BN; BN; NN; NN;
NN; NN; NN; NS; NS; NS; NS; NS; NS; NS; NS; NS; NS; NS; NS; NS; NS; NS;
NS; NS; NS; NS; NS; NS; NS; NS; NS; NS; NS; NS; NS; NS; SS; SS; SS; SS;
SS; SS; SS; SS; SN; SN; SN; SN; SN; SN; SN; SB; SB; SB; SB; SB; SB; SB; SB;
SB; SB; SB; SB; SB; SB; SB; SB; SB; SB; SB; SB; SB; SB; SB; SB; SB; SB; SB;
SB; SB; NB; NB; NB; NB; NB; NB; NB; NB; NB; NN; NN; NN; NN; NN; NN;
NN; NN; NN; NN; NN; NN; NN; NN; NN; NN; NN; NN; NN; NN; NN; NN; NN;
NN; NN; NN; NN; NN; NN; NN; NN; NN; NN; NN; NN; NN; NN; NN; NB; NB;
NB; NB; NB; NB; NB; NB; NB; NB; NB; NB; NB; NB; NB; NB; NB; NB; NB; NB;
NB; NB; NB; NB; NN; NN; NN; NN; NN; NN; NN; NN; NN; NN; NN; NN; BN;
BN; BN; BN; BN; BN; BN; BN; BS; BS; BS; BS; BS; BS; BS; BS; BS; BS; BS;
BS; BS; BS; BS; BS; BS; BS; BS; BS; BS; BS; NS; NS; NS; NS; NS; NS; NS; SS;
SS; SS; SS; SS; SS; SN; SN; SN; SN; SN; SN; SN; SN; SN; SN; SN; SN; SN;
SN; SN; SN; SN; SN; SN; SN; SN; SN; SN; SN; SN; SN; SN; SN; SN; SN; SN;
SN; SN; SN; NN; NB; NB; NB; NB; NB; NB; NB; NN; NN; NN; NN; NN; NN; NN;
NN; NN; NN; NN; NN; NN; NN; NN; NN; NN; NN; NN; NN; NN; NN; NN; NN;
NN; NN; NN; NN; SN; SN; SN; SN; SN; SN; SN; SB; SB; SB; SB; SB; SB; SB; SB;
SB; SB; SB; SB; SB; SB; SB; SB; SB; SB; SB; SB; SB; SB; SB; SB; SB; SB; SB;
SB; NB; NB; NB; NB; NB; NB; NB; NB; NB; BB; BB; BB; BB; BB; BB; BB;
BB; BN; BN; BN; BN; BN; BN; BN; BN; BN; BN; BN; BN; BS; BS; BS; BS; BS;
BS; BS; BS; BS; BS; BS; BS; BS; BS; BS; NS; NS; NS; NS; NS; NS; NN; NN; NN;
NN; NN; NN; NN; NN; NN; NN; NN; NN; NN; NN; NN; NN; NN; NN; NN;
NN; SN; SN; SN; SN; SN; SN; SN; SN; SN; SN

Figure 5.10   The final 500 signals in the information genome of the market index for calibration 5[16]

various permutations of encoded information byte signals for both asset (*a*) and representative asset (*i*) and then mapping this against the evolution of the market index. This is precisely what we have done in Figure 5.12. By presenting our results in such a way, we are able to illustrate how the market index actually *evolves* across time in response to various encoded information bytes. We feel that this is quite a useful visualization. That said, does Figure 5.12 represent the breakthrough we are looking for? Does it truly represent the financial market equivalent of Watson and Crick's double-helix

[16] Note the pronounced clustering of similar recommendations thanks to the presence of price inspired feedback influencing individual analyst recommendations. This "collective" impact of feedback driving the behavior of the microstructure of the analyst community is completely sympathetic to the Complexity Theory views espoused in Chapter 3 and indeed, reflects what we observe in the "real world" as well. We will model the microfoundations driving such behavior on behalf of the analyst community in greater detail in Chapter 6.

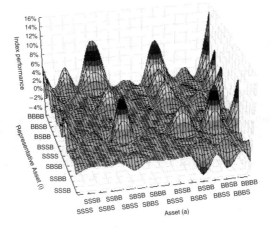

Figure 5.11   The informational impact of memetic information contained within the information genome for calibration 5

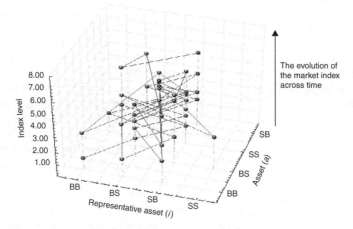

Figure 5.12   The inner core of the informational genome for calibration 5

DNA molecule structure? We think not. For starters, there is no representation of the building of information from byte to meme, meme to theme, theme to overall market sentiment. Further, while definitely elegant, the procedure itself is not scalable as once one expands the representation beyond two assets, you immediately encounter a dimensionality problem in representing your assets graphically on the horizontal *x*/*y* plane. So in short, a 3D trace plot represents a good place to start, but by-no-means does it constitute our final result.

Indeed, to more accurately represent our comprehensive framework for understanding the biological-like molecular building blocks of information, we need to "filter" our informational inputs in some way before presenting them graphically. There are a number of methods by which to do this ranging from relatively simple parametric techniques – such as principle component analysis, factor analysis, multiple correspondence analysis or even discriminant analysis – to the relatively more complex nonparametric methods – which include k-means clustering, EM clustering, regression tree and multidimensional scaling techniques. All can be designed to "filter" the informational genome to look for those memetic strings with the strongest associative asset price impacts. Still, we opted for a far simpler – and more intuitive – technique that takes into consideration what investors are trying to achieve.

Using simple Bayesian trading rule analysis (albeit with a complex overlay of adapted Evolutionary Programming techniques), we were able to detect those memetic sequences that had the largest II upon a chosen asset – or even the impact upon the market index itself. How was this done? By altering various permutations of contemporaneous and latent information byte signals and applying a trading rule "filter" to this information we were able to discern what sequences had the largest "payoff" to an investor attempting to maximize the return on their investment.[17] The obvious beauty of this approach is that it actually mimics "real world" investor behavior – investors receive memetic signals, interpret them and then act accordingly. Further, since our Bayesian trading rules are conditional (by definition), we are able to introduce ancillary constraints to our investment process. Such ancillary constraints in essence augment our algorithmic "key meme" search and include conditions such as maximal drawdown limits, maximal consecutive loss, return volatility minimization, Sharpe ratio constraints, Sortino ratio limits and so on – all of which eventually condition the evolution of the various "structured information" trading rules attached to our key memetic sequence search.

In short, in sympathy with our earlier cited Keynes quote, our investment process at its simplest basically becomes a "beauty pageant" of various structured information memetic-based trading rule strategies (where at times, key memetic sequences can also be used to identify the informational "triggers" forecasting a switch by market participants from one preconceived investment strategy to another).[18] This "switching prediction" capability of our

---

[17] This approach toward key meme identification is not too dissimilar from the genetic mapping algorithms used in genomics which look at various permutations of the "reshuffling" of various bases along a genetic sequence as indicative of the proximity of two genes along a particular chromosome. See Watson (2004) for a highly readable introduction to this field.

[18] It is important to note that our "structured information" approach is a key differentiating feature from previous "Evolutionary Finance" implementations as briefly

Bayesian trading rule (structured information) technique builds upon the earlier mentioned fallibility of most investors simplistic heuristic trading rules inevitably going wrong at varying points in time – as was identified in our Behavioral Finance critique in Chapter 3. As we described there, an "optimal strategy" under such circumstances is one that develops an overarching sequence of strategies that takes into consideration such inevitable "failures" on behalf of the simplistic (typically static) strategies of others and effectively preys upon their shortcomings. To do this in a computationally efficient manner, we had to appeal to a class of nonlinear programming technique known as "Evolutionary Programming" which will be described in considerable detail in Chapter 7. That said, one benefit of our approach that is useful to highlight at this early stage is that the output of our model based investment projections effectively constitute time-series data. Ergo, we are able to plot the evolution of the key memetic sequences along the market's informational genome (and the efficacy of our investment process over time). Needless to say, we will elaborate upon this "structured information" approach toward filtering the market's informational genome and formulating the prediction of (and subsequent evolution of) winning trading rule based strategies in greater detail in Chapter 7, but for now let us reiterate that we found the Bayesian trading rule technique the best place to start in such a process and certainly the most amenable for our purposes. Still, this doesn't tell the reader how such a complex array of information can be efficaciously represented visually and why this concise visualization should be looked upon as such a breakthrough.

Primarily, our challenge from the start has been to find the financial market equivalent of Watson and Crick's double-helix representation of the DNA molecule. In short, we now believe to have found it. Just as Watson and Crick's representation of the DNA molecule had profound implications for the world of science, we feel the following representation of conditional memetic strings which have significant II in terms of asset price growth will have a similar "big bang" effect upon the finance sphere – albeit after the somewhat annoyingly compulsory "take-up lag" that usually accompanies such pronouncements.

described in Chapter 3. In particular, our ability to better understand the "inner fabric" of information – how bytes form memes, which in turn form themes and overall market sentiment – better positions us to actually plot the emergence of winning strategies from a preconceived strategy set. Typically both analysts and investors use such strategies to maximize the return from their research and investments (respectively). Our "structured" approach toward information better understands the true microfoundations of this behavior and thereby differentiates us from simple "random seeding" optimal strategy searches that have characterized past studies under the "Evolutionary Finance" genre. We will contrast our implementation style vis-à-vis those that have preceded us in the Evolutionary Finance field in greater detail in Chapter 7.

So now that we have whetted the reader's appetite, what particular graphical form have we proposed for this all-important representation of the linkage of molecular-like financial information with asset prices? Well, we like to think of it as a "tree-root" type of structure. Small tendrils of radial memetic information inputs feed into the "root core" of the asset price which through time, thanks to these information inputs, grows larger.[19] The memetic inputs are selected via a winning Bayesian conditional strategy regimen with the highest payoff strategy (the "strongest" memetic string) being recorded as a tendril-like extension that has closest proximity to the "root core" of the asset price. Conversely, those memetic strings which have been deemed by our algorithmic process to have been less successful strategies are located the greatest distance away from the asset price "root core." In turn, sub-extensions from the tendrils themselves illustrate the array of individual information bytes feeding into any chosen memetic string. Since these memetic-based winning strategies themselves are in a constant state of evolution, the tendril feeders into the asset price informational core will alter through time – both in their positioning and in outright composition. By comparing these compositional and positioning changes, one is able to discern as to whether there has been a demonstrable change in the informational drivers of the asset in question – and thus, a change in theme. Likewise, by observing the relative proportion of "bullishness to bearishness"/"buy to sell" recommendations in these asset price drivers, one can discern as to whether there has been a change in overall market sentiment. Subsequently, an altering thematic/market sentiment pattern will be represented by a sequence of observable changes in the physical structure of our designated "informational radix."

The true beauty of our depiction of how information comes to affect asset prices – as represented in Figure 5.13 – is that it summarizes a lot of critical concepts in a very concise way. By taking smaller and smaller snapshots of the "informational radix" (as we like to refer to it) it is possible to garner more and more information as to what is actually driving an asset's price at a given point in time. Moving along the structure, one can see what memetic sequences were most important at different points in time and more importantly, how this waxes and wanes as the asset price evolves. Comparing various 2D cross representation snapshots has its advantages too as the "radial clustering" of the key memetic inputs (as determined by our Bayesian conditional trading rule technique) provides information not only as to the relative strength of the key memetic sequences, but also details their composition as well – as shown in Figure 5.14.

---

[19] In this particular example we use an asset's price as the "root core" target variable but as we will explain in Chapter 7, the "root core" can just as easily be the performance of a particular preconceived strategy. Indeed, one can even use moment distributions as a root core.

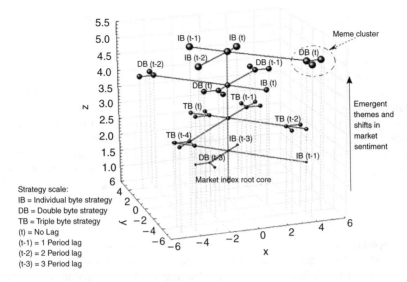

Figure 5.13 The informational radix for calibration 5

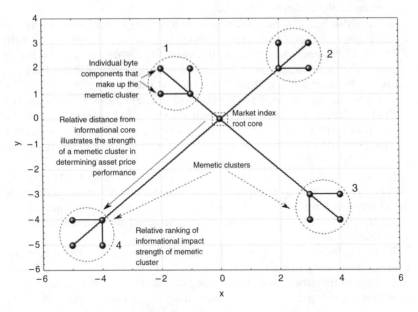

Figure 5.14 A cross-sectional snapshot of the informational radix for calibration 5

Finally, comparing various informational radix structures across differing assets allows one to immediately visualize how various assets "grow" in response to a given informational environ (as set by your choice of the raw information used in constructing the informational genome that is filtered by our Bayesian trading rule technique). Such visual comparisons make the contrasting of highly complex structured information/asset price relationships relatively simple. To illustrate, we undertook yet another calibration of our preferred model – using yet again an associative asset price framework, but this time raising the influence of memetic cluster formation via heightening the externality potential (x) expression for latent information. We did this by altering both Δ (the net sum of latent information enforceable at time t) and Λ (the memetic externality weight placed upon *latent* information) for representative asset (i) – but holding everything else constant as per our previous calibration. The aim of this exercise was to increase the II of latent information and to thus alter both the size and the prevalence of memetic structures as "winning strategy" outcomes within our Bayesian conditioned trading rule screening technique. In particular, our idea was to create an informational radix with discernibly more extensions radiating from its informational core simply because there would be less individual byte (contemporaneous information) winning strategies and more latent information (larger memetic cluster) representations as the preferred outcomes of our memetic string filtering process.

Indeed, this was precisely the result we managed to achieve – as the following figure demonstrates. Simply visually comparing Figure 5.13 with Figure 5.15 presents the reader with a dramatically different sense of the way a given information-set is interacting with the price of two "assets."[20] Mapping the evolution of the way memetic clusters change in positioning and representation between these two graphical depictions across time conveys to the reader how themes are emerging across varying assets. Further, the reader is immediately informed of what information is driving these themes to emerge, its structure, how important particular information sequences are now (and in the past) and even how "informationally intense" a particular asset is. In short, our informational radix representation conveys to the trained eye an immense amount of detail as to the actual fabric and structure of information and how this interacts with the evolution of a given asset's price. It is herein that we feel the greatest potential of this graphical technique truly resides.

To conclude this chapter, we hope we have conveyed to the reader the true beauty of our evolutionary approach toward information structure and in particular, how the study of the biological-like molecular building blocks of information come to influence the evolution of an asset's price. In short,

---

[20] We use the term "assets" in inverted commas here simply because our alternate "assets" are effectively constructed via different calibrations of our preferred model.

Informational radix of the market index with stronger externality potential *(x)* expression

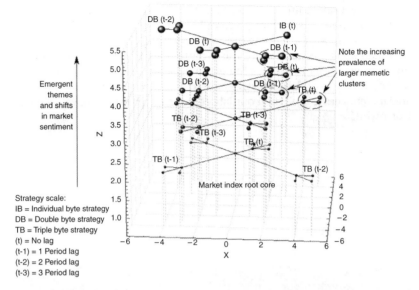

Figure 5.15 Alternate calibration of associative asset prices – representative asset (*i*)
$\Omega = 0.01$, $\Lambda = 0.0107$, $\Delta = 2$, $\omega = 0.01$, $\varsigma_0 = 0.045$, $\theta_0 = 0.01$, $\gamma_0 = 0.01$, $\varsigma_1 = 0.01$,
$\theta_1 = 0.01$, $\gamma_1 = 0.01$; associative asset (*a*) $\Omega = 0.01$, $\Lambda = 0.01$, $\Delta = 1$, $\omega = 0$, $\varsigma_0 = 0.01$,
$\theta_0 = 0.01$, $\gamma_0 = 0.01$, $\varsigma_1 = 0.063$, $\theta_1 = 0.01$, $\gamma_1 = 0.01$

we hope we have managed to put some "flesh on the bones" (so to speak)
of the revolutionary evolutionary informational framework initially mapped
out in Chapter 4. Here in Chapter 5, we have expanded upon our over-
arching philosophy by developing a comprehensive evolutionary model of
the marketplace to give the reader some insight as to how financial markets *really* work. Keystone in this endeavor has been the development of
our informational radix graphical depiction of the way in which structured
information – byte to meme, meme to theme, theme to overall market
sentiment – interrelates with the growth of an asset's price. We feel that this
particular graphical depiction not only summarizes an immense amount of
detail about informational structure to the trained eye, but it also represents a
complete watershed in thinking about the way financial market information
should be assessed.

No doubt, we feel that our "structured information" philosophy will
gradually percolate into a variety of different spheres of influence but one
important port of call before undertaking the *applied* interpretations of our
work will be an examination of the implications of our structured informational approach for distributional form. After all, as was illustrated in
Chapter 2, the Gaussian distributional form has constituted the central

plank with which the traditionalist finance approach has leveraged into an astounding array of financial engineering techniques. For our Evolutionary Finance alternate investment philosophy to prove a worthwhile contender to the crown, we must therefore say something about how a structured informational approach toward the cacophony of information that besieges investors every day influences the distribution of asset prices across time. If we can offer a credible alternative here to "plug into" many an existing financial engineering tool, then our labors will have proven to have been truly worthwhile.

6
The Implications of Our Evolutionary Perspective for Distributional Form

As we have highlighted on a number of occasions throughout this book, distributional form is important. For starters, the ubiquitous Gaussian distributional form underpins the entire edifice of financial engineering techniques that characterize the "traditionalist" thought paradigm – a point already expanded upon in considerable detail in Chapter 2. This distributional form does have its limitations, but as we have also mentioned, it is insufficient to criticize something without at least attempting to offer an alternative. Does our own unique view of Evolutionary Finance provide any insight as to an alternative distributional form? The good news is that "yes it does." The even better news is that this (Game Theoretic based/informational sequence driven) distributional form provides a fundamentally more credible approximation of the intrinsic forces determining the distribution of asset price movements than does the "white noise"/IID (independent and identically distributed) driven assumptions that constitute the Gaussian foundations of the traditionalist perspective.

As we will illustrate in considerable detail in Section 6.1, it is relatively easy to leverage the extensive Game Theoretic analysis that was provided in Chapter 3 to provide a generalized version of a microfoundation model for the emergence of sequence-driven "clumps" of information. This model allows us to more plausibly describe systematic (and at times nonlinear) jumps in asset prices as analysts embark upon different "information byte producing strategies" that intrinsically affect the information format along any particular segment of the market's information genome ($\sum \varphi_{\bar{\imath}}$). Such "information byte producing strategies" are a natural consequence of analysts attempts to produce the various pieces of information necessary to build a certain meme or theme. It will be demonstrated that as these pieces of information become available to the wider market, the market itself typically reacts in a "noisy" – but nevertheless predictable – fashion. Under such circumstances – as we mentioned in Chapter 4 – each information byte is *heterogeneously* determined not homogenously determined. This fundamentally deterministic process is fully explained using Evolutionary Game Theory as a microfoundation. It is not arbitrarily imposed – as has been the case with the IID theoretical underpinning motivating the Gaussian distributional form within the traditionalist thought paradigm. The reward for our efforts? A time dependent Evolutionary Distributional Form (EDF) that responds to the ebb and flow of analyst provided information.

Further, an additional benefit of our labors in motivating our EDF is that we are able to provide an intrinsically better interpretation of the fundamental causes of market "events." Such "events" are an endemic characteristic of financial markets that have (unfortunately) been so poorly described under the auspices of traditionalist theory. The reason for our comparative advantage in this particular field is that we use a Ewens distribution to describe the clustering of analysts into various pure strategy following groupings. It is this "herd like" clustering behavior that at times promotes the emergence of kurtosis in asset prices – hence our model's natural affinity with describing the presence of market "events." The bottom line here is that in Section 6.1 we will put forward a stylized interpretation of the microfoundations of financial markets which we feel will be inherently more appealing to both theorists and practitioners alike. This approach transcends the present (Gaussian directed) industry standard. Such are the motivations behind our EDF.

Following on from this analysis, in Section 6.2 we will focus on the particular Game Theoretic strategies that information producing analysts can adopt. We do this by calibrating our model and observing its behavior. In particular, we use the generalized framework developed in Section 6.1 to describe the "informational leverage" that certain analysts can get from either an adroit positioning of a particular information byte, working cooperatively, or alternately utilizing their stature within the market. By influencing both the flow and format of information in such a way, analysts can generate exceptional return for themselves and in so doing, demonstrably affect the distributional form of asset prices. Such informational leveraging/price-directed strategies manifest as a natural consequence of the interaction between the various members of the analyst community and the wider financial market. Just as certain evolutionary strategies appear to be an endemic feature of any ecology – even artificial ecologies generated within a computer program – so too does it appear that certain analyst "information leveraging" strategies are an endemic feature within the overall financial market ecology.[1] We will focus on but a few of such strategies here but we feel we will leave the reader with a general flavor of how the analyst/market strategy dynamic is integral in explaining some of the more broad-ranging of observed market

---

[1] Indeed, the identification of such strategies within the confines of a purpose built computer program has been attributed to the earlier mentioned "artificial life" computing pioneer Thomas Ray.

phenomena. Yet again, we believe that such analysis will be conceptually more appealing to both theorists and practitioners alike. Especially when compared to the sterile world promised by a slavish adherence to the Gaussian distributional form.

Finally, in Section 6.3 we illustrate the way in which our deterministic foundation/informational sequence driven EDF alters prevailing financial engineering techniques. We will do this by demonstrating how standard financial engineering precepts such as mean-variance optimization, Black–Scholes option pricing and even Value-at-Risk analysis alters under our new interpretation of the linkage between analyst provided information and asset prices. As promised in the introduction to this book, virtually no stone of (traditionalist-based) financial engineering "best practice" is left unturned when it comes to implementing our pioneering new evolutionary approach toward information in financial markets. Given space limitations, we aim to highlight but a few of the more important implications here. That said, as was the objective in Section 6.2, our aim is to leave the reader with an indelible impression of just how far reaching our over-arching philosophy can be. So now, onto the deterministic Game Theoretic microfoundations of our EDF.

6.1 Foundations for an evolutionary approach toward distributional form

To understand the macro, one must at first understand the micro. Physicists have long recognized this problem and indeed have been wrestling for some years with the conundrum of how to unify the principle of general relativity with quantum mechanics.[2] Interestingly, the same fundamental "micro meets macro" principle applies to finance. That is, to truly understand "macro" aspects such as the distributional form of asset prices, one must at first have at hand a firm grasp of the underlying (micro) principles driving the dynamics of such phenomena. This is where the traditionalist thought paradigm has let us down immeasurably. To argue that market information (and therefore asset price movement) is too "noisy" to understand is, we feel, an intellectual dead-end. Furthermore, in trying to defend what was at first an excusable simplifying assumption of IID information arrival (in promoting the emergence of a Gaussian distributional form) with ever more elaborate tautologies of "irrational active investors" and "noise traders," is we feel a regressive step for the industry as a whole. We ask, why devote so much time and effort in defending an assumption that apparently was only inserted

---

[2] In fact, such was the quest for Einstein during the greater part of his latter life. Recent developments in String Theory appear to provide a promising foundation for the development of a singular unifying "theory of everything" – but it is still early days yet. For a very readable exposition on these latest developments, refer to Greene (2000).

in the first place because of the technological constraints imposed by the limitations of 1950s and 1960s computing power in wrestling with complex nonlinear problems? Isn't it time to move on rather than vainly trying to defend the status-quo? Certainly we – along with other adherents to the various "new view" challenges described in Chapter 3 – feel this to be the case.

So if we are to be the agents for change – especially when it comes to understanding the microfoundation drivers of asset price distributional form – where should we start? We feel Game Theory provides a useful framework for modeling the interactions between the various members of the analyst community – especially in terms of the quantum and format of the information bytes they produce. In essence, Game Theory provides a convenient mechanism to put some additional "flesh on the bones" of the types of analyst interaction modeled in the latter stages of Chapter 5. After all, analysts no matter of what ilk, use strategies to guide their behavior and motivate their ideas. The obvious implication of this is that such strategies intrinsically affect the format of the information bytes that analysts produce. Such statements are entirely consistent with our discussion in Chapter 4 as to how we as humans typically interpret information. In short, analysts use various preconceived strategy "filters" to assist them in encoding and categorizing the "data deluge" that besieges the market each and every day. Subsequently, it is possible to categorize analysts into various "warring tribes" in accordance with their preferred strategy filter – fundamental, technical or economic. By doing so, we are able to build up an "ecology" of analyst types within the marketplace.[3]

Consequently, it is through the "strategy conduit" that we will formalize the Game Theoretic microfoundations underlying our evolutionary model of analyst (strategy directed) behavior in determining the actual format of information bytes along any given segment of the market's information genome ($\sum \varphi_I$). Assuming a one-to-one linkage between the format of such information bytes and asset prices (or, in the case of memetic information, to allow for nonlinearities between information format and price to emerge) we can then use the strategy based principles of Game Theory as a basis for explaining some of the more interesting aspects of analyst inspired movements in asset prices and through this, distributional form. So how best

---

[3] The obvious flip-side to this is that investors use similar (if not identical) strategies to implement their investment ideas. Ultimately, these two facets of strategic decision-making are interrelated – analysts quite often are investors and investors quite often are analysts. As a consequence, the two sets of strategies – information production and actual trading – have a tendency to become invariably intertwined. What differentiates the two activities in the minds of most people is that sometimes the "analyst" role is outsourced and therefore they only indirectly benefit from the market generated excess return that their ideas generate – as detailed in Figure 2.3 and modeled explicitly in Chapter 5. As will be seen in Chapter 7, we argue investors and analysts are always the "two sides of the same coin" and model our evolutionary strategy search accordingly.

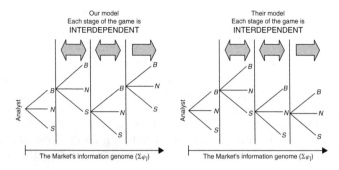

Figure 6.1 An extensive form game of analyst information producing behavior – independent or interdependent?

to implement a Game Theoretic framework within such a context? First, let us begin by way of a visual representation of what we are trying to achieve (Figure 6.1). Here we summarize the basic conceptual Game Theoretic underpinnings of our view vis-à-vis the traditionalist line of thought.[4]

So how to formalize the above schematic into a system of equations? For starters, there is little point in attempting to model the behavior of each

---

[4] Interestingly, as can be readily seen in Figure 6.1, our Game Theoretic framework is amenable to the presence of both interdependent and independent information. For "independent information" in our stylized game you simply assume complete entropy between each sequential t to $t+1$ step of the game. One consequence of this adaptability of our generalized framework is that we are able to describe the traditionalist model as a "special case" of our own. This is an extremely useful perspective when interpreting – as detailed in Chapter 2 – the apparent waxing and waning of support that the various forms of empirical EMH test have given to the traditionalist line of reasoning. For example, if the market sometimes behaves as though information is independent, it is because there is no attempt by analysts (over the specified timeframe of the empirical observation) to *on average* strategically position their information bytes in a manner that takes into account the actions of other market participants. In other words, information bytes are primarily of a singular (Ξ) form. However, if analysts start to think strategically, or there is the publication of an information byte that has significant externality potential (x), then the analyst community may start to produce interdependent information. In this case, information bytes begin to take on a memetic form (M). Under our framework, both such interdependent and independent informational regimes can happily coexist – its simply a matter of distinguishing between the memetic (M) and singular (Ξ) informational format of the various information bytes that constitute the market's information genome ($\sum \varphi_I$) over the time of reference for any particular empirical test. The beauty of such an approach is that it enables us to thereby form an over-arching umbrella across the various empirical EMH studies outlined in Table 2.1 and to explain the alternate outcomes simply via reference to the independence or interdependence of information bytes produced by analysts at a particular point in time along a certain market's information genome ($\sum \varphi_I$).

and every information producing analyst along the near infinite sequence of our extensive form game that represents the market's information genome ($\sum \varphi_I$). Why? Well, in true evolutionary game fashion, there is no guaranteed continuity between the individual analysts represented at each t to $t+1$ step of the game – no matter where t happens to be specified. Rather, different analysts will be entering and exiting the informational sequence continuum at different points in time.[5] This sequential "resampling" of individual analysts into and out of the pool of representative analyst talent at each stage of the game implies that no "reputations" can be built for individual players – at least at the elementary level of the sequential t to $t+1$ iterative steps of information byte production.[6]

So if the analysts in our game are continually changing, how do we model their behavior? The answer is simple – via the payoff function. The specification of the payoff function is the one continuity along our extensive form game that constitutes the market's information genome ($\sum \varphi_I$). Consequently, it is here that our attentions should be directed in building a generalized system of equations for the above schemata. By treating each step along our extensive form game as a distinct (time dimensioned) two person game where each player's actions are motivated by the payoff function, we are able to formalize our Game Theoretic microfoundation structure according to the following generalized form.

A generalized Evolutionary Game format for the market's information genome[7]

Let us primarily concentrate upon specifying a *pure* strategy generalized solution. Assuming two analysts – one at time t and another at time $t+1$ – have the choice of $St = (st_1, \ldots, st_n)$ pure information producing strategies which condition the format of the information bytes they produce, then the payoff to analyst 1 playing $st_б \in St$ is $\pi_{бч}$ while the payoff to analyst 2 playing

---

[5] Such an assumption is entirely consistent with our theoretical model presented in Chapter 5 as different analysts were assumed to produce different information bytes – especially "new theme" generating information bytes over which they possess intellectual monopoly rights. Further, consistent with the above, in Chapter 5 no assumption was made as to the order in which each analyst produces each information byte – rather, the order of each information byte's arrival *itself* was considered as important.

[6] Rather, as will be illuminated in Section 6.2, reputations are built on the basis of the *format* of the analyst's information byte itself – not its positioning. In particular, consistent with our discussion in Chapter 5, the publication by an analyst of an information byte that possesses extremely high externality potential (x) may result in the information producing analyst in question being anointed by the wider market as a "dominant player" in the sense that they become the generator of a new theme. Such a "reputation" will ultimately have its consequences – as will be displayed in Section 6.2.

[7] The analysis presented in this section draws heavily upon the foundations for Evolutionary Games outlined in Gintis (2000).

$st_\text{Ч} \in St$ is $\pi_\text{ЧБ}$.[8] Herein resides the genesis of our "warring tribe" analogy that was motivated in our discussion of artificial models of the marketplace in Chapter 3. While we only assume two types of strategy being pursued at the chosen t to $t + 1$ point in our extensive form game (strategies Б and Ч), our model is sufficiently generalized to allow for a plethora of alternate pure strategy forms. Either Б or Ч could for example be – a technical-based strategy, a fundamental-based strategy, an economic-based strategy and so on. We could even move down into the sub-stratum of each strategy category and specify the actual type of pure strategy – for example, a Fibonacci trading rule strategy. Subsequently, our model is sufficiently generalized to allow for an entire ecology of different pure strategy "types" of analyst to manifest. Ultimately, it is these differences in pure strategy that govern the clustering of analysts into various "warring tribes" and that sets in place the necessary conditions for determining the format of the information byte that a particular "tribal member" will produce when they enter the information byte continuum. Thereby, it is via this process that the format of the market's information genome ($\sum \varphi_\text{I}$) will be determined.

Importantly for our purposes, we also make the distinction between "bullish" ("B" producing) and "bearish" ("S" producing) pure strategy types within each sub-category of our defined analyst "species" that are part of our overall market ecology. The reason for doing so will become all too obvious a little later when we define the conditions for the emergence of an Evolutionary Stable Strategy (ESS) equilibrium and observe the analyst "strategy switching behavior" dynamics of our Evolutionary Game. Suffice to say now, it is important that we draw to the reader's attention to this distinction at the early stages of laying down the generalized format for our microfoundations of analyst behavior.

Furthermore, it is also important to note that we assume our game is symmetric in form – that is, the game is not only symmetric in terms of payoffs but also symmetric in terms of information producing strategies. This implies that analysts can't condition their pure information producing strategy on whether they are Player 1 or Player 2 – an "ordering issue" that we will address in detail a little later in this chapter. Given that this is the case, we can use G to denote a symmetric game matrix at any given representative stage of the game with the vector of Player 1 payoffs within a hypothetically large population of analysts being represented by [Я $= (\pi_\text{Бч})$] – where should more than two players be involved, the variable Ч would represent the strategy choices of the remainder of the analyst group.

Now we add the element of time in the sense that in each period $t = 1, 2, \ldots$ we will assume for simplicity that analysts randomly enter the information producing continuum and produce an information byte (φ) that ultimately

---

[8] Note the notation π for the analysts payoff has been chosen intentionally as it reconciles directly with the profit (π) notation in the previous chapter.

contributes to the overall market's information genome ($\sum \varphi_{\text{I}}$). Remember, we assume each analyst is of a pure information producing strategy type. Subsequently, if the population of analysts following a particular strategy is z at a given point in time, then the analyst population at this time can be represented as $Z = z_1 st_1 + \cdots + z_n st_n$.[9] Under these circumstances, the payoff to an analyst following pure strategy Ђ at the selected time t is

$$\pi_{\text{Ђ}Z} = \sum_{\text{Ч}=1}^{n} \pi_{\text{Ђ}\text{Ч}} z_{\text{Ч}}. \tag{6.1}$$

Now suppose we replace fraction $\omega > 0$ of the analyst population with a "mutant" pure strategy following analyst of type Ч (this could, for example, be a subset of a bullish technical based strategy, a subset of a bearish fundamental based strategy and so on). Still, irrespective of the nature of the "invading mutants" pure strategy, the new state of the analyst population can be specified as

$$\text{Ђ} = (1 - \omega)Z + \omega\, st_{\text{Ч}} \tag{6.2}$$

and the payoff for the randomly chosen nonmutant analyst thus becomes

$$\pi_{Z\text{Ђ}} = (1 - \omega)\pi_{ZZ} + \omega\, \pi_{Z\text{Ч}} \tag{6.3}$$

while the payoff for the mutant strategy following analyst is

$$\pi_{\text{Ч}\text{Ђ}} = (1 - \omega)\pi_{\text{Ч}Z} + \omega\, \pi_{\text{Ч}\text{Ч}}. \tag{6.4}$$

In terms of defining the Evolutionary Stable Strategy, the mutant strategy following analyst can invade the analyst population if $Z = \text{Ђ}$ and (for sufficiently small $\omega > 0$) when

$$\pi_{\text{Ч}\text{Ђ}} > \pi_{Z\text{Ђ}}. \tag{6.5}$$

[9] It is important to highlight that the number of analysts notation is commensurate with the number of informational investors (Z) specified in Chapter 5 – simply because as we have earlier identified, both are effectively the "different sides of the same coin." On the one hand, analysts are "information producers" within our general equilibrium framework described in Chapter 5, on the other hand, they are also "information consumers" (investors). This dual-faceted nature of analyst behavior is entirely consistent with the multiple facets of National Income identities which identify individuals as both producers and investors.

Ergo, as consistent with our definition of an ESS given in Chapter 3, Z is an ESS population of strategies if it cannot be invaded by a mutant strategy following analyst.[10] Interestingly, this relatively simple result lays the foundations for the entire pattern of our market information genome ($\sum \varphi_l$). How is this the case? Well, it basically provides the *generalized* solution for the format of the stream of information bytes produced by a diverse ecology of pure strategy following analysts.

Assuming (as do the traditionalists in their often touted "fair-game" model of market behavior – outlined in Chapter 2) that there is a direct linkage between market information and price, we can now describe in generalized terms how the format of a particular pure strategy motivated information byte affects asset pricing within a relatively simple (deterministic) Game Theoretic process.[11] This is simply because the respective payoffs from following different pure strategies will affect the numbers of analysts represented within each separate pure strategy following faction (or "warring tribe") of our market ecology. It is the total weight of numbers in these relative analyst groupings that determines the most likely outcome for the format of an information byte – given that analysts are assumed to be randomly drawn from the overall analyst population Z. Furthermore, since we have made the distinction between "Bullish" and "Bearish" subcategories of each pure strategy grouping, we are able to generalize swings in overall market sentiment (H) by analyst "herding" into different strategy classifications.

Consequently, in following the lead we established in Chapter 5 and continuing to assume a one-to-one linkage between the format of an information byte and asset prices for all singular information (Ξ) – and a scaled net memetic externality weight in accordance with latent (Λ) and contemporaneous (Ω) effects for all memetic information (M) – we thus have in place a significant portion of the necessary criteria for describing the microfoundations of an EDF. This is done by simply drawing a direct linkage between the predominance of various "clusters" of pure strategy following analysts, the format of the information they produce and the (at times nonlinear) impact that this has upon asset pricing.

So far so good you might say, but how does the above generalized form for an Evolutionary Game across our ecology of information producing analysts (and its implied impact for the pattern of asset pricing behavior) relate back to the series of case study examples presented in Chapter 5 – primarily those presented in Section 5.2? To illustrate this progression from our generalized format Evolutionary Game into the considerably more "macro" in

---

[10] Notice yet again we refer to Z as being a "population of strategies" as we remind the reader there is a one-to-one mapping between individual analysts and pure strategies.
[11] Ultimately this generalized format takes into account the various nonlinear price effects associated with the publication of information bytes that are memetic (M) in form.

perspective dynamic system representation of analyst interaction presented in Chapter 5, we must yet again appeal to the discipline of a generalized form – but this time, for identifying the dynamics of the Evolutionary Game itself.

Our Evolutionary Game's dynamics

As consistent with the earlier, we start by considering an Evolutionary Game where each analyst follows one of n pure strategies and that the t to $t + 1$ sequences of the game are repeated over periods $t = 1, 2, \ldots$. Likewise, we let z be the fraction of analysts adhering to pure strategy $st_Ƃ$ in period t and that the payoff to strategy $st_Ƃ$ is $\pi_Ƃ(z^t)$ where $Z = (z_1, \ldots, z_n)$.

However, in the spirit of our earlier discussion we now explicitly allow for the possibility of a "bull/bear" dynamic to emerge in the sense that the population Z of pure strategy following analysts can be skewed toward either a "bullish" or "bearish" collective mindset. Such swings are entirely consistent with our earlier described process of meme and theme building. Furthermore, we allow for the degree of such a skew to alter across time. To do this we assume that in every time period dt each analyst with probability $ Jbdt > 0$ learns of the payoff of another randomly chosen analyst and changes their pure strategy if they perceive the payoff of the alternate strategy to be higher. That said, we also assume that the information concerning the difference between the two strategies is imperfect, so the larger the difference between payoffs, the more likely it is that strategy switching behavior will manifest. This can be formalized as follows

$$
\mathrm{Pr}^t_{ƂЧ} = \begin{cases} Ƕ\left(\pi^t_Ч - \pi^t_Ƃ\right), & \text{for } \pi^t_Ч > \pi^t_Ƃ, \\ 0, & \text{for } \pi^t_Ч \le \pi^t_Ƃ, \end{cases} \tag{6.6}
$$

which, in effect, denotes the probability $\left(\mathrm{Pr}^t_{ƂЧ}\right)$ that an analyst using strategy $st_Ƃ$ (for instance, a "bullish" GARP strategy) will shift to strategy $st_Ч$ (for instance, a "bearish" DDM strategy) at time t and where the transition factor for analyst switching behavior ($Ƕ$) is sufficiently small so that $\mathrm{Pr}_{ƂЧ} \le 1$ for all $Ƃ, Ч$. As we will demonstrate a little later, the degree of this strategy switching behavior can be influenced by the presence of externality potential (x) promoting the emergence of various pure strategy following "clusters" of like-minded analysts who, in turn, contribute to the process of meme and theme formation. In other words, the degree of externality potential (x) affects the impact of the value of $Ƕ$ upon analyst switching behavior – but more on this later.

For the moment, let us identify that under such conditions the expected fraction of the analyst population using strategy $z_Б$ in period $t + dt$ is

$$Ex\left(z_Б^{t+dt}\right) = z_Б^t - Лbdtz_Б^t \sum_{ч=Б+1}^{n} z_ч^t Hb\left(\pi_ч^t - \pi_Б^t\right) + \sum_{ч=1}^{Б} Лb\,dtz_ч^t z_Б^t Hb\left(\pi_Б^t - \pi_ч^t\right)$$

$$= z_Б^t - Лbdtz_Б^t Hb\left(\pi_Б^t - \bar{\pi}^t\right), \tag{6.7}$$

where $\bar{\pi}^t = \pi_1^t z_1^t + \cdots + \pi_n^t z_n^t$ represents the average return for the analyst population.

Now it is just a simple matter of subtracting $z_Б^t$ from both sides, dividing by dt and taking the limit as $dt \to 0$ to yield the following dynamic equation governing the strategy switching pattern exhibited by the analyst community

$$\dot{z}_Б^t = Лbbz_Б^t\left(\pi_Б^t - \bar{\pi}^t\right) \quad \text{for } Б = 1, \ldots, n. \tag{6.8}$$

A cursory examination of Equation (6.8) in comparison with the types of dynamic equations we have illustrated via our case studies in Section 5.2 demonstrates a distinct similarity in structure (but now we have the added benefit of actually having formalized the microfoundations for the payoffs that "strategy switching" analysts will achieve).

Subsequently, we now know not just the generalized format of the various information bytes that ultimately come to constitute the market's information genome ($\sum \varphi_l$), but we also know the preconditions that motivate changes in analysts pure strategy following behavior. Such microfoundations enable us to drill down even more deeply into the forces constituting a change in the actual fabric of market information given that we can now explicitly identify the reasons why analysts switch from one pure strategy following "cluster" to another. In so doing, we now have a much better appreciation of the forces governing meme, theme and overall market sentiment formation.

Indeed, as we mentioned earlier, such switching behavior appears to be motivated primarily by the presence of externality potential (x) bearing information bytes influencing the perceived payoff for certain analysts to switch strategies. Consistent with our terminology presented in Chapter 5, this externality potential (x) can be either contemporaneous (Ω) or latent (Λ) in form. We will focus upon these aspects of meme, theme and overall market sentiment formation – along with their associated nonlinear asset price impacts – in considerably more detail a little later in this chapter. Suffice to say now, given the above, we now have in place a significant portion of our generalized model for the microfoundation determinants of the fabric of the market's information genome $(\sum \varphi_I)$, and through this, asset prices and distributional form.

The next step in our study of the underlying drivers of analyst decision-making is to introduce the potential for private information. Obviously, the analyst who produces a particular information byte will be privy to such information well in advance of others. This "lead information" can ultimately affect the analyst decision-making process within an extensive form game (this is the "ordering issue" that we drew attention to in the first subsection within this chapter). Subsequently, it is incumbent upon us to introduce some form of generalized formal structure for such a (truly Bayesian) premise within our deterministic Game Theoretic microfoundations of our EDF.

Introducing private information and the Bayesian form[12]

To generalize the games between analysts who possess private information we need to use the concept of "perfect Nash equilibrium in behavioral strategies" in place of standard "Nash equilibrium." What precisely does this new terminology mean? Well, for starters a "behavioral strategy" $st_Ƀ$ for an analyst following pure strategy Ƀ in our extensive form game G is a probability distribution Pr over the analyst's market-wide information set $\in Ï$. Likewise, we use the term "behavioral strategy profile" to represent the actual choice of behavioral strategy made by each analyst. How these concepts relate to the actual concept of Nash equilibrium will become obvious in the following paragraphs.

For example, using the earlier definitions, we can say the behavioral strategy set $St = (st_1, \ldots, st_n)$ is a Nash equilibrium in behavioral strategies if for each analyst, $st_Ƀ$ is the best response to $st_{-Ƀ}$. Unfortunately, while giving us the "solution" to a private information game, this statement provides little to no insight as to the game itself. To circumvent this issue, we must analyze the "out of equilibrium" behavior of the Game Theoretic system.

---

[12] Private information games are ubiquitous in the Game Theoretic literature. A good example of such a game in the applied sense is the often cited "Cobweb" phenomenon – see Chiang (1984) for details. Why the link between "Cobwebs" and private information? Private information games (by default) involve some sort of lag structure in the response function of the game's participants and the Cobweb dynamic explores the cycles that typically manifest as a consequence of this behavior. Likewise, the presence of private information (and the implied lagged response of a game's alternate players that goes hand-in-glove with such a premise) also provides a convenient backdrop for the formalization of such informational principles like "information cascades" and "herding" – both of which have been discussed in Chapters 4 and 5 (respectively). For a comprehensive overview of the herding/information cascade consequences of such games, see Chamely (2004). As a final point, it is interesting that in this discussion Chamely makes the definitional distinction between what precisely constitutes "herding" and what actually constitutes an "information cascade" – a definitional difference often overlooked in much of the present literature.

For example, given the Bayesian foundations of our model it is possible to identify what is known as the "local best response" during each iteration of the game. This, in true Bayesian form, tells us the best response for an analyst following pure strategy Ђ in making a decision at informational node Д ∈ of the game.

To elaborate further, at informational node Д ∈ the expected payoff from using st given st_- is

$$\pi_Ђ(st|Д, st_-) = \sum_{t \in T} \Pr[Д, t|st, st_-]\pi_Ђ(t),$$ (6.9)

where T is the set of terminal nodes, $\pi_Ђ(t)$ is the final payoff to an analyst following pure strategy Ђ at terminal node $t \in T$ and $\Pr[Д, t|st]$ is the probability of reaching a particular point in the game t from Д given (st). Subsequently, if $\Pr[Д|st_-,]$ is the probability of being at node Д, given that an analyst following pure strategy Ђ is attempting to choose the correct path when confronted with , and that (st) is being played, then such an analyst should choose st so as to maximize

$$\pi_Ђ(st|st_-,) = \sum_{Д \in} \Pr[Д|, st_-] \sum_{t \in T} \Pr[Д, t|st, st_-]\pi_Ђ(t)$$ (6.10)

and $\Pr[Д|]$ is given by the Bayesian updating formula[13]

$$\Pr[Д|, st_-] = \frac{\Pr[Д|st_-]}{\Pr[|st_-]}.$$ (6.11)

This gives a much better description of the game as it is sequentially played by our ecology of information producing analysts in true Bayesian form. Why? Well Equations (6.10) and (6.11) identify that if (St) is a Nash equilibrium in our extensive form game G and if is the information set reached at each stage of the game with positive probability Pr, then st is the local best response for each analyst following pure strategy Ђ to respond to st_-. We thus are able to map out each sequential step of our extensive form game under private information conditions.

So now that we have sufficiently generalized the Game Theoretic interaction between the various members of our ecology of information producing analysts (even under private information conditions) for each sequential step along our extensive form game that constitutes the formation of the market's information genome $(\sum \varphi_I)$, does this necessarily mean that our task of formalizing the deterministic microfoundations of our EDF is complete?

---

[13] Note that when $\Pr[|st_-] = 0$ then Equation (6.11) does not make any sense and $\Pr[Д|, st_-]$ would need to be defined arbitrarily.

Unfortunately, the answer is "No." While we do have in place a sufficiently detailed model of the microfoundations that determine the format of each information byte that analysts produce, we still have yet to come to grips with ascertaining some form of generalization as to the impact that this information will have (via asset pricing) upon distributional form.

To do this we could appeal to a number of alternate methodologies, however we have chosen what is known as a "Ewens distribution" – which not surprisingly, given our evolutionary foundations, arises from the genetics literature – as a basis for our EDF.[14] Why this particular distribution? Well, quite simply because a Ewens distribution appears to be a natural extension of the type of analysis we have presented in the preceding subsections. This is because it describes the innate tendency of its underlying components to "cluster." In our case, such micro-component "clustering" refers to analysts crowding into various pure strategy groupings as they gravitate toward where they purview the payoffs to be higher. This, in turn, affects the actual format of the market's information genome ($\sum \varphi_I$) and gives rise to certain memes, themes and causes overall swings in market sentiment to emerge. As will be illustrated in the following subsection, this endemic herding behavior on behalf of analysts – and its concurrent "knock-on" effects regarding the actual fabric of market information – can readily be explained via a convenient marriage between our earlier discussed externality potential (x) and the

---

[14] For example, an alternate distributional form amenable to our purposes is that proposed by Vaga (1991). In particular, Vaga developed a time-variant distributional form based upon how individuals interpret new information as it arrives. In particular, Vaga leveraged the Theory of Social Imitation in discerning the impact of herding on return distributions. Vaga used the following density function

$$f(q) = c^{-1} Q(q) \exp\left[2 \int_{1/2}^{q} \left\{ \frac{K(y)}{Q(y)} dy \right\} \right]$$

where $f(q)$ = probability of annualized return q; $K(q) = \sin h(kq + h) - 2q \cos h(kq + h)$; $Q(q) = 1/n[\cos h(kq + h) - 2q \sin h(kq + h)]$; n = number of degrees of freedom; k = degree of crowd behavior; h = fundamental bias; and

$$c^{-1} = \int_{-1/2}^{1/2} Q^{-1}(q) \exp\left[2 \int_{-1/2}^{q} \left\{ \frac{K(y)}{Q(y)} dy \right\} \right] dq.$$

At $k = 1.8$ and $h = 0$ this distributional form approximates to the normal distribution. At $k > 0$ chaotic behavior eventuates and by altering the value of h, this "random looking" distributional form will be either skewed positively or negatively. See Peters (1996) for more details as to simulating the Vaga distribution. That said, another distributional form possibility that we have considered in our quest is the ubiquitous "Turning Distribution." Here the mathematician Alan Turning (pioneer of some of the fundamental concepts underlying modern day computers) identified that stripe and spot markings in the natural world emerge as a simple "activator/inhibitor" extension of the relatively simple uniform distribution. See Kawczynski and Legawiec (2001) for a recent appraisal of this concept.

clustering processes motivated under a Ewens distribution. So now, over to the basics of this particular distributional form itself.

Formalizing our (deterministic) Game Theoretics into an Evolutionary Distributional Form using Ewens distribution as a guide

As highlighted earlier, the Ewens distribution appears to be a natural extension of the generalized form microfoundations that have been presented in the preceding subsections. Why? Well, the Ewens distribution dates back to the pioneering work done by Ewens (1972, 1979) in describing clustering processes in the biological/genetics realm. Furthermore, recent extensions provided by Aoki (2000, 2002) have popularized the notion in the finance/ economics arena. The Game Theoretic analysis outlined in the preceding pages hinges largely upon the precept that analysts "cluster" into various pure strategy following groupings as they form an overall market "ecology." Dividing these clusters into various fundamentalist, technician, econometrician based sub-stratum allows us to effectively define the numerous factions of the "warring tribe" analogy of an artificial model of the marketplace that we initially motivated in Chapter 3. Furthermore, by dividing each of these strategy sub-stratum into distinct "bull" and "bear" classifications, we are able to model the payoff-driven swings in the analyst population from either a "bullish" or "bearish" pure strategy following mindset. Such shifts *en masse* by the analyst community toward either one of these binary classifications constitute swings in overall market sentiment – as outlined in Chapter 4. Indeed, such cluster based skewing in the collective mindset of the analyst population (which in effect, represents cluster based swings in the distributional form of analyst opinions) is argued to have a commensurate impact upon price activity through a "fair-game" type of informational linkage between market information and asset prices. So to summarize, the Ewens "cluster-orientated" distribution has some nice features which sit well with the Game Theoretic microfoundations outlined so far.

This raises the obvious question – how to generate the Ewens distribution from the analysis that has been presented to date? Appealing to first principles, let us start by assuming there is potentially a large number (St) of types of analysts that constitute our market ecology.[15] As stated earlier, these may be chartists, fundamentalists, economists – even various sub-stratum within these classifications – but importantly each separate sub-stratum is divided up into "bullish" and "bearish" camps. Next, following Aoki (2002), we present a stylized model for the entry and exit into these various analyst sub groupings where for simplicity, we will concentrate upon the transitions between a particular pure strategy following group of "bearish" analysts ($И$)

---

[15] Note that the number of types of analyst within our market ecology is commensurate to the net sum of strategy choices St – as consistent with our earlier discussion.

into and out of the remainder of the overall analyst market ecology (Ж).
Under such conditions

$$\text{Entry}_\text{Ж} = x_\text{Ж} + \text{Ю}_\text{Ж}, \tag{6.12}$$

$$\text{Exit}_\text{И} = \text{Л}_\text{И} z_\text{И}, \tag{6.13}$$

$$\text{Transition}_\text{И} = \text{Њ}_\text{ИЖ} z_\text{И}(x_\text{Ж} + \text{Ю}_\text{Ж}), \tag{6.14}$$

where

$$\text{Њ}_\text{ИЖ} = \text{Њ}_\text{ЖИ} \quad \text{for all } \text{И}, \text{Ж} \text{ pairs}, \quad \text{И}, \text{Ж} = 1, 2, \ldots, St,$$

x_{st} is the externality potential attributable to a given strategy (e.g., И); and z
is the number of analysts adhering to a particular pure strategy.

Equation (6.12) refers to the entry rate into our market ecology of analysts
pursuing strategies (Ж) which are other than our pre-specified bearish pure
strategy (И). Likewise, Equation (6.13) describes the departure rate from such
an ecology by analysts pursuing the pre-specified bearish pure strategy (И).
Importantly, Equation (6.14) refers to the transition by analysts changing
between these representative strategy groupings – which in part reflects an
externality potential (x) driven conversion factor $x_\text{Ж}$ and an additional factor
$\text{Ю}_\text{Ж}$ that reflects the natural proclivity for analysts to be non (И) upon entry.
It is here that one of the serendipitous beauties of the Game Theoretic analysis
we have presented in the preceding subsections and the Ewens distribution
manifests. Think of it in this way, if there has been a "theme forming" information byte published in the past that possesses extremely high externality
potential (x), then it could be exerting an overarching influence upon the
decision by a particular analyst to switch strategies well beyond the given
point in time in which the particular information byte is published. Such
demarcations are consistent with our earlier discussed contemporaneous (Ω)
and latent (Λ) aspects of externality potential as identified in Chapter 5.

In other words, the payoff for switching out of a particular strategy (И) is
contingent upon the net sum of the externality potential (x) "pull" exhibited
across the array of alternate strategies (Ж) at any given point in time, so that

$$z_\text{И} = f(\pi_\text{И}) = f(x_\text{Ж}) = f(\text{Њ}_\text{Ж}). \tag{6.15}$$

Indeed, it is through this equation that we achieve the innate potential of
our Game Theoretic microfoundations for analysts to cluster into various
strategies, for memes and then themes to emerge in a diffusion-like manner,
and for overall swings in bull/bear market sentiment to appear. Why? Well,
as stated earlier, externality potential (x) exerts a "gravitational pull" alluring

analysts to switch strategies – of which our chosen "pure strategy bearish analyst grouping (И) switch into the potential array (Ж)" represents but one generalized form.[16]

That said, what is also readily apparent from the above is that the net balance of externality potential (x) applicable to a particular pure strategy (for instance, И) relative to other pure strategies (for instance, Ж) will constantly be in a state of flux. Why is this the case? Given that there is a continuous stream of information bytes being produced – a fact we established in our general equilibrium framework developed in Chapter 5 – and the fact that some of these information bytes will be "externality potential (x) bearing" implies that the payoff function for adhering to a particular pure strategy will be constantly changing as well. This is evidenced by the presence of a strategy specific transition coefficient Њ in Equation (6.15). This transition coefficient ultimately relates back to our earlier discussed transition payoff specification for analyst switching behavior represented in Equation (6.6). What all of this implies is that we have in place the necessary microfoundations for a model of endemic analyst herding – the relative degrees of which will be determined by the power of externality potential (x) attributable to a particular strategy relative to all other strategies within our overall market ecology – as defined by the strategy set St.

Given these conditions, it is now possible to model the evolution of both (И) and (Ж) switching behavior under our "externality potential (x) driven analyst ecology clustering model" as a Markov process whose steady-state distribution is described as

$$\Pi(z) = \prod_{И=1}^{St} \Pi_И(z_И),$$ (6.16)

where

$$\Pi_И(z_И) = (1 - c_И)^{-Ю_И} \begin{pmatrix} -Ю_И \\ z_И \end{pmatrix} (-c_И)^{z_И},$$

and

$$c_И = \frac{x_И}{Њ_И}.$$

---

[16] In this respect, Equation (6.15) replicates our extensive form game strategy switching condition – as expressed in Equation (6.6). The difference now however is that we place such switching behavior within a distributional form context. We will elaborate upon this in more detail a little later in this chapter.

Assuming for simplicity that our arbitrary constant $c_\text{И} = c$ for all И, then the joint probability distribution is defined as

$$\Pi(z) = \left(\frac{-\sum_z \text{Ю}_\text{ж}}{z} \right)^{-1} \prod_{\text{И}=1}^{St} \left(\frac{-\text{Ю}_\text{И}}{z_\text{И}} \right). \tag{6.17}$$

Introducing the strategy partition vector $St = (st_1, \ldots, st_n)$ where st_z represents the number of pure strategy groupings within our ecology of analysts containing exactly z participants implies

$$\sum_\text{Ђ} st_\text{Ђ} = st_z \leq Z. \tag{6.18}$$

We can thus use this strategy partitioning vector (and adding the simplification assumption that $\text{Ю}_\text{И} = \text{Ю}$ for all И) to enable us to express Equation (6.17) as

$$\Pi(z) = \left(\frac{-St\text{Ю}}{z} \right)^{-1} \prod_{\text{И}=1}^{St} \left(\frac{-\text{Ю}_\text{И}}{\text{И}} \right)^{st_\text{И}}. \tag{6.19}$$

To turn the above into a formal distributional form, let us assume St can become very large and in so doing potentially incorporate a near-infinite range of analyst types within our market ecology of pure strategy following analyst groupings. Next, we keep the innate tendency to cluster into smaller groupings (Ю) very small so that the product of $St\text{Ю}$ approaches a finite positive constant Œ. Now suppose there are $St!/st_1! st_2! \ldots st_z!(St-z)!$ different ways of expressing our strategy partition vector, then it is possible to describe Equation (6.19) as

$$\Pi_z(st) = \left(\frac{-\text{Œ}}{z} \right)(-1^z)\frac{St!}{st_1! st_2! \ldots st_z!(St-z)!} \prod_\text{И} \left(\frac{\text{Ю}}{\text{И}} \right)^{st_\text{И}}. \tag{6.20}$$

Since $[St!/(St-z)!] \times \text{Ю}^z$ approaches Œ^z in the limit of Z becoming infinite and Ю approaching zero, it is possible to express Equation (6.20) in the standard Ewens distribution form

$$\Pi_z(st) = \frac{z!}{\text{Œ}^{[z]}} \prod_{\text{И}=1}^{z} \left(\frac{\text{Œ}}{\text{И}} \right)^{st_\text{И}} \frac{1}{st_\text{И}!}. \tag{6.21}$$

The beauty of the above is that it is described by the single parameter Œ that in turn, is motivated by our externality potential (x). Smaller values of

Œ tend to produce a few large clusters of analyst pure strategy groupings (and the market displays distinct kurtosis type behavior) while larger values of Œ tend to produce a large number of smaller clusters (and the market more resembles a Strong EMH type of outcome). Indeed, as Œ $\to \infty$ and analysts (by definition) become near independent – as each analyst can be regarded as a separate subcategory pure strategy follower – then the stream of information bytes comes more to resemble the IID format "white noise" traditionalist assumption.[17] To contrast, when Œ $\to 0$ the pattern of information byte production is more consistent with analysts displaying a high degree of interdependency.

Importantly, the impact of Œ itself is the converse of our externality potential (x). This is because, as mentioned earlier, externality potential (x) has the effect of binding the distribution of analyst pure strategy groupings together. Higher levels of the variable Ю however – which is the natural proclivity for analysts to want to join a particular pure strategy grouping irrespective of externality potential (x) – has the effect of tearing such larger groupings apart. So it is this relative balance between externality potential (x) and Ю that ultimately determines the order of magnitude of the analyst clustering coefficient Œ.

Subsequently, the fact that the Ewens distribution represented in Equation (6.21) hinges upon a singular parameter Œ – and the fact that the impact of this variable itself hinges upon the relative proportions of our externality potential (x) to Ю – makes such a distribution highly suited for our purposes. Indeed, we feel that an externality potential (x) motivated Ewens distribution represents the best type of distributional form outcome to describe our (deterministic based) Game Theoretic microfoundations for the production of information bytes represented in the preceding subsections. Why? The "endemic herding" exhibited by the analyst community under such conditions is we feel a "natural blueprint" for the foundations of our much desired EDF – that has, in turn, been motivated by our Game Theoretic discussion to date. Indeed, it is this marrying of our information byte motivated externality potential (x) with the generalized clustering type framework described by the Ewens distribution that underlies the entire principle of our own EDF. The advantage that our own EDF possesses is that it is amenable to a whole range of distinct strategic behaviors on behalf of the analyst community. To illustrate, in the following section we will outline a number of different calibrations of our Ewens distribution motivated EDF which are consistent with a range of deterministic Game Theoretic strategic actions exhibited by the analyst community.

---

[17] Ultimately such an outcome is consistent with a very low presence of externality potential (x) – as would be the case if most information was singular (Ξ) rather than memetic (M) in form.

6.2   Analyst/investor strategies and the ecology of the market

The calibration of our Ewens distribution motivated EDF for a range of strategic behaviors exhibited by the analyst community is relatively straightforward. For starters, let us assume our "market ecology" of analysts is populated by just 50 individuals ($Z = 50$) choosing from an identical number of pure strategies ($St = 50$). Given that the entire Ewens distribution is motivated by a single parameter (Œ) – and that this in turn, for EDF purposes, is driven by the relative proportions of externality potential (x) versus the innate desire of analysts to join a particular pure strategy group irrespective of (x) which we have denoted as (Ю) – we can thus use the parameter Œ to describe a plethora of potentially interesting strategic behaviors on behalf of the analyst community. That said, space limitations obviously preclude us from exploring each and every possible permutation of potentially interesting strategic behavior here. Hopefully however, the three examples we have selected will provide the reader with at least some insight as to the capabilities of our time-variant EDF for describing a diverse array of strategic interaction within the analyst community.

A dominant player information producing strategy[18]

In accordance with the underlying principles of our "theme generation" discussion outlined in Chapter 5, let us assume the market has "anointed" a particular seer analyst with a "dominant player" mandate. Such machinations would occur had the analyst produced an information byte (or, in some cases, a series of information bytes) with exceptionally high externality potential (x). Will there be strategic ramifications from such actions? Certainly yes. As will be illustrated next, it is not just a case of "to the victor goes the spoils" in terms of the stream of monopoly profits that a particular theme generating analyst will reap from their ability to "brand" emergent themes within the marketplace – as was specified in Equation (5.21) – but there are also implications from such actions for the distributional form of assets prices.

To illustrate, (see Figure 6.2), detailing the various degrees of pure strategy clustering exhibited by the analyst community for alternate values of Œ. As can be readily seen (and as consistent with our earlier statements), a relatively small value of Œ generates a predominance within the analyst community of a particular pure strategy following cluster, with higher values of Œ implying a much more diverse ecology of pure strategy following groupings.

Undoubtedly under a "dominant player" environ, the degree of herding displayed by the analyst community would be more commensurate with

---

[18] The discussion in this sub-section is consistent with the premise of "reputation based herding." For a detailed discussion of this premise (along with a general overview of the "herding" literature itself) see Bikhchandani and Sharma (2001).

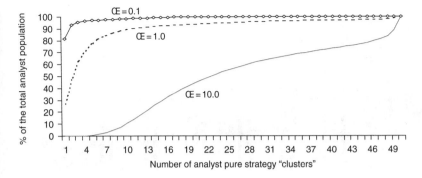

Figure 6.2 The cumulative distribution of pure strategy following clusters for varying levels of Œ – exhibiting alternate degrees of analyst ecology within the marketplace

low values of Œ. Why is this the case? Quite simply if the market has bestowed a "dominant player" title upon a particular analyst, this implies that a particularly powerful emergent theme is already in force.[19] As this theme grows in power – in effect, as the information diffuses within the wider analyst community – there will be (by definition) more converts to the particular pure strategy grouping. Hence, Œ → 0 as the dominant player emerges. Ergo, assuming for simplicity that the nonlinear price impact of the net memetic externality weight across the range of preexisting themes is constant for the duration of our simulation and that there is a one-to-one link between the format of financial information (e.g., each Buy "B," Neutral "N" and Sell "S" recommendation produced by the analyst community) and asset pricing for singular information (Ξ), we can thereby use the various analyst clustering results exhibited in Figure 6.2 to generate a range of time-dependent EDFs. Intrinsically, such time-variance in the EDF must, in turn, depend upon the degree of externality potential (x) that is motivating analysts to cluster and for our "dominant players" to emerge.

To demonstrate this effect, using Œ = 0.1 as a guide, we are able to illustrate the extreme skewness in the distributional form of asset prices under one of the more aggressive dominant player strategic assumptions for a given random seeding of potential "buy/sell" signals at time t across our ecology of alternate strategy groupings (St) (see Figure 6.3).[20] Such skewing could in

[19] That is, unless the dominant player mandate is the result of "reputation" effects from the format of a particular theme producing analyst's previous information bytes. Still, to the extent that the market will be mimicking this analyst's movements, there will still be a predominance of analysts in a singular pure strategy grouping.

[20] These results were generated by assuming 100 distinct randomly seeded "Buy/Sell" signals (constrained within a range $-0.5 \leq P \leq +0.5$) emulating from our various defined pure strategy groupings.

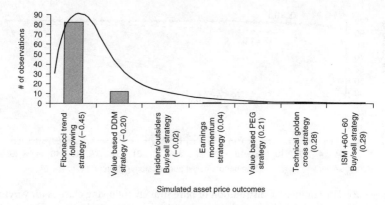

Figure 6.3   The extreme skew in the distributional form of asset prices stemming from the emergence of a dominant player within the marketplace[21]

principle manifest at either the individual asset level (*a*) or, under extenuating circumstances across the overall market index level (Ì) – but more on this later.

Following Equation (6.15) and assuming a direct link between the presence of externality potential (*x*), the strategy transition factor (Hb) and subsequently Œ, we can thus identify direct causality between the actual production of the seminal information byte that bears extremely high externality potential (and, in turn, anoints our "dominant player" within the market), and the concurrent herding by the analyst community into a particular pure strategy following grouping. Such herding, in turn, promotes a marked skewing in the distributional form of asset prices. Consistent with our earlier discussion, these forces are truly time dependent given the incessant nature of information byte – and therefore externality potential (*x*) – production. Ultimately, it is the relative degrees of contemporaneous (Ω) to latent (Λ) externality potential (*x*) effects that will dictate the time dependent evolution of (*x*) relative to IO and through this, the pace of analyst herding, theme diffusion and the skewing of asset prices. Still, the net result remains the same – a distinct shift in the distributional form of asset prices over time which, in turn, enforces a significant departure from the baseline Normal distributional form assumption that underpins the traditionalist literature.

A cooperative clusters information producing strategy

Under a cooperative information producing strategy regimen we witness alternating values of Œ as analysts cooperatively cluster into various pure strategies with the aim of collectively harvesting the powerful monopoly

---

[21] Modeled price impact is given in brackets for each pure strategy specified on the horizontal axis.

rights attributable to the "seer analyst group."[22] It is this seer analyst group that eventually produces the seminal information byte bearing exceptionally high externality potential (x). In accordance with Equation (5.15) the more labor devoted to new theme research, the higher the eventual seminal information byte output, but ultimately such agglomeration efforts are time bounded in terms of their cooperative cluster size simply because of the profit maximizing condition for labor input identified by Equation (5.19). The fact that such cooperative clusters can happily emerge within the presence of singular analyst (preexisting thematic) information producers, and that the success rate of these "new theme seeking" analyst cooperatives may not always be uniform in pattern, implies a waxing and waning of the clustering coefficient Œ as the market gyrates from a near quasi-independent information state to an environ resembling distinct analyst (and therefore information) interdependence.[23] Consistent with our arguments in the previous subsection, obviously the degree of such interdependence skyrockets when a particular cooperative cluster "strikes it lucky" and a new theme is born.

What does all this imply for the distributional form of asset prices? To illustrate, we adopt a time evolution path for Œ in our calibrations later. Here the market is at best characterized as being in a schizophrenic state – alternating between a plethora of various pure strategy following "warring tribe" analyst factions to dramatic "bandwagon jumping" effects. Such "bandwagon jumping" occurs when a particular cooperative cluster develops a new theme generating information byte that possess a high degree of externality potential (x) and singular analysts decide *en masse* to follow the traditional path of information diffusion and switch to the evolving new theme pure strategy.

The "Big Call" information producing strategy

Finally we consider what we label as the "Big Call" information producing strategy. As was briefly discussed in Chapter 5, at times analysts can attempt to brand themselves via a "Big Call" in the sense that the actual format of the new theme generating information byte they produce will be so violently opposed to the existing theme structure that it is likely to provoke a

---

[22] An alternative explanation for such behavior could be motivated by using the theory of social conformity – albeit a shifting one – as developed by Bernheim (1994). Or a model of "stochastic recruitment" as described by Kirman (1993).

[23] The fact that we are implying a certain "lumpiness" in the success rate of a given cooperative cluster tells us that the rate of new theme generation specified in Equation (5.15) should in principle change from a deterministic format to a probabilistic format – if only on a per unit of time basis and only for the exposition of this particular sub-section. That said, such an adjustment should not cause too much disruption to the general equilibrium framework espoused in Chapter 5 as Grossman and Helpman have demonstrated precisely such an extension for this form of model – see Grossman and Helpman (1991) Chapter 4.

significant market response – should such an information byte actually take hold. Ultimately, the "take-up" of such a diametrically opposed opinion to the prevailing market consensus would require considerable additional memetic reinforcing information from extraneous sources – such are the conditions for making a "Big Call" in the first place. That said, the payoff is immense for the enterprising analyst who is willing enough to make such an opinion known to the wider marketplace. This is because such "Big Calls" (by definition) have the potential to invoke swings not just in the theme structure applicable to a particular asset (a) but, given their very nature, also potentially provoke shifts in the overall market index ($\ddot{I}$). Such swings in market sentiment imply even more monopolistic profit for "Big Call" analysts to harvest as their opinions begin to transcend traditional boundaries.

An additional factor to consider in making a "Big Call" is that given their often violent contrast to the ruling market ethos, such calls – on the proviso they are right – can result in a rapid take-up by the residual analyst community once the reinforcing memetic information arrives. Subsequently, in accordance with Equation (5.21) the present value discount factor associated with making such a call is not so much of a drag on the net return to the analyst at time t – so in short, this implies more money up-front to the "Big Call" analyst.

All this adds up to a significant incentive for analysts to embark upon such a strategy but – as always – there is the potential for downside. How does such a downside occur? Quite simply in the sense that if one cries "wolf" too often and gets it wrong then they could suffer some form of reputational loss. Given that our extensive form game microfoundations governing analyst behavior are truly evolutionary in foundation and that there are (by definition) no t to $t+1$ reputations being built, such a "reputational drag" can only manifest within the context of the *format* of each analyst's information byte provision. So in short, if an analyst's strike rate in making a "Big Call" accurately in the past is quite high, they will typically require less ancillary supportive information before other analysts (and thereby investors) convert to the new theme and the "Big Call" analyst starts reaping monopolistic reward. Obviously, if such an analyst's strike rate is quite low, then it may take a considerable amount of supportive memetic evidence before the wider market begins to believe them.

What does this behavior imply for the distributional form of asset prices? Basically, both the pattern of analyst clustering formations and the potential for diametrically opposed skews in asset prices will be even more violent than those expressed in Figure 6.4. To demonstrate, we have calibrated our Ewens distribution motivated EDF for such "Big Call" swings and have demonstrated the results in Figure 6.5.

So how do we move from these types of calibrations of our Ewens distribution motivated EDF toward a better array of financial engineering techniques? Again, space limitations preclude us from exploring each and every possible

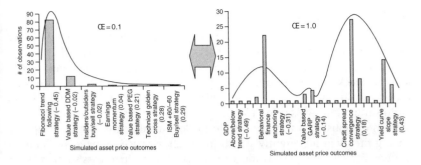

Figure 6.4 Swings in the distributional form of asset prices as cooperative analyst clusters form, become successful and then their influence wanes

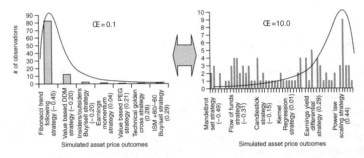

Figure 6.5 Violent swings in the distributional form of asset prices as enterprising analysts embark upon a "Big Call" information producing strategy

permutation. That said, in the following section we will outline at least some of the more significant findings illustrating how our EDF can significantly alter one's precepts of standard financial engineering "best practice" as it is known and applied today.

6.3   Some implications of our results

In the following subsections we detail three of what we believe to be the more significant areas where our evolutionary approach toward information in finance – and through this, the genesis of our EDF – alter standard financial engineering "best practice" precepts in a significant way. In particular, we tackle the issues of option pricing, risk measurement and portfolio construction. Obviously our EDF extends into other areas of financial engineering too – such as fixed-income factor modeling and time-series analysis techniques – but space limitations preclude us from exploring these avenues in greater detail here. That said, we cite these areas as potential fields of fruitful future research.

Implications for option pricing

The implications of our EDF for option pricing are profound. To illustrate why, we must at first provide some background on option price theory as it presently stands. Here we investigate the theoretical underpinnings of the seminal Black–Scholes option pricing equation and numerical techniques such as finite difference methods, binominal/trinomial trees and Monte Carlo simulation. As is often the case in a wide-ranging thesis such as ours, there is obviously no way we can do justice to the entire gamut of option pricing theory as it presently stands in the short space we have available to us here. For such insights, readers are referred to seminal texts such as Haug (1997), Hull (1997) and Wilmott (1998). Still, we hope the box below will

---

Options Pricing 101[24]

The Black–Scholes option pricing method[25]

The Black–Scholes option pricing formula is the benchmark for standard traditionalist-based financial engineering "best practice" as it applies to the difficult area of option pricing. The most simple method to derive the Black–Scholes option pricing equation is through the conduit of delta hedging. To begin with, let us use Π to denote the value of a portfolio composed of one long (European) option position and one short position of the underlying asset. We use Δ (delta) to denote the quantity we are short the underlying asset so that

$$\Pi = V(S, t) - \Delta S; \qquad (6.22)$$

where $V(S, t)$ is the option; and S is the underlying asset.

Now, as consistent with traditionalist theory, we assume asset prices follow the ubiquitous random walk principle. That is

$$dS = \mu S dt + \sigma S dX. \qquad (6.23)$$

This explicitly allows us to introduce the element of time, so our portfolio will now (at least in principle) evolve according to

$$d\Pi = dV - \Delta dS. \qquad (6.24)$$

---

[24] Note that the variable labeling in this shaded section follows standard option pricing terminology and as a consequence the nomenclature may differ a little from the remainder of the book. The reason for this is that a standardized set of variable labels has evolved for this particular branch of financial engineering – the "Greeks" for instance ($\Delta, \Gamma, \Theta, Vega, \rho$) – which are used as an effective shorthand to describe various elasticity responses to an option's price.

[25] See Black and Scholes (1973) – although the derivation of this formula can be found in many derivatives texts.

Appealing to Itô's lemma (which in essence, represents a Taylor's series expansion for stochastic rather than deterministic problems) we know that dV will evolve in a pattern according to

$$dV = \frac{\partial V}{\partial t} dt + \frac{\partial V}{\partial S} dSt + \frac{1}{2} \sigma^2 S^2 \frac{\partial^2 V}{\partial S^2} dt. \tag{6.25}$$

Substituting Equation (6.25) into Equation (6.24) gives

$$d\Pi = \frac{\partial V}{\partial t} dt + \frac{\partial V}{\partial S} dSt + \frac{1}{2} \sigma^2 S^2 \frac{\partial^2 V}{\partial S^2} dt - \Delta dS. \tag{6.26}$$

Now, we explicitly delta hedge by choosing

$$\Delta = \frac{\partial V}{\partial S} \tag{6.27}$$

this important step effectively eliminates the "random" component of the random walk. Therefore Equation (6.26) can now be expressed as a portfolio that is wholly deterministic in form with

$$d\Pi = \left(\frac{\partial V}{\partial t} + \frac{1}{2} \sigma^2 S^2 \frac{\partial^2 V}{\partial S^2} \right) dt. \tag{6.28}$$

Next, we introduce the following no-arbitrage principle to "zero in" on the correct price for the option

$$d\Pi = r\Pi dt. \tag{6.29}$$

Finally, by substituting Equations (6.22), (6.27) and (6.28) into our no-arbitrage principle – Equation (6.29) – and dividing by dt, we are able to ascertain the Black–Scholes option pricing equation

$$\frac{\partial V}{\partial t} + \frac{1}{2} \sigma^2 S^2 \frac{\partial^2 V}{\partial S^2} + rS \frac{\partial V}{\partial S} - rV = 0. \tag{6.30}$$

Obviously in the above, the crucial element is the initial assumption of the "random walk" that underlies the entire principle of the evolution of the underlying asset's price. This is because the "randomness" in dS allows us to delta hedge and convert a stochastic option price formula into a deterministic one. As we have stressed on in a number of occasions in this book, the Game Theoretic microfoundation principles generating our EDF are primarily deterministic in nature – not stochastic – and especially not Gaussian in form. As a consequence, we are precluded from attempting any form of theoretical type delta hedging solutions for modeling the evolution of option prices as "risk" effectively fails to cancel in any manipulations of our results. In short, there is effectively no "closed form" solution for an EDF inspired option price. Thankfully, there is an alternate route – the principles of which we will elaborate upon next.

Numerical methods

There is another route toward pricing options other than purely closed form theoretical expositions. Here computer intensive numerical methods reign supreme. These techniques generally aim to time-step the potential array of (probability driven) asset price permutations via a highly iterative technique with the objective of approximating the various required coefficients for solving the delta hedging/no-arbitrage conditions of an option's price (which, along with the "random walk" assumption, formed the foundations for the solution of the Black–Scholes model cited earlier).[26]

The most frequently used numerical methods are finite difference methods, binomial/trinomial trees and Monte Carlo simulation. That said, we will not dwell too long on these particular solution techniques here as we will actually use a numerical method to solve the necessary "optimal option price" conditions under our Game Theoretic motivated EDF conditions a little later in this subsection.

That said, what we should point out at this stage is that we are not the only authors to adopt numerical methods to describe the derivation of option prices that have been generated from a Game Theoretic microfoundation. Shafer and Vovk (2001) for example, have illustrated the use of numerical techniques to solve various option pricing problems that have *independent information* Game Theoretic foundations as their primary mechanism for generating the "observed market price." To illustrate the applicability of their approach, we have simulated an independent information iterative game between two information producing analysts following the framework generalized in Figure 6.1 and where the format of each information byte is assumed to have a one-to-one link with (a normalized) representative asset (i). As can be seen in Figure 6.6, the price output of such a game does produce highly acceptable index values whose movements do appear to rapidly converge to the targeted Gaussian form.

Given the obvious Gaussian outcome of such a game – an outcome alluded to in our discussion of Game Theory in Chapter 3 – there is little surprise at the relative ease with which these purpose built Game Theoretic foundations can be used to support the traditionalist cause. Importantly however, it is essential to remember that such a game was founded on the principles of independent information – not, as in our case, interdependent information.

---

[26] Finite difference methods are perhaps the closest in structure to the original Black–Scholes option pricing technique with Binomial/Trinomial techniques relying only indirectly on Black–Scholes analysis. Indeed, for Monte Carlo simulations the underlying process can be described in even simpler terms of "attempting to ascertain the expected present value of the payoff of an option." Finally one should note that some of these methods are particularly useful for modeling American options where early expiry is an issue and traditional closed form theoretical modeling has proven ineffectual.

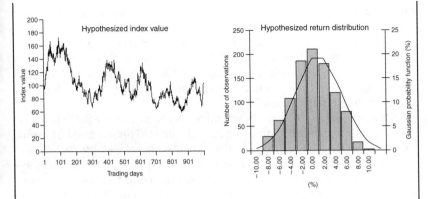

Figure 6.6 The index values and distributional form outcome of an independent information game between two competing analysts

Recent advancements in option pricing theory

Still, the proverbial "thorn in the side" for the traditionalists has been the steadfast refusal of the various applied outcomes of the traditionalist (Gaussian based) option pricing techniques to adhere to their theoretical pure pricing perfection. Nowhere is this better displayed than the ubiquitous presence of volatility smiles.[27] Another major problem from a practitioner's standpoint is that the Black–Scholes option pricing formula assumes delta hedging is a continuous affair whereas real world asset markets do not allow this to be the case.

The fact that such departures from confines of the traditionalist stringent assumption set are so commonplace that it has resulted in a recent explosion of literature attempting to derive alternate option pricing methods that take into account (amongs other things) such features as the ubiquitous kurtosis of asset markets. Some of these studies have already been outlined in Chapter 3 – especially those generated from within the Econophysics paradigm.[28] Still, there are a number of such studies that are not so radically opposed to traditionalist thinking and instead attempt to adapt traditionalist methods so as to make them more applicable to the inevitable realities of applied asset markets. Some references of note operating more within the standard confines of the traditionalist literature are for example Boyle and Emanuel (1980), Leland (1985), Hull and White (1987), Hoggard *et al.* (1994), Avellanda *et al.* (1995), Wilmott *et al.* (1997) and Rebonato (1999).

---

[27] Indeed, the pervasive presence of such "smiley" options markets was briefly discussed in our outlining of the various "attacks" upon traditionalist based assumptions of market behavior – see Chapter 4.

[28] See for instance Bouchaud and Sornette (1994), Bouchaud *et al.* (1996) and Voit (2000). For a perspective on excess kurtosis in financial markets – see Bouchaud and Cont (1998). As for the pricing of options in an "adaptive system" excess kurtosis financial market – see Potters *et al.* (1998).

provide our readers with the necessary criteria for understanding at least some of the basics of this highly complex (and rapidly developing) area of financial engineering. In adopting such a didactic path, our primary aim is simple – that is, to lay the foundations for illustrating why our EDF fundamentally challenges the basic principles underlying standard traditionalist-based financial engineering "best practice" with regard to option pricing (at least as it is implemented in its "purest" form).

So how do we move from these principles to our own evolutionary purview on option pricing? To motivate our EDF directed option pricing formula we appeal to the implied trinomial tree method developed by Yamada and Primbs (2002). Implied trinomial trees have been around for some time – the introduction of the method was first developed in a series of papers by Dupire (1994), Derman and Kani (1994) and Rubinstein (1994). Such methods have the distinct advantage of being able to incorporate real (rather than hypothetical) market data.[29] The beauty of the Yamada and Primbs (2002) extension of the implied trinomial tree method is that it allows for a generalized array of distributional forms – not just the Gaussian distributional form – to be adopted in the algorithmic interpolation of implied probabilities when modeling the price movements of the underlying asset. Theoretically therefore, we are able to "plug in" our own EDF into the Yamada–Primbs algorithmic sequence and "back-out" an optimal option price. We feel this to be a nice result and hence have adopted the Yamada–Primbs procedure.

The general principle of the Yamada–Primbs approach – like all implied trinomial trees – is to construct a series of fitted price steps for the underlying asset and, in accordance with the principle of delta hedging and the no-arbitrage condition, to algorithmically ascertain the present value of the payoff of an option once the potential array of price movements for the underlying asset has been displayed. As the name "trinomial" would imply, such price step movements for the underlying asset are assumed to take on one of three potential directions at any node t to $t + 1$. That is, the price of the underlying asset has the potential to move up, down or sideways.[30] Such an assumption sits perfectly with our hypothesized interdependent information extensive form game between analysts – as represented in Figure 6.1. This is because a "Buy" recommendation is assumed to result in an upward price movement, a "Neutral" recommendation results in a lateral movement and a "Sell" recommendation results in a downward movement – thus providing a one-to-one fit with the trinomial method in general.

---

[29] Indeed, it is this "real versus hypothesized" distinction that differentiates implied trinomial trees from standard trinomial trees as developed by Boyle (1977).

[30] This contrasts the Binomial method – as developed by Cox *et al.* (1979) and Rendleman and Bartter (1979) – where only an "up/down" movement is allowed.

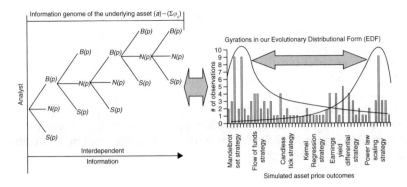

Figure 6.7 A trinomial tree extension to our interdependent extensive form game and the linkage that this has with our Evolutionary Distributional Form

As a consequence of this apparent "natural extension" of the Game Theoretic principles that have been presented in this chapter thus far, we are able to represent the generalized principle of the information driven trinomial tree as shown in Figure 6.7.

So how does one move from this lattice based assessment of an incomplete market model (it is "incomplete" as it hypothetically allows for non-Gaussian distributional forms to which, as discussed earlier, there is no closed form theoretical solution) toward actually solving the optimal hedging problem for a chosen option? The answer is represented in the following relatively simple objective function

$$\text{Minimize} \quad \ddot{E}(\text{Call}_T - \Psi_T | S_0),$$
$$\text{subject to} \quad st_{tr} \in St_{tr}, \tag{6.31}$$

where
Call_T is the value of the call option at terminal time T;
Ψ_T the value of the portfolio of the underlying asset at terminal time T;
S_0 the strike price for the call option;
st_{tr} the optimal trading strategy path given the array of price distributional probabilities so that the final payoff of the call option replicates the value of the underlying portfolio; and
St_{tr} is the entire trading strategy set.

The above represents the objective function for modeling the delta hedged/no-arbitrage condition of a European Call option. Simply by choosing the optimal trading path – given the array of (EDF determined) underlying price probabilities represented on the trinomial tree – we can obtain an outcome at terminal node T where there is little to no difference between the payoff of the call option and a prespecified portfolio (Ψ) of the underlying.

This effectively solves our "optimal" option pricing problem.[31] The fact that we are using a trading strategy (st_{tr}) to derive such a solution sits nicely with the information producing strategies outlined in the previous section – especially given that such an optimal trading path has a classic "Buy," "Neutral" and "Sell" trinomial foundation. Hence, the natural fit between the Yamada and Primbs approach and our Game Theoretic microfoundations that have been methodically built up over the preceding pages.

So how to solve for this optimal trading path given the lattice probabilities determined by our EDF? Quite simply the result is obtained by an iterative algorithmic technique that:

1 assigns the probabilities to the various price steps of the underlying asset (a) in accordance with our earlier identified EDF;
2 determines the probabilities on the edge of the terminal node T by solving the network flow optimization; and
3 computes the initial price of the call option (C_0) by dynamic programming across the trinomial lattice once the necessary probabilities have been obtained in Step 2.

It is in Step 1 that the greatest potential for our approach resides. Why? Well, in accordance with the very nature of our EDF, we are not only able to give "real" approximations of the lattice probabilities stemming from some static (historical) distribution of asset prices, but we are also able to offer the concept of an *evolving* distribution as well. That is, by ascertaining the underlying memetic structure that drives the emergence of certain "winning strategies" (which analysts and investors use to guide their behavior) from a preexisting strategy set, we are able to interpolate the signals from such strategies – along with using a proprietary epoch matching technique – to give a better approximation of *ex ante* asset price movements and through this likely directional swings in a given asset's distributional form.

For example, by identifying that a particular valuation based strategy is in the early stages of forming into a market "theme," we can interpolate the current signals from such a strategy to give a better approximation of the average forward price movements of both "growth" encoded and "value" encoded assets. Further, by matching various epochs in the past when this particular valuation based strategy was emerging in dominance

---

[31] Even though we have used a European Call option by way of example in Equation (6.31), the underlying methodology applies just as well to Put options, American options, barriers and other exotics. See Yamada and Primbs (2002) for more details.

and observing the various moment distributions of the selected underlying asset, we are able to build up a profile of not just the likely mean but also the various additional moments (variance, skew, kurtosis) of our tailored EDF. In other words, we are able to "bespoke" an *ex ante* EDF to suit any particular underlying asset for which the option in question is attempting to replicate.

Ergo, at each t to $t + 1$ node of our Game Theoretic trinomial lattice we can provide a much more accurate representation as to the likely probabilities of any given up, down or lateral price movement of the underlying asset (a). Obviously the further we step ahead in time, the less robust such forecasts become but ultimately our forward looking constantly evolving EDF approach which is tailored to the particular asset in question represents a distinctly better approach than a "one size fits all" generalized (real data) distributional form assumption that is traditionally rigidly applied at each informational node.

Precisely how we are able to estimate such probabilities – and by default, plot the time path evolution of our EDF as a particular emergent strategy begins to take hold – will be discussed in greater detail in Chapter 7. Suffice to say here, it is for the simple reason of providing a better assessment as to the true probabilities of underlying asset prices at each step of our implied trinomial lattice technique that we feel our EDF inspired approach toward option pricing has much to offer in the field of financial engineering "best practice" as it presently stands.

Implications for value-at-risk analysis

Can we utilize our innovative memetic based approach for predicting the time-dependent evolution of our EDF to give a better approximation of portfolio risk? The answer most certainly is "yes." Present industry standards call upon the Gaussian based Value-at-Risk method (or "VaR" as it is commonly known) to approximate the maximal possible drawdown for a given portfolio – subject to a specified timeframe and confidence criteria. As we will demonstrate in the box next, the principles underlying VaR are extremely straightforward. That said, simplicity does come at a cost as VaR at times has failed to provide adequate forewarning as to the potential for portfolio drawdown exposures that manifest when theoretically highly improbable market "events" have occurred – sometimes with surprising ubiquity as our discussion of "volatility clustering" in Chapter 3 intimates. Such is the nature of the kurtosis beast. As will be seen, this unfortunate fact of market life has lead to an array of adapted VaR measures to emerge within the literature – some of the basic principles of which we ourselves will draw upon in our own Evolutionary Value-at-Risk (EVaR) technique.

Value-at-Risk (VaR) for Beginners

The basic principle

As highlighted earlier, the premise underlying VaR is extremely simple. As Wilmott (1998) stated "Value at Risk is an estimate, with a given degree of confidence, of how much one can lose from one's portfolio over a given time horizon."[32] The "estimate" itself is derived from applying the ubiquitous Gaussian distributional form to the asset or portfolio in question – typically via utilizing a Monte Carlo or bootstrapping technique (conditioned of course for the user's objective timeframe) – so as to obtain a range of hypothetical percentage return movements for the underlying. Then it is simply a case of applying a specified confidence interval – typically set at 95, 97.5 or 99 percent – to ascertain the maximum potential loss (drawdown) given the "worst possible" situation. Unfortunately, as we have recounted upon at various times in this text, the Gaussian purview of "worst possible" does not necessarily always hold true.

Figure 6.8 shows just how simple VaR actually is, below we demonstrate (with the aid of a graphic) a hypothetical VaR calculation.

To ascertain VaR from the above it is thus a simple case of taking into consideration the quantity of an asset held, the duration over which the maximum loss condition is to be calculated (one's time horizon) and the present price of the underlying asset (a). Ergo, VaR for a zero drift asset in such circumstances is

$$\text{VaR} = -\sigma_a Q_a P(a)(\partial t)^{1/2} \text{Norm}_{\text{INV}}(1 - c); \qquad (6.32)$$

where

σ_a is the standard deviation of the returns for the underlying asset (a);
Q_a is the quantity held of the underlying asset (a);
$P(a)$ is the price of the underlying asset (a);

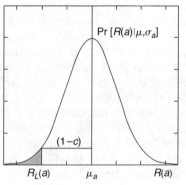

where

$c =$ specified confidence interval;
$R(a) = \%$ return for asset (a);
$\mu_a =$ mean return for asset (a); and
$R_L(a) =$ projected loss for asset (a).

Figure 6.8   VaR at its simplest

---

[32] See Wilmott (1998, p. 547). For further details on the VaR technique, refer to Jorion (1997).

$(\partial t)^{1/2}$ time horizon;

Norm_{INV} the inverse of the cumulative Normal distribution (in standardized form); and

c is the confidence interval.

However, taking into consideration drift (a positive or negative mean return for the underlying asset) Equation (6.32) becomes

$$\text{VaR} = Q_a P(a) \left[\mu_a \partial t - \sigma_a (\partial t)^{1/2} \, \text{Norm}_{\text{INV}} (1 - c) \right]; \tag{6.33}$$

where μ_a is the mean return of the underlying asset (a).

The calculation of VaR for a portfolio is a little more difficult as such calculations need to take into consideration the covariance between assets. Still, the formula is relatively straight forward with VaR being calculated as

$$\text{VaR} = -\text{Norm}_{\text{INV}} (1 - c)(\partial t)^{1/2} \sqrt{\sum_{\mathbf{v}=1}^{\mathbf{B}} \sum_{n=1}^{\mathbf{B}} Q_{\mathbf{v}} Q_n \sigma_{\mathbf{v}} \sigma_n \text{Corr}_{\mathbf{v}n};} \tag{6.34}$$

where

$Q_{\mathbf{v}}$ is the quantity held of the $\mathbf{v}$th asset;

Q_n is quantity held of the nth asset;

$\mathbf{B}$ the number of assets held in the portfolio;

$\sigma_{\mathbf{v}}$ the standard deviation of the $\mathbf{v}$th asset;

σ_n the standard deviation of the nth asset;

and $\text{Corr}_{\mathbf{v}n}$ is the respective correlation between each $\mathbf{v}, n$ asset.

For derivatives, the calculation of VaR is one step further along the road to difficulty as one must take into account the nonlinear nature between a change in the price of the derivative and a change in the underlying asset. Still, for very short time horizons the general approach is not too difficult with a simple delta approximation of Equation (6.34) generating meaningful results[33]

$$\text{VaR} = -\text{Norm}_{\text{INV}} (1 - c)(\partial t)^{1/2} \sqrt{\sum_{\mathbf{v}=1}^{\mathbf{B}} \sum_{n=1}^{\mathbf{B}} \Delta_{\mathbf{v}} \Delta_n \sigma_{\mathbf{v}} \sigma_n \text{Corr}_{\mathbf{v}n},} \tag{6.35}$$

where

$\Delta_{\mathbf{v}}$ is the delta of the $\mathbf{v}$th option; and

Δ_n is the delta of the nth option.

Pitfalls and adaptations to VaR

As we have been at pains to point out upon numerous occasions, Gaussian based methods do have their inherent shortcomings and VaR is no exception. These obvious pitfalls have lead to a number of adaptations to the theory. Some like Wilmott's (1998) Crashmetrics seek to extend the generalized principle with no assumption

---

[33] For longer time horizons, VaR calculations for a portfolio of derivatives becomes discernibly more difficult with a Delta – Gamma approximation providing at least one avenue for solution. See Wilmott (1998) for details.

at all as to distributional form. Others, such as Bouchaud and Potters (2003) aim to adopt an entirely new distributional approach. As will be seen, our EVaR method moves even further away from the static distribution fitting technique by adopting an entirely new memetic string based algorithmic search that is distributional *ex ante* rather than *ex post*.

So how does our method of EVaR actually work? Well, frankly the approach is not too dissimilar from the tact we adopted in our EDF option pricing technique. That is, we:

1 adopt an algorithmic memetic string search matching various preconceived strategy performances vis-à-vis the appearance of particular information bytes, memes and even themes;
2 we then use these results to match against the given informational environ at any given time t to ascertain which particular preconceived strategy is in the emergent stage and is thus (thanks to the strength of payoff function "pull") likely to see a number of analysts switching (herding) into this particular preconceived strategy;
3 we then interpolate the present signals from such a strategy (along with an epoch matching mechanism) to build up the various "moments" of our EDF for a given selected asset (or portfolio of assets); and finally
4 we then use these moments to provide a better approximation as to the potential for any given drawdown over a specified investment holding period.

There are a number of inherent beauties within our EVaR approach. For starters, it is nonparametric – thus circumventing the excess kurtosis problem that plagues the Gaussian based methods. Further, our approach is distributional predictive rather than backward-looking and thus gives the user a better approximation of the true risk of holding an asset over their specified (by definition) forward-looking investment period. Finally, once setup, our approach is relatively simple to use as it makes no distinction between singular assets or entire portfolio's of assets, and no distinction between asset type – with equity, bond and derivative being treated in similar fashion. In effect, all one needs is a time-series of the particular asset's (or portfolio's) returns, a set of preconceived strategy – valuation based, technical based, economic based – performances, and the market-wide information set that is to be used to garner a series of predictive memetic strings for such strategy performances. From this, using the algorithms we have developed in Chapter 7, it is relatively straightforward to build an *ex ante* projection for our EDF and from this compute the probabilities of certain pre-specified drawdowns – it's really that simple.

Implications for portfolio theory

Does our EVaR principle (and its subsequent dependence upon our EDF) have anything to offer in the field of portfolio theory? Again, the answer is absolutely "yes." Markowitz mean/variance portfolio optimization has been around since the 1950s – 1952 to be precise – and like VaR, is now so commonplace that it is generally taken for granted as being part of any financial engineer's "best practice" toolkit. But is there a better way to build an "optimal" portfolio? Part of the problem with principles such as VaR and Markowitz portfolio theory (also known as Modern Portfolio Theory – or "MPT") is that they are so fundamentally simple to use that many do not question the plethora of stringent underlying assumptions that constitute their foundations. As a consequence, quite often there is a serious disconnect between the theory and the applied – which is quite often to the chagrin of many practitioners who fail to realize the stringency of such assumptions until after some unexpected (generally unpleasant) market "event."

As we have done in each of our previous case studies illustrating where our pioneering approach toward EDF stands to upset the reigning status quo of financial engineering "best practice," below we outline the basic principles of the Gaussian based theory of Modern Portfolio Theory as it presently stands. As we have already detailed earlier, the basic principles of MPT are hardly what one would label to be "rocket science" – but within such inherent simplicity reside some very limiting assumptions which, at times, stand to trap unwary market players.

Portfolio Theory in a Nutshell[34]

Basically, the entire thrust of MPT can be summarized with the following objective function

$$\text{Max} \frac{R_\text{P}}{\sigma_\text{P}}, \tag{6.36}$$

where

R_P is the portfolio return over a specified timeframe on ʙ risky assets; and σ_P the standard deviation (risk) of the portfolio.

That is, the objective function of portfolio theory is simply to maximize the return on a group of assets per incremental unit of risk.[35] Underlyingthis premise, is

---

[34] For a comprehensive overview of MPT as it presently stands, see Markowitz (1991) and Sharpe (2000). For an illustration of the resampling technique, see Michaud (1998).
[35] Obviously this is done by varying a series of asset weightings that satisfy certain constraints. The constraints imposed vary as per the requirements (mandate) of the user but they generally include – summation to 100 percent, a holding of >0 percent for "long only" portfolios, and (at times) pre-specified tracking error minimization weighting bands for individual assets.

a utility function that assumes investors prefer less volatility (risk) in their portfolio returns with such a utility function being described as

$$U(w); \quad U'(w) \geq 0 \quad U''(w) < 0, \tag{6.37}$$

where U represents investor utility and w represents investor wealth. Since it is assumed rational investors prefer more wealth to less, their utility function is said to be strictly concave with the absolute risk aversion expressed as

$$A(w) = -\frac{U''(w)}{U'(w)}. \tag{6.38}$$

Note that return and risk observations can be expressed either in *ex ante* or *ex post* sense. When both risk and return are expressed in an *ex ante* sense and return is measured as excess to the risk-free rate, then the maximization condition expressed as Equation (6.36) constitutes the Sharpe (1963) ratio. That said, in many instances historical risk and return variables are taken as a guide in forming an opinion of the most likely future value of the risk/return relativity.

Covariance: The foundation for deriving efficient portfolios

It is really the degree of covariance among assets that drives the rationale behind portfolio theory.[36] Adding assets to a portfolio that have a less than perfect positive correlation with the existing assets within the portfolio (preferably that is, with a *negative* correlation) results in an overall lowering of σ_P and helps lift the overall Sharpe ratio for the portfolio as a whole. In general, covariance is expressed in standardized form – the correlation coefficient ($Corr_{\psi n}$). The $Corr_{\psi n}$ for any two singular assets ranges between $-1 \leq Corr_{\psi n} \leq +1$. Given that this is the case, the range of efficient frontiers generated by an optimized version of MPT for the given range of $Corr_{\psi n}$ values can be concisely represented by Figure 6.9. Obviously "optimized" is the operative word in the previous sentence. Now that we have spelt out the basic principles of MPT, the next question is to ask is – how is the efficient frontier actually formed? In other words, how is an optimization of the maximization condition represented by Equation (6.36) actually conducted so that a set of "efficient portfolios" is effectively obtained? To answer this, we must understand a little about the mechanics of optimizers themselves.

A final word on optimization techniques

Before starting, it is important to realize that in the theoretical world one has the luxury of placing certain limitations on a particular model's form so as to ensure only "well-behaved" functions are optimized. In our case, this implies continuous convexity to guarantee that a local minimum risk per unit of return equates

---

[36] Indeed, the estimation of the covariance matrix for portfolio construction has now become an entire subject in its own right – see Litterman (2003) Chapter 16 for details.

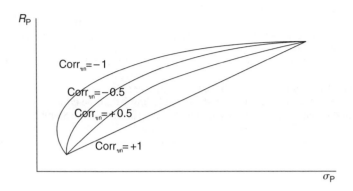

Figure 6.9 The hypothetical efficient frontier for various gradings of Corr$_{\psi n}$

to a global minimum provided $F'(x) = 0$ and $F''(x) > 0$. We can thus be sure that our efficient frontier represents the only possible maximum return (minimum risk) solution. However, the nonlinearities that evolve under constrained portfolio construction in the *applied* world tell us that well-behaved functional forms are generally not guaranteed. In most cases, rational investor behavior implies a well-behaved functional form but it needs to be pointed out that this is an *assumed* starting point, it is not a mathematical fact.

So what of portfolio optimization techniques *per se*? In most cases, portfolio optimization coding involves some form of quadratic programming technique (minimization of a quadratic function subject to linear constraints). Quite often the Simplex Method is used for quadratic programming. The Simplex Method is a recursive program that sequentially deviates from a "best guess" local minima position by advancing if the next variable is less (or if maximizing, more) than the existing or retreating if the opposite is true. As identified earlier, this approach cannot unequivocally guarantee a global maximum/minimum is obtained. To help overcome this deficiency a Simulated Annealing process is used (adding or subtracting a random number to the advance/retreat step function). This provides a useful means of testing the robustness of the local minimum solution. Still, it must be re-iterated that even with Simulated Annealing there are no absolute guarantees that a global minimum (or maximum) has been achieved.[37]

A relatively new approach toward optimization are Interior-Point methods. These emerged from the publication of a paper by Karmarkar in 1984 (Dantzig developed the Simplex Method in the 1940s). Of the class of Interior-Point methods, Primal–Dual algorithms have risen to be the most important. With regard to these algorithms, most Interior-Point software written since 1990 has been based on Mehrotra's predictor–corrector algorithm. What this method does is that rather

---

[37] One way of testing for global minimum is by testing if the Hessian matrix is positive definite. If this is the case, then a global solution is found. If the Hessian matrix is indefinite, a Kuhn–Tucker point (that is usually a local solution) is found.

than moving around the boundary of the solution set from vertex to vertex as Simplex Methods do, this class of Interior-Point method takes an entirely different approach by optimizing both the objective (primal) function and the order of magnitude of the local minimum deviation step (dual) equation simultaneously. One benefit of this approach is that this class of algorithm tends to converge much faster than Simplex algorithms. The approach initially uses the constraint vector as the initial interior point from where to start the local maximization/minimization search. The combination of the speed of this iterative process and the nonrandomness of the local minimum testing facility imply that the Primal–Dual Interior method is generally regarded as mathematically superior to the Simplex Method – even with Simulated Annealing.

So how does our evolutionary approach toward portfolio theory (EPT) actually work? Again, the premise is very simple. Using a fundamentally similar technique to our EVaR for portfolios, we basically conduct a memetic string search matching various preconceived strategy performances vis-à-vis a given memetic information-set. We then match this information-set relative to our present time t and to the predicted (biological-like) byte to meme, meme to theme information building network that is assumed to underlie the predictive emergence of a particular dominant winning strategy. It is through this that we are able to plot the future path of our EDF. Optimal weights are then garnered – subject to minimal drawdown conditions – via an iterative process for each asset within our selected asset universe by examining past epoch performances when such an emergent strategy was growing in dominance. Time is explicitly taken into account in our formulations simply because we plot the forward time path of the emergent strategy's growing dominance – as consistent with our concept of theme diffusion discussed in Chapter 4. Subsequently, through this we are able to generate a set of time evolving optimal portfolio weights for each asset within our selected asset universe.

Therefore, like our EVaR methodology, the general principle of our EPT is intrinsically appealing to the practitioner. For starters, again it represents a nonparametric method that is fundamentally *ex ante* rather than *ex post* in foundation. This is truly a differentiating feature when it comes to risk assessment as most MPT methods are not only Gaussian based but also take a backward looking approach when it comes to quantifying risk. This is a severely limiting assumption when constructing a portfolio as investment decisions (by definition) are forward-looking. Having the benefit of an *ex ante* perspective in the way that not only portfolio return but also portfolio risk is set to evolve going forward in response to the unrelenting flow of information – as typified by the market's information genome ($\sum \varphi_l$) – is an invaluable guide to the practitioner. Furthermore, since our approach is fundamentally algorithmic based as it conducts its iterative conditional search, a plethora of ancillary conditions (other than simple drawdown) can therefore be taken into account within our EPT directed portfolio construction. Finally, because our EPT portfolio weights are time evolving in response to the predicted emergent information flow, they are fundamentally superior to most static approximations of MPT – even with the addition of the periodic rebalancing that some practitioners insist upon. Still, we will dwell on these details more in Chapter 7.

So to conclude, all up we feel our memetic string based algorithmic evolutionary methods – as embodied in our critique of Evolutionary Option Pricing, Evolutionary Value-at-Risk and Evolutionary Portfolio Theory – represent a complete watershed in thinking as to financial engineering "best practice" as it is both known and practiced today. All of these methods have as their foundations the fundamental premise of our strategy directed EDF which uniquely takes a microfoundation informational perspective as to the likely evolution path of both risk and return. Still, all this has not yet demonstrated how our revolutionary evolutionary methods necessarily work in practice. To do this, we turn to Chapter 7 where we outline some of the outstanding results that we have achieved by utilizing such methods.

7
Evolutionary Finance – an Applied Perspective

This is where things get interesting. So far, this book has been laboriously outlining an entirely new theoretical thought paradigm for assessing how the microstructure of information interacts in a biological-like/evolutionary fashion to "build" asset prices. This extensive discussion was a necessary evil for what is about to follow. In this particular chapter we will outline our experience with putting the theoretical principles of Evolutionary Finance (as espoused in Chapters 1–6) into action using our proprietary software platform Natural Selection™. This advanced software package has been purposefully built to demonstrate how an optimal array of risk/return combinations can be constructed for a variety of absolute return mandates using bioinformatics based "structured information" principles. By doing so, like every great conclusion to a good story, we assiduously seek to weave the various strands we have presented in our theoretical foundations thus far into the algorithmic foundations of our Natural Selection™ model. For example, Natural Selection™ is founded on:

- how humans intrinsically categorize and encode interdependent information;
- how preconceived strategies are the most obvious embodiment of such behavior in the financial realm; and
- how it is possible to move "upstream" in this "winning strategy" theme diffusion process by forming an understanding of the early stage signaling of byte to meme, meme to theme, theme to overall market sentiment formation.

By adopting such a tact we will illustrate how our comprehensive investment process can in principle be labeled as being "adaptively predatory." Why is this the case? In a sense, by using upstream memetic informational indicators to provide early stage warning signals as to the take-up (our "longs") or failure (our "shorts") of a particular investment strategy by the wider financial community we are able to prey upon other investors fallibilities. Such incumbents typically are either slavishly adhering to a static investment philosophy (a fatal error in a truly dynamic marketplace) or alternately placing too much emphasis on a simplistic heuristic rule of thumb (strategy) for investing that is not "rich" enough in an informational context to process all the necessary key memetic information signals across the market's information genome ($\sum \varphi_i$). Generally, these latter investors are the late stage entrants to a mature "winning strategy" theme diffusion. By positioning early and selling into this "bid market" we are able to generate exceptional returns for our clients.

So where to start? Well before beginning let us at first, by way of an aside, draw to our readers attention a definitional issue whose demarcation was a necessary evil in the theoretical underpinnings to this book but whose treatment is nevertheless somewhat different here in the applied world. What we are referring to is the definition of an "analyst" and the concurrent definition of a "strategy." As we have highlighted upon numerous occasions so far, in the applied sense we treat analysts and investors as one and the same. That is, investors typically conduct some sort of analysis as part of their necessary due diligence prior to investing. Indeed, this is the case even if investors choose to "outsource" the analyst role as they must at first "analyze the analysts" from whom they seek advice.[1] Subsequently, trading strategies/investor strategies/analyst information producing strategies all fold into the one comprehensive strategy set under the auspices of our applied interpretation of Evolutionary Finance principles. Differentiating between the various trading/investor/analyst substratum was a necessary evil for developing a comprehensive model of marketplace ecology – as we have presented in Chapters 5 and 6. Now however, we are once more able to fold all such strategies into the same preconceived strategy universe so as to give a much more comprehensive range of strategy options from which to "prey" upon.

Still, all this is not necessarily telling you how Natural Selection™ actually works in such an "adaptively predatory" fashion. To do this, we will start

---

[1] That said, as for as those investors who internalize the analyst role, they still essentially reveal their coveted private information to the wider market as they "signal" through the price mechanism – as consistent with the theoretical modeling by Hayek (1995) and more recently by Green (2004) and Chakravarty *et al.* (2004). Indeed, Brommel (2003) has argued that small privately informed traders can improve their returns by spreading rumors after they have established a position. Such actions augment their initially muted asset price "signaling impact." In fact, most applied investment styles – like our own "adaptively predatory" method – rely upon an essential second phase of others rushing into a particular investing (strategy) thematic after you are already set. We however use upstream memetic signaling to determine the extent to which a particular investment theme (the actual embodiment of which is represented by a "winning strategy") is in its ascendant or maturing phase.

by presenting a crash course in Evolutionary Algorithms. Why you might ask? Well, as anyone who has even the slightest amount of tuition as to the basics of Number Theory will soon testify, what we are proposing by way of implementation of our Evolutionary Finance principles (at least from the *applied* perspective) is not an easy task. Take for instance the dimensionality of our information categorization and sorting problem. This is a massive undertaking if we are to truly ascertain the key memetic sequences that determine every potential strategy's performance across a potentially large preconceived strategy universe. This is simply because there will be a near infinite array of possible informational sequences presented to us for each and every strategy's performance within such a universe from the continuous stream of financial information that besieges the market every day.[2] This "dimensionality problem" is made even worse when one considers that we are (somewhat ambitiously) about to embark upon a full decade's worth of out-of-sample backtesting across nearly one hundred highly liquid assets as screened by 36 preconceived strategies and over the very fine granularity of a daily data periodicity. Such are the onerous requirements to display the overall efficacy of our method.

To put the magnitude of this potential problem into context, one should remember that decoding the human genome – a problem of somewhat commensurate size – took a considerable portion of the world's computer processing power many years to unravel. Obviously, we do not have the same resources (or time) at our disposal here so there is no way that we can feasibly undertake an identical process for divining the memetic code for upstream indicators of "winning strategy" emergence across a comprehensive range of assets. Subsequently, it was incumbent upon us to develop a few "tricks" along the way to cut down both computational time and expense in making any implementation of our Evolutionary Finance principles economically feasible.[3] Ergo, it is for this reason that we dive headlong into the relatively difficult area of Evolutionary Algorithms in the first section of this chapter – if only to provide the reader with the necessary background for what is about to follow.

In particular, the assumed prerequisite knowledge level for our discussion in Section 7.2 – where we provide some intimation as to our own adaptations to Evolutionary Algorithms – is unfortunately quite high. In this section we

---

[2] For example, given just 20 information "bytes" spread across a specific day (which is a very conservative assumption when running a global portfolio) implies that there are a massive 2,432,900,000,000,000 different memetic permutations from this information alone. Adding but one additional day has a *geometric* rather than arithmetic scalar impact upon this figure so it is easy to see how very soon the dimensionality problem becomes absurdly large when using a Bayesian trading rule technique to determine the key memetic strings governing a particular strategy's performance.

[3] Indeed, solving this riddle is part of the justification for the patent application underlying our new investment process.

outline precisely how our proprietary (bioinformatics based) adaptations to Evolutionary Algorithms provide both a time and cost efficient mechanism to discern the all-important key memetic sequences that, in turn, provide us with leading indicators as to winning strategy emergence across a select range of assets. By way of background, these specific adaptations have been protected via patent and were painstakingly built up within the Natural Selection™ software platform over a number of "man years."

As will be seen in Section 7.2, one of the true beauties of this software's output is that it is able to tell you the fundamental significance of a particular "byte" of information. For example, we all know that a US GDP release is an important piece ("byte") of information for market participants to closely monitor when it is released – but precisely what assets and strategies is it important for and (even more importantly) when does it have more informational impact (II) significance? Is this II significance heightened by the fact there may have been a Consensus analyst upgrade of the GE equity stock a few weeks earlier? If so, what assets does this apparently emergent "winning strategy" meme benefit (your potential longs) and what assets does it impinge (your potential shorts)? What is the particular "winning strategy" itself? What else needs to fall into place for this meme to emerge into a particular "winning strategy" theme and for the all-important "2nd Phase" late cycle entrants to appear on the scene and bid into the market? Indeed, even if such a "winning strategy" theme were to emerge, will it ultimately grow into such a force that via contagion effects it will eventually cause a swing in overall market sentiment? All such crucial (albeit stylized) questions are eternally asked by market participants over the course of any given trading day and (unfortunately) most go unanswered simply because (up until now at least) there has not been a framework for understanding how the various microcomponents of information necessarily bolt together into a structured entity.[4]

As we have laboriously outlined in the preceding six chapters, this is certainly not the case with our highly detailed Evolutionary Finance investment process. Our Natural Selection™ software platform was designed from "the ground up" to specifically leverage from these Evolutionary Finance principles. In fact, one of the pivotal features that we have found in our modeling work to-date has been the software's uncanny ability to bring to light seemingly innocuous information bytes that form a key component of a particularly strong meme that affects not just one, but a plethora of strategies across a vast range of assets. As mentioned at the very start of this book, it is within this erstwhile overlooked (but nevertheless pivotal) information that much

---

[4] Indeed, it is this "structured information" approach that differentiates our bioinformatics-based adaptations to Evolutionary Algorithms from what are known as "data mining" techniques that traditionally have no concept as to overall information structure in their search for (at times nonlinear) associations between various arrays of data.

of the potential for the generation of exceptional returns necessarily resides – especially for absolute return managers (like ourselves) whose mandate allows them to invest (both long and short) across a diverse array of assets.

Indeed, it is toward the pursuit of such exceptional returns under the absolute return banner that forms the basis of our case study example – which happens to constitute much of the empirical analysis that will be presented in this chapter. In particular, we will take the reader through the proprietary asset selection process adopted by Natural Selection™ in its screening of possible investment opportunities within the Global Macro Hedge Fund space – and along the way, we will no doubt highlight some of the capabilities of this exciting new software.[5] To prove how well this proprietary investment process works, we will document a full 10 years of out-of-sample backtesting of the performance of the Natural Selection™ platform's investment recommendations. Before doing so however, we will at first illustrate how Natural Selection™ can be used to identify the constituents of any asset's memetic structure – as represented by an informational radix – as well as how this principle can be used to develop upstream indicators of "winning strategy" rotation. Furthermore, consistent with our discussion in Chapter 6, we will identify the drawdown risk characteristics associated with the appearance of certain predicted "winning strategies" – as developed under our proprietary EVaR technique. Finally, we put all of the above risk/return characteristics together to build an optimal (absolute return directed) evolutionary portfolio for a Global Macro Hedge Fund mandate and, as mentioned above, illustrate 10 years of out-of-sample backtesting of our "adaptively predatory" investment style. Needless to say, all this would not be possible were it not for innovative investment process detailed extensively in the preceding Chapters 1–6 and our library of patented algorithms that we have developed within our Natural Selection™ software platform. So, without further adieu, over to the basics of these particular algorithms.

7.1   A primer on Evolutionary Algorithms

As stated in our introduction to this chapter, Natural Selection™ draws heavily upon the class of solution procedure known as "Evolutionary

---

[5] In terms of our selected macro asset database for our chosen simulations, we opted for 50 of the world's most liquid futures contracts – as illustrated by daily open interest – covering equity indices, bond indices, interest rates, hard commodities and FX. Likewise, we chose the most liquid Exchange Traded Funds (ETFs) – as demonstrated by a daily turnover in excess of $US 5 million – which covered specifically equity indices (both sector and country allocations). In total, our asset universe comprised 84 macro assets across a diverse geographical array of exposures from Asia to North America to Europe and the Emerging Markets. In general, such a database should be looked upon as being consistent with most "Global Macro" Hedge Fund mandates – with the obvious exception of having a bias toward the "equity" asset class.

Algorithms" to discern the memetic structure of the market's informational genome. Obviously, this raises the issue of precisely what are "Evolutionary Algorithms" *per se* (at least for those as yet uninitiated in the black art of neural network modeling). Without overwhelming the reader too much by way of literature overview, we thought it appropriate to provide at least a cursory introduction to this body of thought before we outline the bioinformatics-based algorithmic reasoning underlying the Natural Selection™ software platform itself.

So what are "Evolutionary Algorithms"? Evolutionary Algorithms are non-linear programming techniques based upon the principles of nature – in particular: natural selection; evolution; and genetic principles.[6] Typically, two questions arise when market practitioners are at first confronted with the concept of an "Evolutionary Algorithm."

1 Why are such algorithmic processes typically referred to as being "nonlinear"?
2 Second, how does the "Evolutionary" component within an Evolutionary Algorithm actually work?

We will answer both questions respectively. To start with, why are Evolutionary Algorithms typically nonlinear? Well, such nonlinearities traditionally stem from the presence of *feedback* between dependent and inde-pendent variables as an Evolutionary Algorithm actively "searches" for an optimal solution. As discussed in Chapter 3, the study of Complexity Theory has highlighted that many of the dynamics exhibited by financial markets can be argued to be fundamentally deterministic/nonlinear in foundation. Indeed, our own concept of memetic information (M) as opposed to sin-gular information (Ξ) and the entire premise of informational impact (II) is founded on the presence of such ubiquitous nonlinearities. Evolutionary Algorithms offer a nice solution mechanism for attempting to identify such principles as they require little by way of prior knowledge of the precise functional form of such relationships from the users initial standpoint.[7]

In the past, it has been the realization of the pervasive presence of empirical nonlinearities in the financial arena that (although a realistic base-case assumption) has presented many a problem for applied finance practitioners. For example, the link between the equity market and the larger economy appears to be a classic nonlinear form. In short, to model the economy

---

[6] The link between biology, mathematics and even computer science has a surprisingly long history. For example, Wolfram (1986) discussed the principles of cellular auto-mata and Complexity Theory. Indeed, von Neumann (1966) discussed self-replicating biological-like systems of artificial life inside a computer.

[7] This in particular, is an endearing feature of this class of algorithm as it in part overcomes some of the earlier stated implementation problems surrounding Chaos Theory – see Chapter 3 for more details.

you need a purview on the equity market but before you can form such an opinion, you need an opinion on the economy. Catch-22? Precisely. In fact, even if you were to undertake a pre-emptive strike against such recursive logic by forming a "base-case" economic scenario and (hopefully) empirically iterating your way out of trouble you would still encounter problems. For example, "what is the correct specificity of the nonlinear functional form for your model," "what is the relative importance of various economic indicators within such a context and how should this relative importance evolve through time," even notions as simple as "what are the best indicators to choose" can at times comprise a major headache for quantitative modelers in this extremely problematic branch of empirical analysis known as "nonlinear econometrics."[8] Thankfully, Evolutionary Algorithms have managed to overcome many of these difficulties as they (adaptively) make many of these difficult decisions upon their own accord – thanks to their innovative (feedback directed) algorithmic structure. Subsequently, quite often an Evolutionary Algorithm represents a far more suitable nonlinear foundation for model building (at least from the practitioners perspective) simply because it is purposefully designed around the premise of (generally nonlinear) feedback to ascertain an optimal solution to a given problem. This represents a "nice fit" with the theoretical foundations of Evolutionary Finance that we have provided to date.

Next, how does the "Evolution" component within Evolutionary Algorithms actually work? As stated earlier, Evolutionary Algorithms have at their heart the principles of recursion and feedback but hand-in-glove with these principles go the concepts of "natural selection" and "survival of the fittest" too. Why? It is the presence of such principles – at least, in a mathematical form – that acts as an iterative screen for the most suitable (generally nonlinear) solution to a given problem. In other words, "natural selection" and "survival of the fittest" are the problem solving "filters" embedded within any Evolutionary Algorithm. It may be that these general principles date back to the 1800s – 22 November 1859 to be precise – but their logic is so powerful that they constitute a convenient way of deriving an appropriate response to pressing present day problems that investors regularly encounter (especially, when such logic is coupled with modern computing power).[9] By mathematically formalizing a natural selection process of various "mutated" possible solutions, Evolutionary Algorithms are able to repetitiously search for the best possible outcome (thanks to the principle of the "survival of the fittest") to many an applied finance conundrum – irrespective of its functional

---

[8] For an extensive discussion of these methods, refer to Mills (1993), Alexander (2001) or the thoroughly comprehensive Hamilton (1994).

[9] 22 November 1859 was when Charles Darwin's "Origin of Species" first went on sale. 1250 copies of this first edition were printed and were almost entirely sold on the first day of release for the princely sum of 15 shillings each.

form, indicator set or even the periodicity of the underlying relationship. Needless to say, there is little wonder as to why such algorithmic processes are now being exploited by an extensive array of proprietary trading desks and quantitative-based absolute return managers from around the globe.

So now that we have sold *you* on the idea, what are the various forms of Evolutionary Algorithms from which you can choose? Below we briefly detail a number of the more common Evolutionary Algorithm classifications and, where appropriate, place them within the context of our own Natural Selection™ software platform. Still, given space limitations and the specialist nature of this field, our coverage of the specifics of these particular algorithms will be by no means comprehensive. Interested readers are therefore referred to the ample number of excellent texts in the area – such as Goldberg (1989), Beale and Jackson (1990), Caudill and Butler (1992), Holland (1992), Peretto (1992), Muller *et al.* (1995), Refenes (1995), Beltratti *et al.* (1996), Mitchell (1998) and Coley (1999). All these references will provide an excellent embellishment to the various principles that will be presented here.[10] One caveat however is that readers wishing to delve further into this field should be aware of the at times confusing definitional basis within the literature at hand. This is especially the case when it comes to differentiating between the various sub-stratums of the Evolutionary Algorithm family. Indeed, even we – with considerable background in the area – found this definitional inconsistency confusing at first. Still, we have done our best to decipher most of the commonalities between contrasting definitions so hopefully the brief summary next will provide readers with a good grounding as to the basics of this exciting emergent modeling field.

Evolutionary Programming

This particular facet of Evolutionary Algorithms focuses most intently on the "natural selection" aspects of the evolutionary process. A nice example of these algorithms at work is the Adaptive Belief System (ABS) proposed in a series of papers by Brock and Hommes (1997a,b, 1998, 1999) – which also happen to fall under the bailiwick of "artificial models of the market" as described in Chapter 3.[11] In short, Evolutionary Programming is

---

[10] Also, should one want an introduction to Evolutionary Algorithm and Neural Network code examples, we recommend readers see Rao and Rao (1995) and Steeb (1999).

[11] Indeed, it could be argued that these models could also be classified as being sympathetic to the Behavioral Finance literature as Brock and Hommes Adaptive Belief System (ABS) is consistent with Sargent's (1993) earlier theoretical work on the concept of "bounded rationality" – and indeed Arthur's (1990) even earlier computational representation of this concept. Ultimately, underlying the principle of the Adaptive Belief System is a "minority game" where individuals are assumed to improve their decision-making by learning the outcomes of their past actions. It is important to recognize that such Game Theoretic foundations are not consistent with our earlier

an iterative procedure where alternate solutions to a (generally nonlinear) problem are sequentially put forward to some time-variant (adaptive) screening procedure – a simple example of which is an "error correction" mechanism. It is by making successive changes to the format of this adaptive filter (which, importantly for our purposes, can be Bayesian in foundation) that the best possible solution to a given problem is found.[12] Hence the strong affinity between Evolutionary Programming and the process of "natural selection" as in effect, "Mother Nature" makes successive changes to the "adaptive filter" of an organism's genetic code in the refinement of species.

When conducting Evolutionary Programming, it is important to note that the input parameters fed into the adaptive filter are typically encoded in binary form with each possible "solution" represented as a binary string. This encoding – which happens to be thoroughly sympathetic to the encoding of information into "manageable" signals as described in Chapter 4 – substantially speeds up the computation time associated with ascertaining the best possible solution. Furthermore, such encoding is a "natural fit" with the principle of Bayesian trading rule analysis – which again, suits our purposes immeasurably.

Needless to say, given our chosen nomenclature for our proprietary software platform – Natural Selection™ – which embodies a series of algorithms that have been designed specifically to decipher the market's information genome, we somewhat unsurprisingly draw heavily upon both encoding and Evolutionary Programming principles in our bioinformatics-based approach toward financial markets. That said, unlike most past approaches in the Evolutionary Programming field, we use a far more comprehensive (micro informational building block) "structured" encoding of information in our own proprietary search procedure which – importantly – is fundamentally "out-of-sample" *ex ante* rather than *ex post* in perspective. Furthermore, we

discussed Evolutionary Stable Strategy (ESS) based Evolutionary Games as minority games (at least in part) rely upon reputations being built between t to $t + 1$ whereas within Evolutionary Game principles the changing "rules of the game" preclude such reputations being built – that is with the exception of our own Evolutionary Game format where "reputations" are formed via the actual format of an analyst's information byte production.

[12] An alternate tact to this is by successive changes (mutations) to the format of the sequence of the input parameters themselves and holding the filter constant. Ultimately the choice of which direction to proceed depends upon the structure of the problem at hand. We generally find that altering the filter (rather than the binary string "solution") works best for testing the Bayesian foundations of memetic strings that help predict the emergence (and subsequent maturing) of preconceived strategy sets. Earlier work on Evolutionary Programming – see for instance Fogel *et al.* (1966) – used a random mutation process as part of the solution procedure. However we generally find such approaches unnecessary given the foundations of the out-of-sample memetic string search problem.

also explicitly consider the signaling from the wider information-set along with the *order of arrival* of each information byte as an important factor in discerning the key memetic sequences along the market's informational genome. We feel these adaptations to traditional Evolutionary Programming techniques makes our approach much more robust and certainly given our *ex ante* perspective, considerably more easy to implement from the practitioners standpoint as we rely less upon backward-looking "error correction" in making our adaptive filter adjustment and more on the forward-looking "structured information" fabric of the current signaling from the market's informational genome in deriving an "optimal solution" investment projection. As a consequence, portfolio drawdown is typically less using our "adapted version" of Evolutionary Programming principles than is traditionally the case with standard Evolutionary Programming techniques and this is one major appeal of our proprietary approach.

Evolutionary Strategies

Important too for the final recommendations of Natural Selection™ are the principles of Evolutionary Strategies. These algorithmic code structures contain most of the underlying elements of Evolutionary Programming but here the focus is mostly on using preexisting sequences of information (in our case, preexisting strategies) rather than the (potentially randomized) mutated search for any particular arbitrarily defined binary string solution to a prespecified problem.[13] Indeed, this particular facet of Evolutionary Algorithms fits nicely with the Evolutionary Stable Strategy identification and subsequent Evolutionary Portfolio Construction attributes of our Natural Selection™ software package. How is this the case? Well, for starters we are using preexisting strategy sets common throughout the marketplace – such as valuation strategies, economic strategies, technical strategies and behavioral strategies – as part of our raw information input.[14] Next, the binary encoded inputs that are loaded into our *ex ante* adapted version of an Evolutionary Program are in effect "strategies" themselves – at least in the sense that we use the payoffs from the various permutations of Bayesian trading rule strategies to discern the key memetic sequences along the market's informational genome. Once

---

[13] Note, this is not to say binary strings are not used for deriving the solution to Evolutionary Strategies. Indeed, the strategies themselves are typically encoded as binary strings – so an "evolutionary strategy" is simply a time-variant version of a binary string. Rather, the original format of the string itself is generally prespecified. This illustrates just how useful binary notation can be from a computer programmer's perspective – especially in cases such as ours where it has proven to be extremely flexible in specifying the format of both the "information structuring" problem and subsequent solution to this problem as well. For a discussion of the various methods of information encoding using binary strings, refer to MacKay (2003).

[14] See our Glossary in Appendix 1 for an exposition of some of the more frequently encountered forms of these strategies.

these key memetic sequences are found, they can then be easily used to identify the predictive success (and failure) of any preexisting strategy so that an overarching sequence of preexisting strategies can be put forward as a "solution." This "sequenced strategy" solution approach is generally consistent with Maynard-Smith's concept of an ESS. It is the representation of this ESS as a (purposefully mutated) binary string – which interestingly feeds off the signaling from the memetic encoded binary strings – that makes our general approach consistent with Evolutionary Strategy/Evolutionary Programming principles.

As we stated in Chapter 6, both investors and analysts typically use preexisting strategies to implement their ideas – but unfortunately most such strategies are static. By "moving upstream" in the information flow and identifying the early stage information byte signals that highlight the formation of a particular meme or theme that will cause a particular preexisting strategy to fail while for another to succeed we can effectively "prey" upon the fallibilities of others. We label this combined Evolutionary Programming/Evolutionary Strategy approach as "adaptively predatory." Our objective in adopting such a stance is simple – we aim to position "long" into preexisting strategies at the early stages of their inevitable theme diffusion, while at the same time looking to "short" those strategies at the mature stage of their life cycle which have a high probability of "strategy failure."[15] To do this, we use the key memetic strings identified by our adapted Evolutionary Programming Bayesian "strategy" screening technique to identify a particular preexisting strategy's informational drivers. Doing so helps us to build up a much better picture of the overall informational environ – including importantly, other investors responses to such an environ.

Subsequently, Natural Selection™ is able to give its investment recommendations not just in terms of a simple "buy" or "sell" signal for a given asset but also in terms of which preexisting strategy from a given strategy set – valuation, economic, technical or behavioral – is likely to be the key strategy driver determining this asset's performance over a given investment horizon. How do we determine which assets to choose from the array of recommendations put forward by Natural Selection™? In part this relies upon a proprietary "strength of signal" screening device to determine which memetic signals we have the most confidence in, however it also relies upon how each asset interrelates within an overall portfolio

---

[15] It is a well-known fact that no alpha-generating strategy works forever as incumbents soon "bid the alpha" out of a given winning strategy idea. Commensurately, certain "winning strategies" tend to comef in and out of favor over the course of time – style investing (growth versus value investing) being a natural case in point. Natural Selection™ is able to take into account such inevitable strategy longevity in forming its investment recommendations through a combination of its "structured information" approach toward discerning the market's informational genome and the fact that it uses preexisting strategies as part of its raw data input.

structure – as determined under the auspices of our Evolutionary Portfolio Theory (EPT) technique.

As stated earlier, by taking such an evolutionary perspective on the emergence, maturation and failure of successive "winning strategies," our approach is highly sympathetic to the concept of an Evolutionary Stable Strategy (ESS) as espoused by John Maynard-Smith (1982). That is, we develop an overarching "sequence of strategies" perspective which are fundamentally conditional upon the responses of others. Indeed, it is from such ESS formulations that Natural Selection™ is able to give an approximation as to what is the optimal forward looking set of time-variant weights that needs to be applied to each asset within a select universe under EPT principles. Such principles necessarily apply as we are taking into account the inevitable "theme diffusion" of a particular "winning strategy." Furthermore, such a facility also helps to minimize drawdown risk as we are able to identify those key memetic sequences that are likely to result in "strategy failure." This "strategy failure" can at times be the catalyst for market "events" – as evidenced by the demise of the Yen carry trade and the LTCM Crisis. Consequently, this approach (while complex) is far superior, we believe, to most static "forward looking" portfolios construction/risk management techniques that are presently being used today.

Genetic Algorithms

Like Evolutionary Strategies, Genetic Algorithms at times contain some of the "adaptive" aspects of Evolutionary Programming but the core component of any Genetic Algorithm remains its focus on the "seeding" of random inputs into each generation (iteration) of solutions in an attempt to select a "best of breed" group of survivors – and subsequently "crossing over" these survivors at each generation by the "interbreeding" winning pairs.[16] It is this Darwinian "survival of the fittest" selection mechanism – as hybrid "best" solutions effectively compete against each new generation of randomized "mutant" inputs – that forms the basis of a Genetic Algorithm's solution to a particular problem. One appealing by-product of this structure is that unlike Neural Networks (see the following box) and (to a lesser extent) Evolutionary Programs, Genetic Algorithms have less potential to be convergent toward any singular local solution to a particular problem. This is primarily because of the random seeding of alternate solutions embedded in each iteration (generation) within the algorithm effectively knows no bounds and is thus less dependent upon the "initial conditions" of a given solution that has already been achieved by the algorithm. This "initial condition" problem is endemic in recursive programming techniques – as discussed in our critique

---

[16] This particularly useful insight as an adjunct to the algorithmic solution search process has been attributed to the earlier work by John Holland – see Holland (1974).

of optimizers in Chapter 6 – and has (at times) been a major hindrance limiting the uptake of nonlinear programming techniques by practitioners in the past.[17] Indeed, to their distinct advantage, quite often a Genetic Algorithm will suggest multiple solutions to a given problem that matches the multiple existence of localized maxima or minima – an outcome that most traditional recursive nonlinear solution procedures find hard to achieve.[18]

That said, there are drawbacks in using Genetic Algorithms that practitioners should consider. First and foremost is the fact that random mutation and hybridization of possible solutions can at times result in nonsensical results – especially when one is using preexisting strategies as a component of raw data input. Hybrid strategies such as $\frac{1}{3}P/E$ Relative valuation/$\frac{1}{8}$ Head & Shoulders technical/$\frac{13}{24}$ US GDP economic factor weighting may have little *a priori* rationale as a clear driver of a certain asset's performance.[19] Furthermore, like standard Evolutionary Programming techniques, Genetic Algorithms tend to be backward looking in the sense that they attempt to fit the best solution (via the mutation of various possible outcomes) at each time t and then extrapolate this solution to time $t + 1$. Implicit in this is the belief that what works today, will work tomorrow whereas those who have been at the coalface of global markets know, such assumptions are at best "heroic" and at worst downright dangerous. Given this circumstance, the random mutation and hybridization of possible solutions at each time t may actually contribute to instability in the model's prediction output – thus resulting in excessive transaction costs. Indeed, our own proprietary work in this area proved this to be exactly the case – hence our preference for the more stable *ex ante* (forward looking) adapted version of an Evolutionary Programming/Evolutionary Strategy algorithm for our bioinformatics-based approach toward financial markets.[20] In fact, it is

---

[17] Further, Genetic Algorithms also have less potential to be affected by the presence of outliers that tend to pull Neural Network solutions away from what is ultimately the correct steady-state solution path.

[18] This allowance for multiple local minima (or maxima) in Genetic Algorithms does enable the better approximation of catastrophic localized events that tend to get "averaged out" in a Neural Network solution – a nice attribute not only for biological modeling but financial modeling as well.

[19] This is even more the case when one is using pure index price series as raw data inputs as here the problem becomes one of "spurious technicals" – a mélange of varying momentum observations – driving the "solution" to asset price performance prediction. The danger in these circumstances is that the procedure itself may become little more than a data fitting exercise and one can place little faith in the out-of-sample forecasting ability of the Genetic Algorithm.

[20] Indeed, the presence of excessive transaction costs is a criticism often leveled at earlier empirical attempts under the "Evolutionary Finance" banner as adopted by Hens *et al.* (2002) and Dempster and Jones (1999a,b,c, 2000, 2001) – see the next subsection for more details. That said, our own purview on Evolutionary Finance overcomes the excessive transaction cost shortcoming through our comprehensive modeling

primarily this forward-looking "structured information" adaptation toward Evolutionary Programming/Evolutionary Strategy techniques that differentiates our own implementation of Evolutionary Finance principles from the largely Genetic Algorithm/Genetic Programming based "Evolutionary Finance" studies of the past.

Genetic Programming

Genetic Programs are inherently similar to Genetic Algorithms in that they focus on the random "seeding" of different possible solutions into each successive population of possible outcomes (with only the best results surviving into the next iteration) but importantly they also differ from Genetic Algorithms in one key respect. Genetic Programs allow the sequence (or "string") of possible solutions to vary in number. This may not sound like much but ultimately it adds yet another dimension to the range of possibilities in the search for the best possible solution to a particular problem. For example, should one rely upon four (binary) signals or five from a given population of inputs? Such a question is typically answered with sophisticated filtering or autoregressive techniques in standard econometric analysis but they are also nicely covered by Genetic Programming too. In fact, as mentioned earlier, a considerable portion of the earlier work done under the "Evolutionary Finance" banner utilized this particular form of nonlinear modeling – see for instance Dempster and Jones (2000).

As cited upon in numerous occasions in this book, we have always sought to leverage from the principle of "standing upon the shoulders of giants" and in this case part of our own adaptations to the Evolutionary Programming/Evolutionary Strategy class of algorithms explicitly allows for the binary string of any proposed "solution" to vary in length – hence (at least in part) following in the footsteps of "Evolutionary Finance" Genetic Programmers of the past. Indeed, this feature was crucial in allowing the explicit consideration of the *order* of arrival of information as part of our overall solution mechanism. Without such an appendage, we would not have been able to bring our own adaptations to Evolutionary Programming/ Evolutionary Strategy algorithms that much closer to the type of code structures used in the bioinformatics sphere and thereby move the concept of "Evolutionary Finance" closer to the "structured information" realm.

of the microfoundation building block structure of financial information – with the most readily identifiable embodiment of this in the *applied* sense being – the *ex ante* adjustment appendage to existing Evolutionary Programming/Evolutionary Strategy techniques. Here, we have the advantage of not only explicitly considering how the various subcomponents of information necessarily "bolt together" but also we take into account the order of arrival of information as an integral feature of our solution procedure.

Memetic Programming

Finally, let us briefly touch upon Memetic Programming techniques – one of the more recent introductions to the emergent field of Evolutionary Algorithms. In essence, Memetic Programming adds yet another dimension to Genetic Programming in the sense that it explicitly allows for separate local randomly seeded searches to be conducted in parallel before the various "populations" of "optimal solution" results are then recombined and a final "best of breed" is selected for each new generation.

Here – consistent with our earlier discussion of Professor Richard Dawkins (1976) concept of memes – it is the *association* between binary inputs that is explicitly considered as an additional input into the "best of breed" solution screening process. In short, those local "optimal solution" results with the best associations with other local "optimal solution" outcomes generally have the best chance of being accepted as the "best of breed." Subsequently, their survivorship through each of the successive generations (subject to the randomized onslaught of the seeding of alternate solutions) is generally greatest. While on the surface this approach may appear to be ideally suited to our own "structured information" based purview of Evolutionary Finance, one of the major drawbacks of this particular algorithmic structure from an implementation standpoint is that (like Genetic Algorithms and Genetic Programming) such algorithms periodically result in excessively "noisy" output. Yet again, this fallibility is instigated by the random seeding of possible solutions at each new generation and translates into excessive transaction costs for the practitioner. That said, further refinements to this relatively new type of solution procedure should be closely monitored as the overall structure of the approach looks quite compelling – indeed, a nice example of this form of iterative solution technique can be found in Hart (2003).

Neural networks: The most ubiquitous form of nonlinear programming technique

Neural networks: The basics

When most people think of nonlinear programming techniques, Neural Networks generally spring to mind. So what are they precisely and how do they differ from Evolutionary Algorithms? In short, Neural Networks are mathematical models developed to replicate the information processing of the human brain. Intuitively these models are appealing from the financial modelers perspective because they process raw information in much the same fashion that the human brain responds to a given set of stimuli. What precisely does this mean? Well, at first there is a cognitive search by a Neural Network algorithm (using encoding and categorization) for associative patterns within interdependent information. Then, the Neural Network typically uses a trial and error type of process (mathematically, this is known as backpropagation) to search for an optimal solution to a given problem.[21] This sequence of information processing replicates the pattern adopted by the human brain as it seeks to determine the most appropriate response to a given set of information stimuli (such is the process of "learning" as we have described in Chapter 4).

The beauty of Neural Networks from a market practitioner's standpoint is that not only do they provide a framework that allows for investors to effectively identify what are the important factors in driving a particular asset's price performance, but they also allow for such a system to be adaptive. In other words, the functional form of a solution to a Neural Network problem is nonconstant over time. In particular, the weighting applied to alternate information sources can change between the months and years over which a Neural Network is run. Even the specificity of the constituent information sources that comprise the "optimal solution" to a Neural Network problem can alter across time.

This contrasts with traditional nonlinear time-series econometric modeling where the approach is to at first assume what the key factors are in driving a particular asset's price performance and then to assume what is the most appropriate function form for the econometric model to take. Neural Networks typically require no such assumptions. Instead, they allow the data to "speak for itself." In this sense, Neural Networks:

> are non-parametric, nonlinear estimators of financial data that require little to no prior assumption about functional form, sign or the weighting that should be given to a set of independent variables in determining a dependent variable

These are fundamentally nice attributes of the Neural Network modeling technique.

That said, strong models that make strong assumptions about the interaction between dependent and independent variables are generally superior to Neural Network forecasting when the underlying relationship between a set of variables is unquestionably known. However, strong models also suffer from the significant fallibility of inducing a considerable degree of bias into the modeling process when the underlying relationship between the dependent and independent variable is less well understood – as is often the case in financial markets. In this sense, the data driven foundations of Neural Network techniques are perfectly suited to the ever-changing nature of an adaptive investment process where the relationships between dependent and independent variables are constantly evolving thanks to the relentless flow of market information.

Neural network design

To start off with, let us concentrate on the physiological aspects of the human brain's information processing. As partly mentioned earlier, upon receiving an information "signal" the human brain immediately seeks to encode all

---

[21] Whether this trial and error process is formalized in the form of a sigmoid function (as is generally the case in backpropagation) or alternately is simply the outcome of the training process (which is the case when no explicit backpropagation layer is present) depends upon the whims of the programmer.

interdependent information and then to categorize it. This task is performed (at the smallest operational scale) by neurons. Neurons consist of: (1) a cell body; (2) branching extensions for receiving information (called dendrites); and (3) an output channel (called an axon).

Information (in the form of electrical impulses) are collected by the dendrites and channeled to the synapses – if you like, a "bridge" over which information deemed to be relevant (i.e., if the information signal is of sufficient strength) will flow. It is the task of the synapse to effectively sum the information and if the combined strength of the information exceeds a certain threshold a reaction signal is on-sent to the axon via a chemical/electro transmission mechanism. In doing so, the original information undergoes a type of transformation as what is sent along the axon is an *interpretation* of the original information provided by the dendrites. If this interpretation is deemed to be "appropriate" or "correct" the chemical weights applied to the original information at the (synaptic) threshold level are memorized and the brain has effectively rote "learnt" a new signal/response. This is human learning at its most primal physiological/neurological level.

So how does all this relate to Neural Network algorithms? In short, it is useful to think of a Neural Network algorithm as essentially replicating this three stage process of information transformation. A group of interdependent data is amassed by the prospective modeler in much in the same way that dendrites collect signals within the human brain. However, for our purposes, the final transformation of this information is already known – at least in an *ex post* sense – as it is represented by the market price of a chosen asset. It is therefore the task of the Neural Network algorithm to solve stage 2 of the process – input is known, output is known but how to determine the appropriate threshold weights so that only the correct information signals are identified as to the true determinants of an asset's price?

What a Neural Network algorithm does to overcome the problem of appropriate threshold weights is to use an error correction mechanism to determine the most effective set of weights to apply to the raw data input so that the forecast output price is as close as possible to what is actually observed in the real market. In doing so, the weights applied to the particular data in question must be sufficient enough so as to solicit an appropriate response when the market price changes but not to solicit a response when the market price is static.

Mathematically, the objective of the Neural Network algorithm is to minimize the total error in the predictive capability of the threshold response function over a chosen timeframe of data. As displayed in Figure 7.1 it does this by iteratively setting alternate weights to the raw data input using a "trial and error" type of

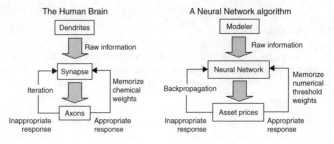

Figure 7.1 Information processing by the human brain and by a Neural Network Algorithm

process (or backpropagation as it is known in computing lexicon). Neural Network models then choose as a "solution" the set of weights with the minimum total error value.[22] It is this "adaptive filtering solution" technique that makes Neural Networks highly sympathetic to Evolutionary Algorithms (especially Evolutionary Programming/Evolutionary Strategy) procedures.

Neural Networks and Evolutionary Algorithms: Commonalities and differences

Neural Networks differ from Evolutionary Algorithms primarily in terms of algorithm design. Both Neural Networks and Evolutionary Algorithms use a recursive error correction mechanism to discern a total minimum error solution to a given informational problem, but where these approaches differ is basically in terms of how this total minimum error solution is found.

- For Neural Networks, the total minimum error solution is found by setting the correct levels for the threshold function – which effectively "filters" raw input signals for the correct (minimum error) response output and sets the input weights accordingly.

To contrast, Evolutionary Algorithms use a variety of (evolution based) principles to discern the total minimum error solution to a given problem.

- For Genetic Algorithms/Genetic Programming this entails randomly "seeding" alternate (weighted information input) solutions at each iteration of the algorithm and then selecting the "best of breed" (those solutions with the lowest total minimum error) as "survivors" to the next iteration.
- For Evolutionary Programming/Evolutionary Strategy selection, the contrast with Neural Network techniques is much more subtle in the respect that no explicit threshold function is set under Evolutionary Programming principles. Rather, the adaptive error correction mechanism itself "evolves" across time – both in terms of input signal weights and in terms of overall specification. Here, the number of iterations (rather than the prespecification of the level of the threshold function) plays the predominant role in the accuracy of the final solution.

That said, both Evolutionary Programming/Evolutionary Strategy as well as Genetic Algorithm/Genetic Programming principles do still have much in common with Neural Networks in the sense that they all use a form of raw information input/output error correction mechanism in their attempts to ascertain an optimal solution. Where these programs differ however, is in precisely how the "optimal solution" is filtered at each stage of the iteration process. It is this key factor that is the differentiating feature between these various sub-stratum of the nonlinear programming family of algorithms.

---

[22] Crucial to the Neural Network approach is the choice of the weighting applied to the threshold function. If the threshold is set too low, the model will find too many solutions. Alternately, set the threshold too high and the model will find no solution. Herein resides the "art" in Neural Network modeling.

So how best to put our proprietary adapted version of nonlinear Evolutionary Programming/Evolutionary Strategy techniques into action? Well, before starting it must be said that – in sympathy with our comments made in Chapter 3 on the implementation of Chaos Theory – nonlinear programming has been a siren song beckoning onto the rocks many a would-be investment guru in the past. What makes our approach so different? Primarily it is because of our "structured" interpretation of how the microcomponents of information come together to form larger and larger entities in a classic biological/evolutionary fashion. Having spent such a large part of this book getting the theory "right" one would hope there would be some benefits to reap from this unique perspective for the *applied* realm as well. Indeed, this appears to be the case. To demonstrate this, below we sequentially build up a case study that ultimately culminates in a comprehensive out-of-sample backtest of our proprietary Natural Selection™ software platform over ten years of data. This backtest applies to using our "adaptively predatory" principles for a Global Macro Hedge Fund mandate.

That said, this is not to say such principles (which leverage our proprietary theoretical "structured information" investment process detailed exhaustively in Chapters 1–6) are necessarily confined to the "Global Macro" absolute return space. Indeed, depending upon the dataset used and the constraints set upon the Bayesian trading rules that are directed toward discerning the key memetic sequences along the market's informational genome, Natural Selection™ could just as easily identify investment recommendations consistent with a long/short equity fund mandate, a statistical arbitrage absolute return mandate, short only, CTA, even event driven strategies. Basically, our adaptively predatory structured information approach toward investing in financial markets is intrinsically applicable to a whole range of absolute return investment styles and is thus (importantly) scalable across a variety of alternate mandates. Readers should make note of this.

7.2   Evolutionary asset selection

So where to start in our Global Macro Hedge Fund mandate case study? Our first port of call will be identifying the key memes that constitute both the asset price and "winning strategy" drivers of each asset within our selected asset universe. Obviously space limitations prevent us from displaying each and every result from this comprehensive study here so by way of example we decided to choose an asset that most investors should be familiar with – the short-term futures contract on the S&P500.

An example of the identification of key memes in an asset's information genome

Following our lead established in Chapter 5, it is possible to use Natural Selection™ to discern the key informational drivers of a particular asset's performance that are already in place. To do this, the software must inexorably discern an asset's *informational genome* and (as we stated in Chapter 5) the most concise way to visually represent this overwhelming array of information is by appealing to our aforementioned "informational radix" structure.

The following figure illustrates the informational radix for the S&P500 short-term futures contract – as has been ascertained by using an adapted (forward looking) form of bioinformatics based Evolutionary Programming for structured information search within our Natural Selection™ model. To do this, as mentioned in Chapter 5 and in the preceding section, we appeal to Bayesian trading rules ("strategies") as the criteria via which the relative strength of alternate memetic sequences are assessed. It is the relative payoff performance of these various memetic-based Bayesian trading rules (as independently defined using our bioinformatics-based Evolutionary Program within Natural Selection™) that identifies the key memetic sequences that have the largest impact upon the S&P500's short-term futures contract (Figure 7.2).[23]

In interpreting Figure 7.2, as discussed in detail in Chapter 5, it is those memetic sequences with the strongest impact upon the S&P500 future contract's price performance at a given point in time that are recorded as clusters closer to the "inner core" of the informational radix. Conversely, those memetic clusters with less of a bearing on this particular asset's price performance are identified more toward the periphery. Time is measured by an upward movement along the vertical axis while the waxing and waning of emergent themes or even a change in overall market sentiment itself is illustrated by a demonstrable change in the actual structure of the informational radix.

For our given Global Macro Hedge Fund asset universe, Natural Selection™ can identify a similar type of structure for each and every asset. It does this by identifying what proportion of a given asset's key memetic drivers are present at any given point in time and also makes a prediction for the formation of certain memes in the future given:

1 the relative percentages of as yet unformed memes already in existence at a given point in time; and
2 using an epoch matching formula, identifying the probabilities of certain meme forming bytes manifesting in the near future given the similarities between the present epoch and past epochs.

Still, Natural Selection™ does not rely upon simple asset price projections to form its investment recommendations. Why? To do so would fail to bring

---

[23] Note that in Figure 7.2 we have constructed our assessments on an annual basis from June 2000 onward.

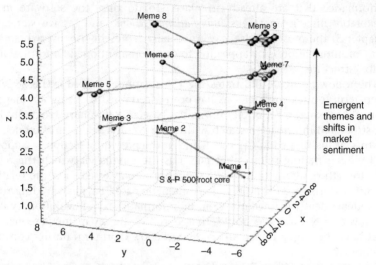

Meme Legend:[24] Meme 1: Euroland CPI (t-1) 1, IBM 12mth Fwd Earnings Revision (t-2) -1, Nokia 12mth Fwd Earnings Revision (t-5) -1, Dell 12mth Fwd Earnings Revision (t-4) -1. Meme 2: Euroland CPI (t-1) 1, Citigroup 12mth Fwd Earnings Revision (t-6) -1, Nokia 12mth Fwd Earnings Revision (t-6) -1. Meme 3: US PPI (t-3) 1, IBM 12mth Fwd Earnings Revision (t-3) -1. Meme 4: US PPI (t-4) 1, Japan CPI (t-3) -1, Euroland CPI (t-2) 1, Coca-Cola 12mth Fwd Earnings Revision (t-2) 1, Bank of America 12mth Fwd Earnings Revision (t-1) -1, Pepsico 12mth Fwd Earnings Revision (t-5) 1. Meme 5: Japan CPI (t-3) -1, Euroland CPI (t-1) 1. Meme 6: Euroland CPI (t-3) 1. Meme 7: Euroland CPI (t-2) 1, Citigroup 12mth Fwd Earnings Revision (t-5) 1, Procter & Gamble Dividend Yield (t-4)-1, Shell P/E Ratio (t-2) -1, Pepsico 12mth Fwd Earnings Revision (t-5) 1. Meme 8: Time Warner P/E Ratio (t-4) 1. Meme 9: Euroland CPI (t-2) 1, Citigroup 12mth Fwd Earnings Revision (t-5) 1, Procter & Gamble Dividend Yield (t-4) -1, Shell P/E Ratio (t-2)-1, Pepsico 12mth Fwd Earnings Revision (t-5) 1.

Figure 7.2   The price-based informational radix for the S&P500 futures contract[25]

[24] Note that consistent with our discussions to date, each information byte (φ) is encoded. To do this, we took the running five-year (out-of-sample) average of the time-series for each information byte and compared each respective observation (φ) at time t relative to its average to "normalize" all our observations. Subsequently, the encoded outcomes for the normalized data were as follows: $0 < \varphi < 0.5 = 1$, $0.5 < \varphi < 1.0 = 2$, $1.0 < \varphi < 1.5 = 3$, $1.5 < \varphi < 2.0 = 4$, $2.0 < \varphi < 2.5 = 5$, $2.5 < \varphi < 3.0 = 6$ and $\varphi > 3.0 = 7$. With the converse applying for the left hand (negative tail) of the distribution.

[25] Something to look for in making any casual observation of an informational radix is the frequency of triple byte memes. Triplet bases form the foundation for the genetic code underlying messenger RNA so the presence of triple byte memes is automatically analogous to the types of key sequences we observe in nature. Yet again, it appears that nature is providing a "guiding light" to our evolutionary interpretation of markets as we have observed a surprising number of triplet memes in our studies to date. Furthermore, some information bytes seem to appear on a reoccurring basis – irrespective of the

into consideration the explicit reactions of other investors in responding to specific (memetic) information signals. As stated in the introduction to this chapter, the *raison d'être* of Natural Selection™ is to be "adaptively predatory" upon other investors fallibilities. Since such investors typically use heuristic rules of thumb (i.e., simple strategies) to encode and categorize the vast amount of information presented to them each and every day, then Natural Selection™ needs to explicitly consider such strategies as part of its "raw data" information-set. Doing so enables us to form a far more comprehensive purview of the actual drivers of asset performance and also allows us to develop a range of comprehensive "2nd Phase" strategies that prey upon the inevitable fallibilities of standard preexisting strategies as they emerge, mature and fail in a manner predicted by our biological-like/evolutionary information "building block" principles. We feel this byte to meme, meme to theme, theme to overall market sentiment approach to the "winning strategy" diffusion process across the wider market is considerably more sophisticated than many investment styles undertaken by market practitioners in the present day. Subsequently, we turn our attentions to implementing information radix techniques for forming a better understanding of the "winning strategy" life cycle.

An example of evolutionary strategy selection

As stated earlier, it is insufficient for Natural Selection™ to make an investment recommendation on singular predictive asset price performance alone. Somewhere explicit in our analysis must be an interpretation of investor action (in other words, the greater market's response) to specific information signals. As investors typically use preconceived strategies to guide their behavior in response to such signals, then it is incumbent upon us to build as many of these preconceived strategies as possible into the raw dataset of Natural Selection™.[26] Doing so adds an all-important "2nd Phase" perspective to the investment recommendations of Natural Selection™ in the sense that we can identify whether a particular preconceived "winning strategy" is in the early stages or at the mature stages of its inevitable theme diffusion path in response to the presence of supportive ancillary

price series used. Obviously, these information bytes are deemed to have a wide significance by market participants across all asset classes and the externality potential (x) of such bytes is exceedingly large. For a comprehensive discussion measuring the frequency of such phenomena along very long sequences, see Allouche and Shallit (2003).

[26] Some of these preconceived strategy formulae are represented in Appendix 1.

memetic information.[27] In short, by considering such strategies as part of its overall dataset, Natural Selection™ is automatically able to identify the "greater herd" investor response to the appearance of specific memetic sequences and this enables us to develop an appropriate "predatory" strategic response.

For example, a particular memetic sequence may be commensurate with the outperformance of "high Beta" assets such as Emerging Market equities or High Yield bonds. This memetic sequence may also be indicative of investors focusing on certain (largely momentum based) strategies for their investment timing decisions into and out of these assets. In marrying the two together – the preferred asset class along with the investment strategy governing the timing of participation into and out of this asset class – Natural Selection™ is able to signal whether to "Buy" into this strategy (that is, assuming the particular momentum strategy is in the early stage formation of its theme diffusion process) or alternately whether to adopt a contrarian standpoint and "Sell" into this strategy as it is near the end of its life cycle and our memetic-based signaling is predicting a near-term strategy failure.[28] The beauty of such an approach is that not only does it give us an investment recommendation but also a "timing" recommendation as governed by the month to month vagaries of the signaling to the wider market that is originating from the "winning strategy" at hand (which in the example earlier was a momentum-based strategy). All up, we feel this is a far more comprehensive approach toward absolute return investing than is the case with a simple (predictive) asset price investment style.

To illustrate the efficacy of our earlier "strategy orientated" approach toward investing, later we illustrate how a memetic string-based information radix can be constructed for a particular preconceived strategy (in this case, a technical-based seasonality strategy) for our chosen short-term S&P500 futures contract.[29] The differentiating feature between Figure 7.3 and Figure 7.2 is that in the information radix below a movement up the vertical scale is represented by the payoff function from a slavish adherence to this

---

[27] Obviously by taking into consideration such inevitable "winning strategy" life cycles, the investment philosophy provided under the auspices of Natural Selection™ is consistent with Maynard-Smith's premise of an ESS.

[28] In such cases, the "winning strategy" that is in the early stages of forming is a "short momentum" strategy. Likewise, such "long to short" strategy swings are inevitably accompanied by the emergence of other "winning" long strategies. So the investment recommendations from Natural Selection™ are typically neither 100 percent long or 100 percent short – even though theoretically this is possible.

[29] For a detailed exposition on the various methods available to construct such a seasonality strategy, see Kaufman (1998).

particular strategy. Those memes with the most significant influence upon the payoff function are located in clusters closer to the "inner core" while those memes with less of an impact are located more toward the periphery. We illustrate both complementary and contrarian standpoints in an attempt to ascertain the key "long" strategy signaling memes and "short" strategy signaling memes.

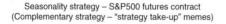
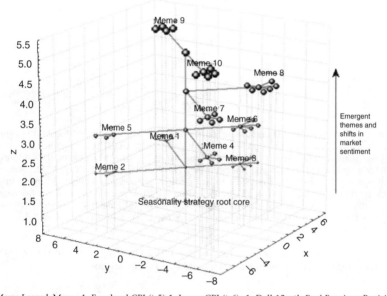

Meme Legend: Meme 1: Euroland CPI (t-5) 1, Japan CPI (t-6) -1, Dell 12mth Fwd Earnings Revision (t-4) -1. Meme 2: Euroland CPI (t-5) 1, Oracle 12mth Fwd Earnings Revision (t-1) -1. Meme 3: US Fed Funds Rate (t-3) 1, US CPI (t-5) 1, Euroland CPI (t-4) 1, Citigroup 12mth Fwd Earnings Revision (t-1) 1, IBM 12mth Fwd Earnings Revision (t-5) -1. Meme 4: Japan CPI (t-6) -1, Euroland CPI (t-5) 1, IBM 12mth Fwd Earnings Revision (t-4) -1, Dell 12mth Fwd Earnings Revision (t-4) -1. Meme 5: Japan CPI (t-6) -1, Euroland CPI (t-5) 1. Meme 6: US PPI (t-2) 1, Euroland CPI (t-1) 1, Japan CPI (t-2) -1, GE 12mth Fwd Earnings Revision (t-6) -1, IBM 12mth Fwd Earnings Revision (t-5) -1. Meme 7: Japan CPI (t-6) -1, Euroland CPI (t-5) 1, IBM 12mth Fwd Earnings Revision (t-4) -1, Pepsico 12mth Fwd Earnings Revision (t-3) 1. Meme 8: Japan CPI (t-6) -1, Euroland CPI (t-5) 1, Johnson & Johnson 12mth Fwd Earnings Revision (t-6) -1, Time Warner 12mth Fwd Earnings Revision (t-3) -1, Procter & Gamble 12mth Fwd Earnings Revision (t-4) 1, Novartis 12mth Fwd Earnings Revision (t-5) -1. Meme 9: Euroland CPI (t-5) 1, IBM 12mth Fwd Earnings Revision (t-4) 1, Vodafone 12mth Fwd Earnings Revision (t-1) -1. Meme 10: Japan CPI (t-6) -1, Euroland CPI (t-5) 1, Time Warner 12mth Fwd Earnings Revision (t-3) -1, Pepsico 12mth Fwd Earnings Revision (t-5) 1.

(Figure 7.3 continued)

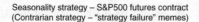

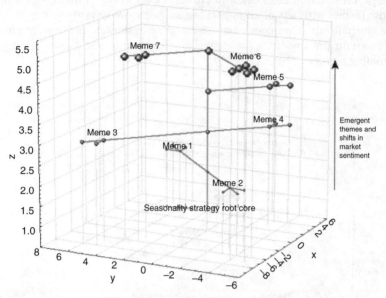

Meme Legend: Meme 1: OECD Leading Indicator (t-5) -1, Novartis 12mth Fwd Earnings Revision (t-2) -1, Nokia 12mth Fwd Earnings Revision (t-4) -1. Meme 2: OECD Leading Indicator (t-2) 1, Walmart 12mth Fwd Earnings Revision (t-6) 1, Time Warner P/E Ratio (t-4) 1. Meme 3: Shell 12mth Fwd Earnings Revision (t-4) 1, Bank of America Shell 12mth Fwd Earnings Revision (t-1) 1. Meme 4: Time Warner P/E Ratio (t-4) 1, Toyota 12mth Fwd Earnings Revision (t-5) -1. Meme 5: Time Warner P/E Ratio (t-4) 1, Toyota 12mth Fwd Earnings Revision (t-5) 1. Meme 6: Mexican Stock Exchange (t-1) 1, IBM 12mth Fwd Earnings Revision (t-6) 1, Time Warner P/E Ratio (t-4) 1, Altria 12mth Fwd Earnings Revision (t-6) 1. Meme 7: Time Warner P/E Ratio (t-4) 1, Toyota 12mth Fwd Earnings Revision (t-5) 1.

Figure 7.3 The information radix for a seasonality strategy approach to the S&P500 futures contract – both a complementary and contrarian viewpoint

By comparing such information across a range of different assets and a plethora of different preconceived investment strategies, Natural Selection™ is then able to provide a specified number of its strongest signal recommendations across a selected asset universe. Still, such a "best of breed" set of investment recommendations must sit well together within an over-all portfolio structure. Which leads us nicely into the next stage of the algorithmic structure of Natural Selection™ – how to use these investment recommendations to build an evolutionary portfolio which is consistent with the principles outlined in our critique of Evolutionary Portfolio Theory (EPT) in Chapter 6.

7.3   Evolutionary Portfolio construction

What about risk? Return projection – no matter how comprehensive – represents just one side of the coin when it comes to prudent investing. Obviously, to minimize portfolio drawdown potential we must also incorporate some aspect of risk management into our ruminations. Thankfully, Natural Selection™ – with its barrage of patented bioinformatics-based adapted Evolutionary Algorithms – does this with comparative ease. As was explained in Chapter 6, using an epoch matching technique to examine the various "moments" of selected assets when a particular "winning strategy" (either complementary or contrarian) is garnering wider market dominance underpins the very foundations for our Evolutionary Value-at-Risk (EVaR) technique.[30] Further, by matching (strength of signal adjusted) projected "winning strategy" returns vis-à-vis projected drawdowns as each emergent "winning strategy" within our "best of breed" asset universe progressive matures, it is possible to iteratively assign a series of time-variant weights for a truly evolutionary portfolio – as consistent with the underlying premise of our EPT. Both these elements to our comprehensive investment approach will be discussed in due course.

An example of EVaR principles at work

Again, space limitations preclude us from outlining an EVaR assessment for each and every asset within our selected asset universe across our chosen case study of a Global Macro Hedge Fund mandate. Still, consistent with the earlier EVaR assessment, we have opted for an illustration of this technique using an asset that most investors should be familiar with – the short-term futures contract on the S&P500.

Below we illustrate the time-variant probability attached to a 5 percent drawdown on a S&P500 futures contract position as determined by a series of out-of-sample tests of our Natural Selection™ platform. These results were constructed for a one-month investment horizon and by obtaining a projection for a particular "winning strategy" for the S&P500 at each time t. These strategy projections – together with the ancillary data within the overall dataset – were then (on an out-of-sample basis) progressively matched against past epochs to build up a purview of the EDF for the S&P500 at each time t. It is from this EDF that we obtained the representative probability of a 5 percent drawdown on our selected asset – as conditional upon a one-month investment horizon. What is immediately obvious is the significant degree of volatility in such projections. Obviously such

---

[30] Which, in turn, is driven by our concept of an Evolutionary Distributional Form (EDF).

volatility would not occur had one assumed a static (traditionally Gaussian-based) distributional form to estimate drawdown potential (Figure 7.4).[31]

But still, the above does not provide us with any insight as to how to best use this information. To do so, we must amalgamate both risk and return aspects of the overall investing conundrum. This is done in the following sub-section where we discuss the application of Evolutionary Portfolio Theory (EPT) to our Global Macro Hedge Fund case-study.

Putting it altogether using Evolutionary Portfolio Theory as a guide

Amalgamating both the risk and return aspects of our revolutionary evolutionary approach to information in financial markets is the charter of our EPT technique. In doing so, we combine both the ESS/investment

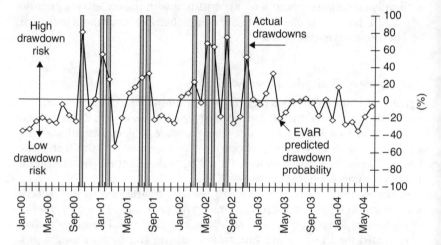

Figure 7.4 A 5 percent drawdown probability assessment for the S&P500 futures contract – using our proprietary EVaR principles as a foundation[32]

---

[31] That said, some volatility in such assessments does occur when "static distribution" techniques are used. Still, given the degree of overlap in the out-of-sample "rolling window" over which such distributions are fitted, the degree of volatility is considerably less marked than that displayed in Figure 7.4.

[32] Obviously a negative probability for a 5 percent drawdown implies our EVaR assessment is approximating a mean positive (rather than negative) return. Ultimately, such predictions marry in nicely with our ESS formulations as both risk and return facets of our evolutionary approach toward investing are effectively amalgamated under our Evolutionary Portfolio Theory (EPT) banner.

projections obtained under the auspices of our Evolutionary Asset Selection, along with the projected drawdown assessments compiled under our EVaR technique to provide a comprehensive list of each asset's projected risk and return characteristics (over a specified timeframe) across an entire asset universe. Importantly, our return assessments are formulated on the basis of past (epoch matched) experience where each asset's predicted "winning strategy" is forecast to emerge in dominance over a specified timescale – as conditioned by the expected pace of meme formation over one's chosen investment horizon. Furthermore, these return assessments are then weighted by our proprietary "strength of signal" assessment that takes into consideration the confidence attached to each risk/return forecast.

Using this "raw data" input Natural Selection™ is able to iteratively build a sequence of optimal time-variant portfolio weights (for a chosen number of assets) that takes into account both the risk and return attributes of a particular asset as well as the offsetting potential for overall portfolio drawdown that certain asset combinations have (in other words, their covariance structure). Importantly, Natural Selection™ does this by factoring in the fundamental byte to meme, meme to theme, theme to overall market sentiment evolutionary path of each "winning strategy" projection for the selected assets within the portfolio. So does the system work? We let the results speak for themselves in the following section.

7.4 Does it work? the results of ten years of out-of-sample backtesting for the investment recommendations from Natural Selection™

Without doubt, the proof of the pudding is in the eating so what happens when we put together the various mechanisms separately identified in the previous section to develop a truly evolutionary portfolio? Figures 7.5–7.9 and Tables 7.1 and 7.2 illustrate the performance of Natural Selection™ for the past ten years of out-of-sample backtesting in generating exceptional returns across our selected asset universe within a Global Macro Hedge Fund mandate. As can be seen, the results are quite impressive.[33]

---

[33] As always, the normal disclaimers apply in the sense that past performance is not always a reliable guide as to future performance. Further, these results apply to a specific asset universe over a specific epoch of time and should be taken as broadly indicative only. Note: all performance estimates are in NAV terms – net of 2 percent management fees and 20 percent performance fee above the Federal Reserve Funds Rate. Transaction costs and market impact ratios are included in these estimates. The portfolio is denominated in $US terms.

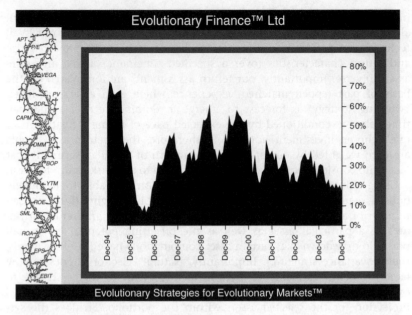

Figure 7.5   Rolling 12-month performance

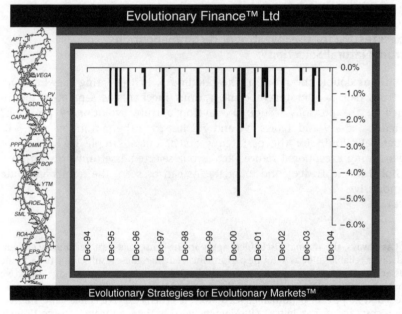

Figure 7.6   Drawdown distribution

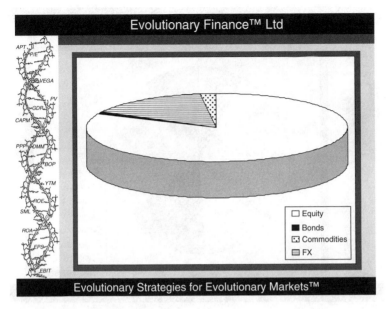

Figure 7.7   Asset class preference

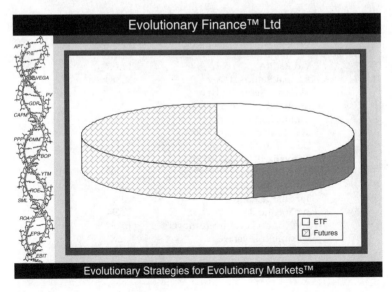

Figure 7.8   Instrument preference

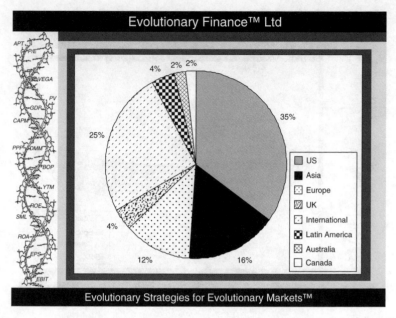

Figure 7.9 Geographic diversification

Table 7.1 Key performance indicator summary

Indicator	Performance
Ytd Return	17.05%
Annual Return St Dev	4.38%
Monthly Return St Dev	1.82%
Annual Return Sharpe Ratio	3.50
Annual Sortino Ratio	2.14
Max Drawdown Since 2000	−4.75%
Down Mths Since 2000	25.00%
Annualized 3yr Return	23.23%
3yr Annual St Dev	6.20%
3yr Monthly St Dev	2.75%
3yr Sharpe Ratio	3.47
3yr Sortino Ratio	3.94
Aug 12mth Portfolio Turnover	665.32%
Maximum Leverage	2X

Table 7.2 Monthly performance profile since 2000 (%)

Year	Jan	Feb	March	Apr	May	Jun	Jul	Aug	Sep	Oct	Nov	Dec	Total
2000	1.37	14.69	−1.94	5.29	4.08	1.84	−1.55	1.60	1.43	2.10	10.91	0.53	38.50
2001	6.97	−4.75	7.60	−2.27	−1.15	1.01	4.20	3.15	16.74	−2.63	8.38	2.57	37.36
2002	1.63	−1.12	−0.54	−1.14	0.88	5.76	1.06	−1.47	8.38	1.86	10.01	−1.64	22.63
2003	0.79	2.78	5.72	2.20	2.04	2.61	1.14	2.27	−0.47	5.38	0.38	5.79	30.01
2004	1.49	2.11	−1.60	−0.31	4.58	−1.28	2.55	0.43	1.77	2.85	2.12	2.65	17.05

Hopefully the above provides some insight as to the capabilities of our revolutionary new investment process and its most obvious physical embodiment – the Natural Selection™ software platform. As we mentioned earlier, even though the above results refer primarily to a Global Macro Hedge Fund mandate, this particular case study was chosen for illustrative purposes only. Indeed, Natural Selection™ could just as easily be used to identify investment opportunities consistent with long/short, statistical arbitrage, even event driven strategy absolute return mandates. We leave such examples as grist for the mill in future embodiments of our work – see Appendix 3 for details. Suffice to say here, we are extremely pleased with the results to date and the future looks bright for Evolutionary Finance both on a theoretical and applied basis.

8
Future Directions – The Path Ahead for Evolutionary Finance

Let us begin our conclusion to this book by way of summary of what we have presented. In Chapters 2 and 3 we outlined the foundations of both "traditional" and "new" views on the mechanics underlying information and asset price dynamics in financial markets. Without doubt, our own purview of financial market information forming together in a manner consistent with biological/evolutionary principles falls firmly under the "new" view umbrella.[1] Indeed, we directly incorporated some preexisting "new" view principles into our own model of information mechanics presented in Chapter 4. This, "Evolutionary Finance" perspective on markets began with the simple premise that investors interpret interdependent information using a process of encoding and categorization. This fundamentally all-too-human response toward informational structure can be readily described by a byte to meme, meme to theme, theme to overall market sentiment "information building" mechanism where the order of information arrival is deemed to be important. Still, in undertaking such behavior investors typically use heuristic rules of thumb (simple strategies) to assist with the encoding and categorization process. Unfortunately, as we have seen, such rules have a tendency to go wrong.

In Chapter 5, we heightened the degree of rigor by developing a comprehensive model of intertemporal information consumption/production. We used this model to establish the economic foundations for the publication

---

[1] Our appeal to biology as a means of explaining observed market phenomena should not come too much as a surprise. Mankind has repeatedly referred to nature for inspiration. For example, the Eiffel Tower's crisscross pattern replicates that of spongy bone in the femur, Alexander Bell's studies on the ear lead to his ideas for the microphone in the 1870s, Georges de Mestral – a Swiss engineer – developed Velcro in the late 1940s after observing how burrs stuck to his dog after a country walk. See Thompson (1942) classic *On Growth and Form* for a comprehensive overview of some of the mathematical inspirations one is able to draw from "solutions" to certain problems that nature has ascertained – thanks to the refining comb of natural selection.

of an infinite stream of information bytes by the analyst community. The establishment of such foundations were a necessary evil for contrasting our perspective vis-à-vis the long-standing traditionalist view that such information "manufacture" is of little economic value in a "fair-game" marketplace. By establishing an economic justification for a continuous stream of analyst produced information – the effective constituents of the market's information genome ($\sum \varphi_{\bar{i}}$) – we were then able to develop a series of appendages to our model that displayed how the format of information affected market price dynamics. This was done via a classic information binding process where bytes form memes, memes form themes and themes form even broader swings in market sentiment. Finally, we presented our preferred model in the final stages of Chapter 5 – a model that was able to be sufficiently generalized to display a plethora of interesting Evolutionary Finance principles simply through its ability to encode the infinite stream of analyst output into a market information "genotype" ($\sum \varphi_{\bar{i}}$) that, in turn, determined an array of interesting asset price "phenotypes." Such asset price time-variant phenotypes were displayed at the broad market index level using our innovative informational radix technique.

In Chapter 6, we again opted for a rigorous approach but this time to describe the actual microfoundation principles governing analyst choice as to the format of the information bytes they produce. Here Evolutionary Game Theory was the chosen medium of formalization and through a variety of generalized models we were able to describe how the relative payoff from pursuing alternate "information producing strategies" (which determine information byte format) will dictate the appearance of endemic "strategy herding" within the analyst community. Ultimately, this "strategy herding" behavior is a necessary evil for the appearance of theme formation in a classic diffusion-like pattern but one consequence of this inevitable analyst clustering is a marked skewing in the distributional form of asset prices. We were able to demonstrate the link between analyst "strategy herding" and the time-variant movements in asset prices under a variety of alternate strategy regimens. Further, we were able to illustrate how such a time-variant principle as our Evolutionary Distributional Form (EDF) will affect financial engineering's "best practice" in the fields of option pricing, Value-at-Risk, and mean-variance optimization.

Finally, in Chapter 7, we put all the theoretical principles developed in the preceding chapters to work in demonstrating how our revolutionary evolutionary perspective on information within financial markets can be used to garner exceptional returns across an array of absolute return strategy mandates. Using a Global Macro Hedge Fund investment style as our chosen forum for case study, we illustrated how our proprietary bioinformatics adapted Evolutionary Programming/Evolutionary Strategy software platform – Natural Selection™ – was able to discern a series of key memetic strings for "upstream" winning strategy prediction from a preconceived strategy set. Using our unique "adaptively predatory" approach we were then able to discern whether to "Buy" into a newly emergent "winning strategy" or adopt a contrarian stance and "Sell" into a mature "winning strategy" – which ultimately stood a good chance of an eventual wider investment market realization of a "strategy failure" over our specified investment horizon. Further, using our proprietary Evolutionary Value-at-Risk (EVaR) and Evolutionary Portfolio Theory (EPT) techniques developed in Chapter 6, we were able to place such exceptional return investment recommendations within the context of a drawdown risk adjusted evolutionary portfolio whose time-variant weightings reflected the expected pace of winning strategy theme diffusion for each of our selected assets. All up, a very comprehensive and sophisticated approach.

So where to from here? What is the path ahead for Evolutionary Finance? Without doubt, the future looks rosy for our revolutionary evolutionary perspective on information within financial markets. That said, to provide some brief insight as to how we expect our approach to evolve in the years to come, we have divided our thoughts along the lines of the theoretical and the applied.

8.1 Future directions for Evolutionary Finance – the theory

How is the theory of Evolutionary Finance likely to "evolve" going forward? One item of particular interest from a theoretical perspective briefly touched upon in the early stages of this book is the symbiotic relationship between information orientated institutions and the actual fabric and texture of information itself. Looked at through a certain lens, much of the institutional structure that constitutes what we today perceive to be the "market" is actually involved in some way with an informational role. Investment banks produce research, so too do fund managers, news organizations report this research, regulators monitor the dissemination of this research, and investor's interpret this research and adjust asset prices accordingly – but in doing so generate an informational feedback that eventually gets embodied in new research. And so goes the financial market information cycle.

The analysis presented in this book was focused purely on establishing a better understanding of the information cycle itself via examining information not as some textureless generic concept but rather as a diverse array of forms – forms that have an immediate analogous reference to the way information is encoded in the evolutionary/biological sphere. But there is another form of evolutionary feedback that we did not have the space to consider here but nevertheless is crucial to forming a better understanding of the information/market price dynamic. This evolutionary feedback is between information and the "market" itself. How do institutions form to enhance the evolutionary principles of information dispersal? Consistent with the work of Watts (1999) and Bonabeau *et al.* (1999) we know that complex networks are formed from relatively simple principles. So does the institutional structure of what we observe as the information orientated marketplace actually constitute such a "complex network" and if so, how does this institutional structure actually affect the fabric of information itself? Certainly both entities – the institutional structure of the market *and* information itself – consistent with the principles of North (1990), Mainzer (1997) and Olsen (1982) – are evolving through time. We would argue that – in the spirit of Ayres (1994) – both are evolving symbiotically. We simply cite this institutional/informational evolution aspect to our approach as an area of potentially fruitful future research under the Evolutionary Finance banner.

8.2 Future directions for Evolutionary Finance – the practice

But what of Evolutionary Finance from an applied sense? Certainly we – as we will outline in Appendix 3 – will be ardent supporters of the principle of Evolutionary Finance in pursuing exceptional absolute returns across a variety of investment mandates for our clients. But what of the actions of others? Well, in part in response to its proprietary nature and in part is response to the overall complexity of the approach, we feel the take-up of Evolutionary Finance principles by the wider investment community will be relatively slow. This is not such a bad thing as it provides us with more opportunities to "prey" upon those who are either slavishly following a static preconceived investment strategy or alternately are following very rudimentary heuristic investing rules of thumb that have neither the depth nor the breadth of our overarching investment philosophy.

Which ultimately brings us to another point. Some have argued that by writing a book about our proprietary investment process we are opening ourselves up to others mimicking certain aspects of our investment style. This, they argue, in true evolutionary fashion would progressively "bid the alpha" out of our unique investment philosophy. To these critics we have two responses. First and foremost is the fact that our investment process is patent protected at the broad business method level and is thus proprietary and defensible. Second, even if future aspirants were to attempt to mimic our approach, our underlying principles are adaptive enough so that if such moves had a marked impact upon market dynamics our "adaptively predatory" approach would take into account such behavior within our preconceived strategy set.

As a consequence, consistent with Keynes' beauty pageant analogy cited in Chapter 3, we would not only concentrate on "2nd Phase" investor response behavior but also (where appropriate) "3rd Phase" investor response behavior and the hunters would effectively become the hunted. In our particular field having first mover advantage is a key principle to success so given that we have already spent many man years refining and developing the key principles of our revolutionary new investment philosophy, the future looks very bright for Evolutionary Finance Ltd indeed.

Appendix 1

A Glossary of Investment Terms

Active Holdings: Percentage holdings of assets above or below a specified Benchmark. See Benchmark.

$$h_{\text{Active}} = h_{\text{Portfolio}} - h_{\text{Benchmark}},$$

where

$h_{\text{Portfolio}}$ = vector of total portfolio holdings; and
$h_{\text{Benchmark}}$ = vector of benchmark holdings.

Active Management: The pursuit of investment returns in excess of a specified Benchmark. See Benchmark.

Active Mode: The operation of pursuing portfolio construction (and performance) relative to a specified Benchmark. See Benchmark.

Active Return: Portfolio return in excess of a specified benchmark. Referred to as "Alpha." Generally forward-looking – *ex ante* – rather than historic – *ex post*. Historic Active Return is known as Residual Return. For example, if a portfolio's Total Return is 12 percent, and the specified Benchmark Return is 7 percent, then the portfolio's Active Return is 5 percent. See Benchmark Return.

$$R_{\text{Active}} = h_{\text{Active}}^{T}(R_{\text{Asset}}) = \alpha,$$

where

h_{Active} = vector of active holdings; and
R_{Asset} = individual asset return.

Active Risk: The risk (as measured by standard deviation) of the active return. Also known in the finance profession as "tracking error."

$$\sigma_{\text{Active}} = \sqrt{h_{\text{Active}}^T V h_{\text{Active}}},$$

where

h_{Active} = vector of active holdings; and
V = covariance matrix of portfolio of assets.

Alpha: The expected (or *"ex ante"*) residual return. Alpha is sometimes referred to as the "expected exceptional return." See Active Return, Residual Return.

American Option: An option which allows exercise prior to maturity. See European Option.

Analyst: Any individual who contributes to the stream of market information. Typically categorized on the basis of the type of strategy they pursue – technical, fundamental, economic and so on. Role is at times outsourced to professionals but it is important to remember that most investors are also "analysts" as they undertake some form of analysis as part of their due diligence before investing. See Strategy.

Anchoring: Facet of Behavioral Finance that concerns the power of "suggestion" in the sense that investors use heuristic "rules of thumb" to form judgments but at times such rules become "anchored" to a particular belief structure.

Arbitrage: The "risk-free" pursuit of profit in buying and selling comparable assets across markets where price disparities exist.

Arithmetic Return: The discrete return on an asset or a portfolio. Where possible, most return measurements are made using the less volatile Geometric Return method. See Geometric Return.

For an individual asset, the historic Arithmetic Return is calculated by

$$\text{HR}_{\text{Asset}} = \left[\frac{(R_{\text{Asset}})_t}{(R_{\text{Asset}})_{t-k}} \right] - 1,$$

where

R_{Asset} = asset return; t = present time; and k = starting time.

Asset: Any instrument considered in a portfolio that has an expected return and associated risk. A time series of returns is helpful for any asset to be considered as a potential candidate for inclusion within a portfolio as it allows the investor to form some judgment of *ex ante* return and risk. "Assets" can be as diverse as Real Estate, Vintage Wine, Art, Equities, Bonds, Options, Collectibles and Commodities.

Asset Return: Return for an individual asset. Can be expressed in *ex post* of *ex ante* sense. See Active return, Alpha, Benchmark Return, Excess Return, Exceptional Benchmark Return, Benchmark Return, Systematic Return, Residual Return, Total Return.

Asset Universe: The number of individual assets (n) considered for a portfolio. This may be entirely at the investor's discretion – Total Mode. Or, specified by a Benchmark – Active Mode. See Total Mode, Active Mode.

Attractor: The equilibrium state in a nonlinear dynamic system.

Autoregressive Process: A simple time-series process in which past values for a given variable are related to present value of this variable (also known as "AR").

Autoregressive Conditional Heteroskedasticity (ARCH): A nonlinear stochastic time-series process in which variance is time-variant and conditional on past variance.

Behavioral Finance: Behavioral Finance is all about the fact that investors make mistakes. They do not possess complete information, they are not equal in financial ability, and their wants and desires alter. In other words, it treats investors as though they are human.

Behavioral Strategy – Downside Risk Aversion: A cross asset class strategy that takes into account the Behavioral Finance observation that investors typically display an asymmetric response function to market volatility – they do not like large negative returns. To give some insight as to how extreme a particular measurement of downside risk actually is, it is useful to compare an observation in the present period vis-à-vis a long-run average. In the formula below we measure 12 month rolling downside risk and then compare this to the long-run average to generate our "buy" and "sell" signals. See Downside Risk.

$$
\text{Downside Risk} =
\begin{cases}
\text{Asset a "Sell" if } \left[\sqrt{\sum_{i=1}^{12} (R_i - \mu)^2} \Big/ n - 1 \right] > \text{Long-run Avg.,} \\
\text{Asset a "Buy" if } \left[\sqrt{\sum_{i=1}^{12} (R_i - \mu)^2} \Big/ n - 1 \right] < \text{Long-run Avg.,}
\end{cases}
$$

where, $R_i = 0$ if > 0.

Behavioral Strategy – P/E Momentum (Anchoring): An equity strategy that is based on the Behavioral Finance observation that investors can incorrectly "anchor" a valuation at extreme levels for short periods of time in moments of market excess. In the formula below we measure the "valuation creep" of a given asset by taking the rolling six month change in the P/E for the given (equity) asset. See Valuation Strategy – Simple P/E Relative.

$$P/E \text{ Momentum} = \begin{cases} \text{Asset d "Sell" if } \left(\sum_{i=0}^{-6} PE_i/6 \right) \Big/ \left(\sum_{i=-7}^{-12} PE_i/6 \right) < 0 \\ \text{Asset a "Buy" if } \left(\sum_{i=0}^{-6} PE_i/6 \right) \Big/ \left(\sum_{i=7}^{-12} PE_i/6 \right) > 0 \end{cases}.$$

Benchmark: A "yardstick" portfolio that an active manager's performance (in the sense of percentage return over a given time) is assessed against. Common examples are MSCI indices, S&P500 and FTSE indices. The ultimate goal of an active manager is to achieve "positive alpha" – a return in excess of their specified benchmark. See Active Return, Benchmark.

Benchmark Return: The aggregate of benchmark asset holdings multiplied by the expected return for each individual asset. This gives an estimation of the expected performance of the Benchmark portfolio. For the individual asset, Benchmark return is made up of the Risk Free Rate of Return, the Risk Premium, and the Market Timing Return. Can be expressed in either *ex ante* or *ex post* sense. See Asset Return, Benchmark, Risk Free Rate of Return, the Risk Premium, and the Market Timing Return.

$$R_{\text{Benchmark}} = h_{\text{Benchmark}}^T (R_{\text{Asset}}),$$

where

$h_{\text{Benchmark}}$ = vector of benchmark portfolio holdings; and
R_{Asset} = individual asset return.

Benchmark Risk: The risk (as measured by standard deviation) of the Benchmark Return. See Benchmark Return.

$$\sigma_{\text{Benchmark}} = \sqrt{h_{\text{Benchmark}}^T V h_{\text{Benchmark}}},$$

where

$h_{\text{Benchmark}}$ = vector of benchmark holdings; and
V = covariance matrix.

Beta: The volatility of a portfolio (or asset) relative to its benchmark. Beta is simply the ratio of the Covariance of an asset with its benchmark relative to the overall variance of the benchmark. A Beta of 1 implies commensurate volatility of the asset (or portfolio) with its specified benchmark. See Covariance, Variance.

For an individual asset,

$$\beta_{Asset} = \frac{\text{Cov }(R_{Asset}, R_{Benchmark})}{\text{Var}(R_{Benchmark})},$$

where

R_{Asset} = asset return; and
$R_{Benchmark}$ = benchmark return.
The portfolio beta is

$$\beta_{Portfolio(Asset)} = \frac{V h_{Benchmark}}{\sigma^2_{Benchmark}},$$

where

V = covariance matrix;
$h_{Benchmark}$ = vector of benchmark holdings; and
$\sigma^2_{Benchmark}$ = benchmark variance.

Bera–Jarque Test: Tests for the normality of a distribution of asset returns by simultaneously analyzing both third and fourth moments of the distribution. See Kurtosis, Skew.

$$\text{Bera–Jarque} = N \left[\frac{S^2}{6} + \frac{(K-3)^2}{24} \right],$$

where

N = number of observations;
S = skew of distribution; and
K = kurtosis of distribution.

Bifurcation: When a nonlinear dynamic system develops twice the solutions it had past some critical level. Also known as "period-doubling."

Binomial Method: Numerical time-step technique for option pricing. Relies upon simulated expansion over specified interval where the price of the underlying asset is assumed to move in a probabilistic determined fashion either upward or downward. See Trinomial Method.

Black–Litterman: An input conditioning mechanism developed by Fischer Black and Bob Litterman that ensures the linearization between an investor's *ex ante* return and the historical risk of assets within a portfolio. This ensures spurious Efficient Frontiers are not generated as a result of optimization. See Efficient Frontier.

$$\text{Total Mode: } BL(R_{\text{Asset}})_{(N\times1)} = \left[\Lambda^{-1}_{(N\times N)} + \left(\varphi_{(1\times1)} \, V_{(N\times N)} \right)^{-1} \right]^{-1}$$

$$\times \left[\Lambda^{-1}_{(N\times N)} R_{\text{Asset}}_{(N\times1)} + \left(\varphi_{(1\times1)} \, V_{(N\times N)} \right)^{-1} \text{HR}_{\text{Asset}}_{(N\times1)} \right],$$

$$\text{Active Mode: } BL(R_{\text{Asset}})_{(N\times1)} = \left[\Lambda^{-1}_{(N\times N)} + \left(\varphi_{(1\times1)} \, V_{(N\times N)} \right)^{-1} \right]^{-1}$$

$$\times \left[\Lambda^{-1}_{(N\times N)} \text{IMP}(R_{\text{Active}})_{(N\times1)} + \left(\varphi_{(1\times1)} \, V_{(N\times N)} \right)^{-1} \text{HR}_{\text{Benchmark}}_{(N\times1)} \right],$$

where

Λ = diagonal $(N \times N)$ matrix indicating investor uncertainty.

Note: Default set at the standard deviation of each asset.

$$\Lambda_{(N\times N)} = \begin{bmatrix} (SD_n)_{11} \ldots\ldots 0 \\ 0 \ldots (SD_n)_{22} \ldots\ldots 0 \\ \vdots \\ \vdots \\ 0 \ldots\ldots\ldots (SD_n)_{NN} \end{bmatrix}.$$

V = covariance matrix;
φ = scalar of confidence for expected returns $(0 < \varphi \leq 1)$;
R_{Asset} = individual asset expected return;
HR_{Asset} = historical asset return;
$\text{IMP}(R_{\text{Active}})$ = implied active return; and
$\text{HR}_{\text{Benchmark}}$ = historic benchmark return.

Black–Scholes: Theoretical method for option pricing.

Bootstrapping: Iterative interpolation technique common used for determining forward interest rates.

Bounded Rationality: Field of thought which recognizes that even a truly "rational" investor can only make decisions on the basis of what they already know – or in modeling jargon, the information-set available at time t. Hence, the "bounded" component to bounded rationality. Sometimes known as "quasi-rationality."

Brownian Motion: See Random Walk.

Capital Asset Pricing Model (CAPM): An equilibrium-based asset pricing model developed independently by Sharpe, Lintner and Mossin. The simplest version states that assets are priced according to their relationship to the Market Portfolio via individual asset Beta's. Assuming a number of stringent conditions such as rationality on behalf of investors, costless information (or costless arbitrage) and highly liquid markets, CAPM's most obvious (and perhaps most useful) foundation is the requirement for a direct relationship between risk and return. See Market Portfolio, Beta.

Chaotic Systems: Nonlinear dynamic systems that are primarily deterministic rather than stochastic in foundation, but display non-replicating/unstable behavior dynamically.

Cluster Analysis: Statistical technique based upon the idea of measuring distances, subject to a given metric, between points in a multidimensional space. Successive levels of aggregation allow for a "cluster hierarchy" to be built and represented graphically. Typical approaches include k-means clustering and EM clustering.

Complex Systems: Nonlinear dynamic systems that display self-organizing behavior and are primarily deterministic rather than stochastic in foundation.

Constraints: Percentage limitations under which a portfolio optimization is run. Constraints can be arbitrary in nature such as a investor's preference to hold a majority of their assets in a preferred stock, to something a little more rigorously imposed such as a fund manager's mandate not to have more than a certain proportion of a given portfolio in emerging markets.

Cornish–Fisher VaR: An adaptation of standard Value at Risk (VaR) to take into consideration skewness and kurtosis. See VaR.

$$\text{Cornish–Fischer VaR} = Z + \frac{1}{6}(Z^2 - 1)S + \frac{1}{24}(Z^3 - 3Z)K - \frac{1}{36}(2Z^3 - 5Z)S^2,$$

where

Z = VaR critical value;
S = skewness; and
K = kurtosis.

Correlation: The degree to which alternate assets influence each other as expressed by a numeric between -1 and $+1$. Correlation is the standardized measure of Covariance. See Covariance.

$$\rho_{XY} = \frac{\text{Cov}(R_{\text{Asset}(X)}, R_{\text{Asset}(Y)})}{\sigma_{\text{Asset}(X)}\sigma_{\text{Asset}(Y)}},$$

where

$\text{Cov}(R_{\text{Asset}(X)}, R_{\text{Asset}(Y)})$ = covariance between asset X and asset Y;
$\sigma_{\text{Asset}(X)}$ = standard deviation (risk) of asset X; and
$\sigma_{\text{Asset}(Y)}$ = standard deviation (risk) of asset Y.

Covariance: The degree to which alternate assets influence each other as measured by the co-movement of excess returns.

$$\text{Cov}(R_{\text{Asset}(X)}, R_{\text{Asset}(Y)}) = \sum_{\text{all } X} \sum_{\text{all } Y} (R_X - \mu_X)(R_Y - \mu_Y),$$

where

$R_{\text{Asset}(X)}$ = return on asset X;
$R_{\text{Asset}(Y)}$ = return on asset Y;
$\mu_{\text{Asset}(X)}$ = mean return on asset X; and
$\mu_{\text{Asset}(Y)}$ = mean return on asset Y.

Covariance Matrix: A ($n \times n$) matrix representation of the Covariance between all the assets in the specified Asset Universe. See Covariance.

Cumulative Return: Returns calculated as compound returns. See Historic Asset Return.

Currency Hedging: The attempt to optimally allocate the currency exposures associated with cross-border investment. Each foreign asset has an implied currency exposure – assuming an investor can only purchase a foreign asset in its domicile currency. As a consequence, at times it is worthwhile for an investor to use mean/variance optimization in an attempt to ascertain the optimal combination of the basket of associated currency exposures associated with their portfolio. This is done in much the same way as that for the optimal asset basket (portfolio). That is, an expected return for each currency exposure is formulated and then assessed against relative risk within the portfolio of currency exposures.

Currency Return: Currency Return is the difference between the projected and present spot rate for a given currency, taking into consideration the "carry" (short-term interest rate in the domicile country) of holding a particular currency position.

Calculation for Currency Return against $US:

$$R_{\text{Currency}} = (1 + \text{Cur}_{\text{Loc\$US}}(r)) * (1 + RF_{\text{Loc}}) - (1 + RF_{\text{Base}}),$$

where

$\text{Cur}_{\text{Local\$US}}(r)$ = raw exchange rate return;
RF_{Local} = risk-free return for local currency; and
RF_{Base} = risk-free return for chosen base currency.

Currency Risk: The Risk (standard deviation) attributable to currency exposure of cross-border assets.

$$\sigma_{\text{Currency}} = \sqrt{h_{\text{Currency}}^T V_{\text{Currency}} h_{\text{Currency}}},$$

where

h_{Currency} = vector of currency holdings; and
V_{Currency} = currency component of covariance matrix.

Deterministic System: Systems that are non-random in nature – their foundations are mathematical rather than statistical and, as a consequence, pre-ordained in advance.

Dividend Discount Model: A model for asset pricing based on discounting future expected dividends by an interest rate (or "discount factor"). Primarily applicable to the valuation of common stocks in the equity market but can also be applied to aggregate equity indices provided an appropriate "normalization" procedure is used.

Dividend Discount Return: The rate of return which equates the present value of future expected dividends with the current market price of a security.

Dividend Yield: The dividend per share divided by the price per share.

Downside Risk: Estimates the probability of negative return by only considering the negative return tail of the distribution of asset returns.

$$\text{Downside Risk} = \sqrt{\frac{\sum_{i=0}^{n} \min(r_i - \bar{r}, 0)^2}{n - 1}},$$

where

$\bar{r}$ = downside mean;
r_i = downside returns; and
n = number of observations.

Drawdown: A negative return on a selected strategy. See Maximum Drawdown.

Dynamic System: System of equations where the output of one equation is part of the input for another.

Earnings Yield: The earnings per share divided by the price per share. It the inverse of the P/E ratio. See P/E Ratio.

Efficient Frontier: A diagrammatical representation of the relationship between portfolio risk and return as expressed by a plot of the series of "efficient" portfolios in mean/variance space. "Efficient" portfolios are combinations of assets with the highest return for a given level of risk (or vice-versa – lowest risk for a given level of return). Such efficient portfolios are constructed using nonlinear constrained optimization techniques such as the simplex method, downhill simplex method, neural network, and primal dual interior solution algorithms. Efficient portfolios inevitably display a curve upwardly sloping in a northeasterly direction. (Note: risk is traditionally measured on the horizontal axis, return on the vertical axis of mean/variance space). The Efficient Frontier relationship identifies the basic premise of CAPM – higher risk portfolios require higher returns to compensate investors for additional volatility. The nonlinearity (curve) in the Efficient Frontier arises because rational investors invariably require ever-higher increments of return to compensate them for each additional increment of risk they are undertaking. This curvature relationship between risk and return is consistent with the nonlinear representation of the investor's utility function. See CAPM.

Efficient Market Hypothesis (EMH): Classification basis for the way information is distributed within financial markets. Three separate versions, strong, semi-strong and weak. Strong EMH states that asset prices reflect all information from public and private sources at each and every point in time. Weak EMH states that asset prices only reflect relevant asset market information. Semi-Strong EMH encapsulates Weak EMH but adds a time restraint in arguing that asset prices adjust *rapidly* to the release of relevant asset market information. A belief in Active Management implies a belief in either the Semi-Strong or Weak version of EMH.

Equilibrium: The stable-state of a system.

Equity Risk Premium: Excess return above bond yields argued to be demanded by investors for the higher risk associated with holding equities. Can be constructed on both prospective and historical basis. The prospective equity risk premium is measured as present 12 month earnings yield estimates minus the difference between the long-term earnings growth rate (generally assumed to be 3.0 percent) and the real risk free rate of return (for instance, ten year US Treasury Yields minus Annual Headline US CPI). The historical equity risk premium is measured as current 12 month trailing dividend yield minus the difference between the long-term growth rate (again, generally assumed to be 3.0 percent) and the real risk free rate of return (again, for instance, ten year US Treasury Yields minus Annual Headline US CPI). Note the secular decline in dividend payments has resulted in a considerable scalar departure between the two measures with prospective measures of the equity risk premium traditionally being considerably larger than the historical based measures.

European Option: an option that can only be exercised at maturity. See American Option.

Evolutionary Finance: The application of evolutionary/biological principles to the field of finance.

Evolutionary Games: Basically "repeated games" with the added condition that the actual rules of the game are dynamically changing – or *evolving* – through time.

Evolutionary Stable Strategy: The equilibrium state of an evolutionary game that occurs when no individual (Player I) playing one strategy could improve its reproductive fitness by switching to an alternate strategy and no "mutant" (Player II) playing a different strategy could establish itself ("invade") the existing population and so disrupt the equilibrium.

Excess Return: Return above the risk free rate of return. For example, assuming an asset's return is 9 percent and the Risk Free Return is 4 percent, then the asset's excess return is 5 percent. See Asset Return, Risk Free Return.

Factor Model: A computationally efficient way of deriving the Covariance Matrix. Large dimension Covariance Matrices require considerable degrees of freedom (number of observations) to form a robust estimate of true Covariance. At times, the short history associated with the time-series of some asset classes can restrict this requirement. A Factor Model is used to construct the Covariance Matrix in these instances. See Covariance Matrix.

Factor Model $[R_{Asset}]_{n \times 1} = [\phi]_{n \times (t-k)} [F]_{(t-k) \times 1} + [\varepsilon]_{n \times 1}$,

where

R_{Asset} = individual asset expected return;
ϕ = factor matrix;
F = factor variable vector;
ε = error vector;
n = number of assets in universe; and
$(t - k)$ = timescale.

Fair Game: Augment to EMH developed by Fama (1970) to gauge the level of market efficiency and its concurrent impact on price. Under the principles of a "fair game" Fama argued that, market participants of similar financial ability (provided they have equal access to information) cause the market price and the expected price of any asset to converge. In a sense, everyone comes to roughly the same conclusion as to the correct price of an asset when presented with the same information. See Efficient Market Hypothesis (EMH).

Finite Difference Method: numerical method for option pricing. See Monte Carlo techniques.

Fractal Analysis: Studies that typically look for interdependencies (correlation) between datasets in different timescales in attempt to identify the deterministic foundations that make up what we perceive to be overall market dynamics. See Scaling.

Fundamental Information: Information relating to the underlying state of an asset. In stock analysis, fundamental information is related to the cost structure and earnings prospects of a specific firm – in particular, as they relate to the valuation of that asset.

Game Theory: A mathematical representation of the strategic interaction between two or more "agents" – each attempting to maximize a specific utility function and each imparting information to the other through their actions. In many respects, Game Theory is like a mathematical formalization of the game of chess.

Gaussian Form: Also known as the "Normal Distribution." Distributional form that states returns are $N(\mu, \sigma^2)$. Product of the combination of Fair Game principles, IID information arrival and the Central Limit Theorem. Forms the distributional foundation of most traditionalist orientated financial engineering techniques.

Genetic Algorithms: Computer code structure that mimics Darwinian principles in obtaining solutions.

Geometric Return: The continuously compounded return of an asset or a portfolio. Return calculations using this method are less volatile than Arithmetic Returns. See Arithmetic Return.

For an individual asset, the historic Geometric Return is calculated by

$$HR_{Asset} = Ln(R_{Asset})_t - Ln(R_{Asset})_{t-k},$$

where

R_{Asset} = asset return;
t = present time; and
k = starting time.

Granger Causality: Econometric method used to determine existence of causal determination between dependent and independent variables.

Granger Causality is determined by the comparison of residual sum of squares between restricted (Autoregressive with inclusion of independent variables) and unrestricted models (Autoregressive with inclusion of just dependent variable)

$$F_{var,df} = \frac{RSS(Y_{t-1} \to Y)}{RSS(Y_{t-1}, X_{t-1} \cdots \to Y)},$$

where

Y = dependent variable;
X = independent variable(s);
F = critical value; and
RSS = residual sum of squares.

Historic Active Return: Return for the portfolio over a specified time period minus the Historic Benchmark Return. This is calculated in Geometric Average terms. See Active Return, Asset Return, Historic Benchmark Return.

$$HR_{Active} = Ln(R_{Active})_t - Ln(R_{Active})_{t-k},$$

where

R_{Active} = active return;
t = present time; and
k = starting time.

Historic Asset Return: Return for an Asset over a specified time period. This is calculated in Geometric Average terms. See Asset Return, Geometric Average.

$$HR_{Asset} = Ln(R_{Asset})_t - Ln(R_{Asset})_{t-k},$$

where

R_{Asset} = asset return;
t = present time; and
k = starting time.

Historic Benchmark Return: Return for a Benchmark portfolio over a specified time period. This is calculated in Geometric Average terms. See Benchmark Return.

$$HR_{Benchmark} = Ln(R_{Benchmark})_t - Ln(R_{Benchmark})_{t-k},$$

where

$R_{Benchmark}$ = benchmark return;
t = present time; and
k = starting time.

Historic Total Return: Return for an aggregate portfolio over a specified time period. This is calculated in Geometric Average terms. See Total Return, Geometric Average.

$$HR_{Portfolio} = Ln(R_{Portfolio})_t - Ln(R_{Portfolio})_{t-k},$$

where

$R_{Portfolio}$ = total return;
t = present time; and
k = starting time.

Hurst Exponent: Measure of the bias departure of a given distributional form away from the Gaussian distributional form. Interestingly, this measure arose out of an analysis of frequency of flooding (using an 847 year record no less) of the Nile river.

$$R/S = kT^H,$$

where

R/S = range/standard deviation;
T = index for number (time) of observation;
k = some constant for the time series; and
H = Hurst exponent.

For $H = 0.5$ the series is independent. For $0 \leq H < 0.5$ the series is antipersistent. For $0.5 < H \leq 1.0$ the series is persistent.

Implied Active Return: The Active Return (Alpha) for a specific asset implied from holding that asset within a specified (benchmarked) portfolio. Backed out of the linearity between risk and return implied under CAPM. Note: the Implied Active Return is benchmark-dependent since the benchmark determines the asset universe and therefore the dimensionality of the covariance matrix. See Active Return.

$$IMP(R_{Active}) = 2 * \lambda * (h_{Active}^T V),$$

where

λ = investor risk aversion;
h_{Active} = vector of active holdings; and
V = covariance matrix.

Derivation:
Start with the investor's utility function

$$U = \alpha - \lambda \sigma_{Active}^2, \qquad (A1.1)$$

and expressing this in matrix form

$$U = h_{Active}\alpha - \lambda(h_{Active} \times V \times h_{Active}^T). \qquad (A1.2)$$

We then maximize utility by

$$\text{Max} \quad U \Rightarrow \frac{\partial U}{\partial h^T} = 0.$$

This gives

$$\alpha = 2 * \lambda * (V * h_{\text{Active}}). \tag{A1.3}$$

Implied Risk Aversion: Also known as "lambda," Implied Risk aversion is the standardized measure of Risk Aversion implied under a global CAPM framework. Representing the generic Risk Aversion of the universal investor, this measure – λ – becomes useful for a variety of implied return calculations in portfolio theory. A common value for λ is 0.75. See CAPM, Implied Active Return, Implied Benchmark Return, and Sharpe Ratio.

Starting with the investor's utility function

$$U = \alpha - \lambda\sigma^2; \tag{A1.4}$$

then substituting the Sharpe ratio SR = α/σ using an *ex post* calculation of risk as an approximation of the *ex ante*, it is possible to now express the investor's utility function as

$$U = \text{SR} \times \sigma - \lambda\sigma^2. \tag{A1.5}$$

We then maximize utility by

$$\text{Max } U \Rightarrow \frac{\partial U}{\partial\sigma} = 0,$$

which gives

$$\text{SR} - 2 * \lambda * \sigma = 0. \tag{A1.6}$$

Rearranging gives

$$\lambda = \frac{\text{SR}}{2\sigma}. \tag{A1.7}$$

Using an equity risk premia of around 6 percent as a long-term approximation of α and a historical risk for a global portfolio of around 20 percent, it is possible to express the Sharpe Ratio as SR = α/σ = 6 percent/20 percent = 0.3. Substituting this result into Equation (4) above and continuing to use 20 percent as an approximation for σ gives

$$\lambda = \frac{\text{SR}}{2\sigma} = \frac{0.3}{2 * 20\%} = 0.75. \tag{A1.8}$$

Hence, our estimation of the level of Implied Risk Aversion under global CAPM is

$$\lambda = 0.75. \tag{A1.9}$$

Indexing: Practice of matching portfolio composition as close as possible to that of a specified Benchmark. See Passive Management.

Information Coefficient: The degree to which forecast returns correlate with actual outcomes. Provides a measure of an active portfolio manager's skill. Default in most implied value calculations traditionally set to 0.05.

Information Ratio: The ratio of annualized expected Residual Return to Residual Risk. Not to be confused with the Sharpe Ratio which refers to Active Return (alpha) and Active Risk rather than Residual Return and Residual Risk. The square of the information ratio is directly proportional to the value added by the active manager. See Active Return, Active Risk, Residual Return, Residual Risk.

Jensen's Alpha: Measures the extent to which the performance of a portfolio (or strategy) exceeds the expected return given the amount of risk undertaken by the strategy.

Jensen's Alpha $= r_p - E(r_p),$

where

r_p = portfolio return; and
$E(r_p)$ = expected portfolio return $= r_p + \beta_p(r_m - r_f)$

where

r_f = risk free return;
β_p = portfolio beta; and
r_m = market return.

Kurtosis: Measures the "fatness" of the tails of a distribution of asset returns. A high level of kurtosis implies "fat tails" and therefore a predominance of extreme returns (outliers). Applied to a normal distribution, the raw kurtosis below would equal 3 – hence the normalization subtraction of 3 from the formula below.

$$\text{Kurtosis} = \sum_i \frac{(r_i - \bar{r})^4}{N\sigma^4} - 3,$$

where

r_i = asset return;
$\bar{r}$ = mean asset return;
N = number of observations; and
σ = standard deviation.

Levy Distribution: An alternate distributional form that has been suggested by Complexity theorists as an alternative to the Gaussian form in attempt to better capture the short-term kurtotic behavior of financial markets but also their tendency to approximate to $N(\mu, \sigma)$ as $t \to \infty$. The characteristic "fat-tails" of such distributions arise in response to the presence of an inverse power law.

Lyapunov Exponent: Measures the dynamics of an attractor in various dimensions. See Attractor.

Magical Thinking: Behavioral Finance term which highlights that under conditions of uncertainty investors have a tendency to form associations between non-dependent outcomes. As if by "magic" a particular outcome may become associated with a market event and then it is possible that the two outcomes can become (erroneously) associated within investors heuristic rules – even though there is no economic/finance rationale for doing so.

Marginal Contribution to Return: The additional contribution to the Total Return for a portfolio from an individual asset. Simply a multiple of the asset's weight in the portfolio by its expected return.

$$R_{\text{Marginal Contribution}} = h_{\text{Asset}} \times E(R_{\text{Asset}}),$$

where
h_{Asset} = individual asset holding;
E = expectation set; and
R_{Asset} = returns for individual asset.

Marginal Contribution to Residual Risk: The additional contribution to the Residual Risk of a portfolio from an individual asset. See Residual Risk.

$$\sigma_{\text{Marginal Contribution(Residual)}} = \frac{VR - h_{\text{Portfolio}}}{\omega_{\text{Portfolio}}},$$

where

$h_{\text{Portfolio}}$ = vector of portfolio holdings;
VR = residual covariance matrix; and
$\omega_{\text{Portfolio}}$ = residual risk for portfolio.

Marginal Contribution to Risk: The additional contribution to Total Risk for a portfolio from an individual asset. See Total Risk.

$$\sigma_{\text{Marginal Contribution}} = \frac{Vh_{\text{Portfolio}}}{\sigma_{\text{Portfolio}}^{2}},$$

where

$h_{\text{Portfolio}}$ = vector of portfolio holdings;
V = covariance matrix; and
$\sigma_{\text{Portfolio}}^{2}$ = total risk.

Market: Nomenclature used to describe a group of associated assets. Term is generally used quite loosely to describe a collection of assets in similar class such as equities, bonds, commodities, FX, or futures and options but at times can encompass all assets. A prefix generally helps clarity such as "equity market" or even better, an additional geographic locale such as "US equity market" or trading board such as NASDAQ. The most accurate way to describe the "market" is via benchmark such as S&P500 and this is the definition of the "market" most commensurate with portfolio theory.

Market Events: Substantial volatility in the market – generally over a short duration of time. Traditionally measured as a daily percentage move in the return of an asset in excess of three standard deviations from the historical norm. See Standard Deviation.

Market Portfolio: The portfolio that is assumed to constitute the entire "market" for a given investor. Most commonly refers to a global asset class aggregation of various assets used for CAPM construction. See Market.

Maximum Drawdown: The maximum negative return on a selected strategy (usually applied to Hedge Funds) over a specified timeframe.

Mental Compartmentalization: Refers to the tendency for investors to compartmentalize assets into specific groups based upon superficial attributes.

Modern Portfolio Theory: Refers to mean/variance optimization of portfolios in attempt to ascertain the most "efficient" portfolio structure. See Efficient Frontier.

Moments: The dispersion characteristics that describe a distributional form. Most commonly known are mean (1st moment), variance (2nd moment), skew (3rd moment) and kurtosis (3rd moment). See Variance, Skewness, Kurtosis.

Monte Carlo Techniques: Numerical method for option pricing. Relies upon determining the present value of the payoff of an option via extensive iteration of the (randomly determined) price of the underlying asset. See Finite Difference Method.

Nash Equilibrium: The equilibrium endpoint of a game when the respective players in essence decide (generally, but not always, independently of each other) that they have achieved the optimum outcome. This occurs when no player would be advantaged by making an alternate decision based upon the given array of payoffs.

Neural Network: Computer code structure that mimics the parallel processing conducted in the human brain for obtaining solutions.

Normal Distribution: See Gaussian Distribution.

Ordinal Ranking: A methodology via which Alpha values can be constructed. Users "score" assets using a series of criteria such as fundamental and technical indicators. Typically, scores range between +2 and −2. Alpha values are then constructed by matching the aggregate score with Residual Risk. The process assumes that market efficiency results in a linearization between Residual Return (Alpha) and Residual Risk.

Starting with the investor's utility function

$$U = \alpha - \lambda\sigma^2, \tag{A1.10}$$

where σ_t^2 is the volatility of the investor's portfolio and α is the active portfolio return. Expressing this in matrix form yields

$$U = h_{\text{Active}}\alpha - \lambda\left(h_{\text{Active}} \times V \times h_{\text{Active}}^{\text{Trans}}\right). \tag{A1.11}$$

We then maximize utility by

$$\text{Max} \quad U \Rightarrow \frac{\partial U}{\partial h_{\text{Active}}^{\text{Trans}}} = 0,$$

which gives

$$\alpha = 2 * \lambda * (V * h_{\text{Active}}), \tag{A1.12}$$

where

V = covariance matrix;
h_{Active} = vector of portfolio holding's; and
λ = investor's risk aversion.

Using

$$V = VR + \beta\sigma_{\text{Benchmark}}^2\beta^{\text{Trans}};$$

VR = residual risk covariance matrix;
β = vector of asset beta's;
$\sigma_{\text{Benchmark}}^2$ = benchmark volatility;

we can by aid of substitution give

$$\alpha = 2 * \lambda * VR * \beta\sigma^2_{\text{Benchmark}} \beta^{\text{Trans}} * h_{\text{Active}}. \tag{A1.13}$$

Effectively, certain segments of the RHS of Equation (A1.13) can be broken down into three components: $2 * \lambda$ = measure of information coefficient for individual asset (its simply a scaled risk aversion parameter); UR = residual risk covariance matrix; and $\beta\sigma^2_{\text{Benchmark}}\beta^{\text{Trans}} * h_{\text{Active}}$ = a combination of systematic risk and portfolio holdings (or in other words, the *Score* for individual holdings assuming a linearization between risk and return).

As a consequence, it is possible to express the Alpha of an individual asset as

$$\alpha = \text{IC} * \omega^2_{\text{Asset}} * \text{Score}, \tag{A1.14}$$

where

IC = information coefficient;
ω^2_{Asset} = residual volatility of an asset; and
Score = ordinal ranking.

Overconfidence, Representative Heuristic and Over/Under-Reaction: Refers to the phenomenon that even though they are often confronted with their failure, investors as a class often have a tendency to express excessive confidence in their own judgments. From an Evolutionary Finance institutional perspective this may be a Darwinian attribute of the Market in that "only the confident (or foolhardy) survive."

Passive Management: Also known as Index Management. Structuring a portfolio so that its performance mimics that of a specified benchmark.

Payout Ratio: Ratio of dividends to earnings. Fraction of earnings paid out as dividends.

Persistence: Data series that follows an observable trend.

Phase Space: Chart that depicts all possible states of a dynamic system. Also known as a "Phase Portrait" or "Phase Diagram."

P/E Ratio: (Price/Earnings Ratio). Price of a stock divided by earnings per share (EPS).

Portfolio Return: See Total Return.

Prospect Theory: Field of thought that espouses that consumers specify their value function in terms of wealth rather than utility. See Behavioral Finance.

Random Walk: Also known as Brownian Motion. Argument where the expected change in the value of an asset is unrelated to its past and present changes. Supportive of the Strong EMH viewpoint of market behavior. Acts as a modeling framework for such advanced financial tools as the Black–Scholes option pricing model and various Factor models used in bond pricing. See Efficient Market Hypothesis.

Regret and Cognitive Dissonance: Type of investor behavior that represents a classic case of "living in denial." The psychology of "regret" implies that investors may be reticent to crystallize a loss due to the emotional pain of being wrong.

Residual Return: Asset Return independent of the Benchmark Return. *Ex post* version of Alpha. See Alpha, Asset Return, Excess Return.

Residual Risk: The standard deviation of the residual return.

For a portfolio

$$\omega_{Portfolio} = \sqrt{\sigma^2_{Portfolio} - \left(\beta^2_{Portfolio} \times \sigma^2_{Benchmark}\right)},$$

where

$\beta^2_{Portfolio}$ = squared beta of portfolio;
$\sigma^2_{Benchmark}$ = benchmark volatility; and
$\sigma^2_{Portfolio}$ = total risk.

Note that

$$\left(\beta^2_{Portfolio} \times \sigma^2_{Benchmark}\right) = \text{Systematic Volatility (Portfolio)} = \sigma^2_{SYS(Portfolio)}.$$

For individual asset

$$\omega_{Asset} = \sqrt{\sigma^2_{Asset} - \left(\beta^2_{Asset} \times \sigma^2_{Benchmark}\right)},$$

where

β^2_{Asset} = squared beta of asset;
$\sigma^2_{Benchmark}$ = benchmark volatility; and
σ^2_{Asset} = asset volatility.

Note

$$\left(\beta^2_{Asset} \times \sigma^2_{Benchmark}\right) = \text{Systematic Volatility (Asset)} = \sigma^2_{SYS(Asset)}.$$

Return Ratio: Historical version of Sharpe Ratio in matching return versus risk. See Sharpe Ratio.

Robust Beta: A proprietary method of Beta calculation developed by Mathsoft® that removes extreme values from Beta calculation so as to provide a more precise estimation of an asset's Beta. See Beta.

Risk: The standard deviation of Asset returns (sometimes measured on an annualized basis). Generally denoted with the Greek letter σ. See Volatility.

Risk Aversion: The amount of Risk an investor is willing to tolerate for a given rate of Return. This will vary for each individual depending upon the makeup of their utility function which includes such conditions as present wealth, age, dependents etc.

Risk Free Return: Return achievable with little to no volatility. Typically a short duration Treasury Bills is taken as the effective "risk free" return.

Risk Premium: Excess Return over a specified Benchmark. For equities, traditionally return on excess of Risk Free Return. See Benchmark, Excess Return, Equity Risk Premium, Risk Free Return.

Scaling: Process by which the change in a series is observed as the scale of the measuring device is altered – usually intertemporal correlation. This is typically done by constructing a k-dimensional vector $X(t) = [P(t), P(t+\tau), P(t+2\tau), \ldots, P(t+(k-1))\tau]$ and plotting its trajectory in k-dimensions as t varies. Indeed, one can further formalize this approach by examining Hurst exponents. A Hurst exponent $\neq 0.5$ implies intertemporal correlations and thus fractal dimensions. See Fractal Analysis.

Security Market Line: A foundation of CAPM, the Securities Market Line (SML) is the linear relationship between asset returns and their respective Betas (relative risk). See CAPM.

Sharpe Ratio: The ratio asset/portfolio *ex ante* Excess Return to *ex ante* Risk. Developed by Professor Bill Sharpe, hence the name. See Active Return, Excess Return, Risk.

$$\text{Sharpe Ratio} = \frac{E(r_i) - r_f}{\sigma_i},$$

where

$E(r_i)$ = expected asset (or portfolio) return;
r_f = risk free return; and
σ_i = standard deviation of asset (or portfolio) return.

Skewness: Measures the asymmetry of an asset's return distribution around its mean. "Positive skew" means the return distribution is disproportionately tailed toward positive returns (whereas the converse applies for "negative skew"). Applied to a normal distribution, skewness as measured by the formula below would be zero.

$$\text{Skew} = \sum_i \frac{(r_i - \bar{r})^3}{N\sigma^3},$$

where

r_i = asset return;
$\bar{r}$ = mean asset return;
N = number of observations; and
σ = standard deviation.

Sortino Ratio: Similar to Return Ratio but considers downside risk rather than absolute risk as the denominator to compare to the asset (or portfolio's) numerator. See Downside Risk, Return Ratio.

$$\text{Sortino Ratio} = \frac{r_i - r_f}{DR},$$

where

r_i = asset (or portfolio) return;
r_f = risk free return; and
DR = downside risk.

Specific Return: Portion of Excess Return uncorrelated with the returns of other assets (common factors). Also called the idiosyncratic return. See Excess Return.

Specific Risk: Risk (standard deviation) associated with a Specific Return. See Specific Return.

Standard Deviation: The square root of the volatility of asset returns. Generally denoted by the Greek letter σ. See Risk, Volatility.

$$\text{Standard Deviation} = \sigma = \sqrt{\frac{\sum_{i=1}^{n}(r_i - \bar{r})^2}{n-1}},$$

where

r_i = asset return;
$\bar{r}$ = mean asset return; and
n = number of observations.

Standard Error: Standard Deviation = Standard Error. See Standard Deviation.

Sterling Ratio: Measures return relative to drawdown – either for an asset strategy, portfolio strategy or an overall portfolio. See Sharpe Ratio, Sortino Ratio.

Stochastic Region of Risk: A representation of the total risk of the portfolio where the stochastic region is calculated dynamically in proportion to total risk.

$$\text{Stochastic Region of Risk} = c\sigma_{\text{Portfolio}},$$

where c is the dynamic proportional adjustment factor.

Stochastic System: Systems that are probabilistic in nature – their foundations are statistical rather than mathematical.

Strange Attractor: An attractor where points never repeat themselves or their orbits don't intersect. Usually depicted as a Phase Diagram. See Attractor.

Strategy: Mechanism via which an analyst's or investor's actions are governed. Ranges from the very simple heuristic rules of thumb as identified under the Behavioral Finance literature, to the considerably more complex Evolutionary Stable Strategies championed (but not exclusive to) the Evolutionary Finance literature. See Behavioral Finance, Evolutionary Finance, Evolutionary Stable Strategy.

Systematic Return: Portion of an asset's return that relates to (is correlated with) Benchmark Return. Excess Return is a combination of Systematic Return and Residual Return. See Asset Return, Benchmark Return, Excess Return, Residual Return.

Systematic Risk: Risk (as measured by Standard Deviation) of Systematic Return. See Benchmark Risk, Beta.

Technical Analysis: Body of thought that uses trends in (usually price) series and various filtering techniques to make projections of future movements.

Technical Strategy – Volume Rises/Falls: A cross asset class strategy that simply takes the three month average of number of rises relative to falls within a given index. Can be either a complementary (momentum driven) or contrarian strategy. In the example below, we illustrate the complementary case.

Volume Rises/Falls

$$= \begin{cases} \text{Asset a ``Sell'' if } (\text{Volume Rises/Volume Falls})_{3 \text{ monthly Avg.}} < 1 \\ \text{Asset a ``Buy'' if } (\text{Volume Rises/Volume Falls})_{3 \text{ monthly Avg.}} > 1 \end{cases}$$

Technical Strategy – Price Momentum: A very common cross asset class strategy that simply takes the recent performance of an asset as indicative of its future performance. In the formula below we use the three month average of price change for a given index as indicative of its "momentum."

$$\text{Price Momentum} = \begin{cases} \text{Asset a ``Sell'' if } \left( \sum_{i=0}^{-3} P_i / 3 \right) \bigg/ \left( \sum_{i=-4}^{-6} P_i / 3 \right) < 0 \\[2em] \text{Asset a ``Buy'' if } \left(\sum_{i=0}^{-3} P_i / 3 \right) \bigg/ \left(\sum_{i=-4}^{-6} P_i / 3 \right) > 0 \end{cases}$$

Technical Strategy – Skew: A cross asset class strategy that relies upon the premise of the long-term symmetry of asset returns and the potential for "mean-reversion" for asset classes displaying extreme skew. In the formula below we take the 12 month rolling skew estimate of monthly returns as indicative of an asset's "skew." See Skew.

$$\text{Skew} = \begin{cases} \text{Asset a "Sell" if (skew)}_{12 \text{ month}} < 0 \\ \text{Asset a "Buy" if (skew)}_{12 \text{ month}} > 0 \end{cases}$$

Technical Strategy – Seasonality: A cross asset class strategy that takes the average departure of a selected month's price growth from its trend as indicative of a seasonal pattern. There are a number of methods to calculate "seasonality" ranging from the very simple – a simple arithmetic mean of the average monthly return for a given month over a given timeframe – to the exceedingly complex (e.g., taking into consideration trend departure, differences in working/trading days for a given month etc). In the formula below we opt for a mid-way level of complexity by taking the average trend departure for the past 30 years of a given month – we measure "trend" as a three month trend.

$$\text{Seasonal Effect} = \begin{cases} \text{Asset a "Sell" if } \sum_{i=1}^{30} (P_i - Trend_{3 \text{ month}}) < 0 \\ \text{Asset a "Buy" if } \sum_{i=1}^{30} (P_i - Trend_{3 \text{ month}}) > 0 \end{cases}$$

Total Mode: The operation of pursuing portfolio construction (and performance) without reference to a specified Benchmark.

Total Return: Aggregate return for the portfolio. Is the sum of Active and Benchmark Return. Can be expressed in either *ex ante* or *ex post* sense. See Active return, Benchmark Return.

Total Risk: The risk (as measured by standard deviation) of the aggregate portfolio. Includes both Active Risk and Benchmark Risk but note risk is not additive (due to covariance) so Total Risk is not the sum of Active Risk and Benchmark Risk. Also note, when cross-border assets are held, Total Risk also includes the risk associated with the currency overlay (hedge) to the portfolio. See Active Risk, Benchmark Risk.

Tracking Error: Standard Deviation of a portfolio return away from its specified Benchmark – also known as Active Risk. See Active Risk.

Tracking Error Minimization: Technique that places the desire of the investor to minimize the volatility of the recommended portfolio relative to that of a specified Benchmark. Investors specify the degree of tolerance they are prepared to deviate away from the Benchmark via the tradeoff between their target Active Return (alpha) and their tolerance boundary – tracking error. See Active Risk, Active Return.

Traditional Finance Theory: Body of thought premised on the belief in Strong EMH, Gaussian Form and the Random Walk hypothesis.

Treynor Ratio: Resembles the Sharpe Ratio however uses Beta rather than standard deviation as a measure of risk in the denominator. See Sharpe Ratio.

$$\text{Treynor Ratio} = \frac{E\,(r_i) - r_f}{\beta_r},$$

where

$E(r_i)$ = expected asset (or portfolio) return;
r_f = risk free return; and
β_r = Beta of asset returns relative to some specified benchmark.

Trinomial Method: Numerical time-step technique for option pricing. Relies upon simulated expansion over specified interval where the price of the underlying asset is assumed to move in a probabilistic determined fashion either upward, sideways or downward. See Binomial Method.

Valuation Strategy – Simple *P/E* Relative: An equity strategy that compares present *P/E* relative to its long-run average. If present *P/E* exceeds its long-run average then equity asset is deemed "expensive" (with the converse applying if the present *P/E* is below its long-run average).

$$\text{Simple } P/E \text{ Relative} = \begin{cases} \text{Sell if } \left(\dfrac{P}{E}\right)_t > \left(\dfrac{P}{E}\right)_{\text{Long-run Avg.}} \\[2mm] \text{Buy if } \left(\dfrac{P}{E}\right)_t < \left(\dfrac{P}{E}\right)_{\text{Long-run Avg.}} \end{cases}$$

Valuation Strategy – Extreme *P/E* Relative: An equity strategy that compares present *P/E* relative to its *deviation adjusted* long-run average. If present *P/E* exceeds its deviation adjusted long-run average then equity asset is deemed "expensive" (with the converse applying if the present *P/E* is below its long-run average). See Valuation Strategy – Simple *P/E* Relative.

$$\text{Extreme } P/E \text{ Relative} = \begin{cases} \text{Sell if } \left[\left(\dfrac{P}{E}\right)_t > \left(\dfrac{P}{E}\right)_{\text{Long-run Avg.}} + 1(\text{stdev})\right] \\[2mm] \text{Buy if } \left[\left(\dfrac{P}{E}\right)_t < \left(\dfrac{P}{E}\right)_{\text{Long-run Avg.}} - 1(\text{stdev})\right] \end{cases}.$$

Valuation Strategy – Simple Earnings Yield Relative: An "equity versus bond" strategy that compares bond yields relative to equity yields. If present earnings yield on equities exceeds that of bonds then bonds are deemed "expensive" (with the converse applying if the yield on bonds exceeds that of equities). Note that since equities traditionally require a higher yield (in accordance with the equity risk premium) then it is necessary to compare earnings yield relativities in the present period versus their long-term average to take into account the variant risk premia across assets.

Simple Earnings Yield Relative =

$$\begin{cases} \text{Equities "Expensive" if } \left(\dfrac{\text{Bond Yield}}{\text{Earings Yield}} \right)_t > \left(\dfrac{\text{Bond Yield}}{\text{Earings Yield}} \right)_{\text{Long-run Avg.}} \\[4mm] \text{Equities "Cheap" if } \left(\dfrac{\text{Bond Yield}}{\text{Earings Yield}} \right)_t < \left(\dfrac{\text{Bond Yield}}{\text{Earings Yield}} \right)_{\text{Long-run Avg.}} \end{cases}$$

Valuation Strategy – Extreme Earnings Yield Relative: An "equity versus bond" strategy that compares bond yields relative to equity yields but now only records "buy" or "sell" signals for values in excess of ±1 standard deviations from the long-term mean. If present earnings yield on equities exceeds that of bonds by more than 1 standard deviation from the long-run average then bonds are deemed "expensive" (with the converse applying if the yield on bonds exceeds that of equities by more than one standard deviation from the long-run average). See Valuation Strategy – Simple Earnings Yield Relative.

$$\text{Simple Earnings Yield Relative} = \begin{cases} \text{Equities "Expensive" if } \left(\dfrac{\text{Bond Yield}}{\text{Earings Yield}} \right)_t \\[4mm] > \left[\left(\dfrac{\text{Bond Yield}}{\text{Earings Yield}} \right)_{\text{Long-run Avg.}} + 1\,(\text{stdev}) \right] \\[4mm] \text{Equities "Cheap" if } \left(\dfrac{\text{Bond Yield}}{\text{Earings Yield}} \right)_t \\[4mm] < \left[\left(\dfrac{\text{Bond Yield}}{\text{Earings Yield}} \right)_{\text{Long-run Avg.}} - 1\,(\text{stdev}) \right] \end{cases}$$

Valuation Strategy – 2-Stage DDM: An equity strategy that values an asset on the basis of the expected stream of its future dividends. The "1st stage" takes recent dividend growth (e.g., as observed by the three monthly rate of change in dividend yields) and extrapolates this forward for a finite time period (e.g., five years). The "2nd stage" takes a conservative long-term growth assumption (e.g., 1 percent dividend growth) and extrapolates this over an additional (long) horizon (e.g., 15 years). This future stream of dividends is then discounted to present value. To take into consideration the impact that structural shifts – such as alterations in corporate tax treatment – it is useful to compare each discounted dividend value at time t vis-à-vis a rolling average (e.g., three years).

$$2 \text{ Stage DDM} = \begin{cases} \text{Equities "Expensive" if } PV \text{ (Dividend Stream)}_t \\ \quad < PV \text{ (Dividend Stream)}_{3 \text{ year Avg.}} \\ \text{Equities "Cheap" if } PV \text{ (Dividend Stream)}_t \\ \quad > PV \text{ (Dividend Stream)}_{3 \text{ year Avg.}} \end{cases}$$

VaR: Value at Risk. Gaussian-based measure of risk based upon probability (assuming a normal distribution) of portfolio exhibiting a specified value.

$$\text{VaR} = n\sigma W dt^{0.5},$$

where

n = number of standard deviations;
σ = standard deviation of portfolio;
W = present value of portfolio; and
dt = year fraction.

Variance: The Volatility of an asset. See Volatility.

Volatility: The square of the standard deviation (Risk) of asset returns. Generally denoted by the square of the Greek letter σ^2. See Standard Deviation.

$$\text{Volatility} = \sigma^2 = \frac{\sum_{i=1}^{n} (r_i - \bar{r})^2}{n-1},$$

where

r_i = asset return;
$\bar{r}$ = mean asset return; and
n = number of observations.

Appendix 2
An OLG Form Evolutionary Model of the Marketplace

As stated in Chapter 5, when attempting to formalize a molecular-like building block process toward financial information using infinite horizon/ finite agent techniques we soon encounter a major problem. Why? Simply by doing so, we are essentially saying that the price of assets has the potential to depart from "fundamental" value – otherwise intuitively, there would be little market for the investment research output from the analyst community. Tirole (1982) illustrated that under such conditions – even though they may be realistic – the general equilibrium conditions for infinite horizon/finite agent models will be violated.[1]

We thus faced two alternatives:

1 build a comprehensive information production/consumption model for our evolutionary marketplace which has a finite number of infin- itely lived investing agents but *excludes* informationally imperfect asset prices explicitly from the general equilibrium conditions (rather, the general equilibrium conditions apply only to that segment of the economy responsible for the production and consumption of information and not to the factors underlying the speculative assets themselves); or

2 build an overlapping generations (OLG) model of finite lived investing agents but also include under the one all-encompassing general equi- librium "umbrella" the production and consumption of information/ financial assets/goods and services.

In Chapter 5 we chose to illustrate the former, in this Appendix we choose to illustrate the latter.

---

[1] For the mathematically minded, such a violation arises simply because the "departure from fair value" assumption implies a failure to satisfy the transversality conditions of the general equilibrium solution within this particular framework – see Chiang (1992) for an exposition of these.

A2.1   An introduction to our OLG framework

Our model's foundations follow closely that of Blanchard and Fischer's (1993) interpretation of the model presented by Diamond (1965) who was in fact, building upon earlier work by Samuelson (1958). We will omit certain obvious and tautological elements of this model so that we can focus purely on what is necessary to illustrate an intertemporal framework of information consumption and production – which also allows for the presence of financial "bubbles" within its all-encompassing framework. For detailed aspects of the model that are too extensive to be presented here, readers are referred to Blanchard and Fischer (1993).

To begin with, let us assume the marketplace is composed of just individuals and firms. Individuals are assumed to live for two periods – "young" and "old." A "young" individual born at time t consumes C_{1t} in period t and C_{2t+1} in period $t + 1$ – when they are "old." Their utility function can thus be specified

$$U(C_{1t}) + (1 - \text{Б})^{-1}U(C_{2t+1}), \quad \text{where } \text{Б} \geq 0, \ U'(.) > 0, \ U''(.) < 0. \quad \text{(A2.1)}$$

We assume individuals work only in the first period of their life, supplying inelastically one unit of labor and earning a real wage of w_t. They consume part of this income and save the rest for retirement consumption. Part of this consumption represents standard goods and services, part of it represents financial information.[2]

The saving of the young in period t generates the capital stock that is used to produce this output in period $t + 1$ (in combination with the labor supplied by the young generation at period $t + 1$). Also, it is important to note that embodied in the production function is a relative share of capital and labor that represents the financial analyst "community."

The number of individuals born at time t and working in period t is Z_t. The population grows at rate $\breve{y}$ so that $Z_t = Z_0(1 + \breve{y})^t$. It is assumed firms act under perfect competition conditions and face constant returns technology for their production function $Y = F(\acute{\text{K}}, Z)$.[3] Given the presence of perfect competition, each firm takes the wage rate (w_t) and the price of capital (r_t) as given.

---

[2] For a detailed exposition of the cost/benefit conditions in the purchase of information, refer to Hirshleifer and Riley (1992) Chapter 7.

[3] We also assume $F()$ is a *net* production function with depreciation already accounted for. Further, we assume $F()$ is strictly concave and satisfies what are known as the "Inada" conditions – $F(0) = 0, F'(0) = \infty$ and $F'(\infty) = 0$.

A2.2 Equilibrium for the consumer

An individual born at time t thus has a maximization problem

$$\text{Max} \quad U(C_{1t}) + (1 - Б)^{-1}U(C_{2t+1}), \tag{A2.2}$$

$$\text{Subject to} \quad C_{1t} + S_t = w_t,$$

and

$$C_{2t+1} = (1 + r_{t+1})\emptyset_t.$$

This implies a first order maximization condition of

$$U'(C_{1t}) - (1 - Б)^{-1}(1 + r_{t+1})U'(C_{2t+1}) = 0. \tag{A2.3}$$

Taking into consideration Equation (A2.1), this implies a savings function

$$\emptyset_t = \emptyset(w_t, r_{t+1}) \quad 0 < \emptyset_w < 1, \quad \emptyset_r <> 0, \tag{A2.4}$$

where
$\emptyset_w$ is the savings with respect to wages; and
$\emptyset_r$ is the savings with respect to interest rates.
    It is straightforward that an increase in wages results in increased savings.
That said, should interest rates increase there is both an incentive to:

1 consume more in the second period (and thereby save more in the first period) as it decreases the discount rate on future consumption; but also
2 to save less in both periods as a higher return on savings in both periods raises the overall incentive to consume more.

This is known as the "income" versus "substitution" effect on savings from a change in interest rates – hence the ambiguity in sign in the side condition to Equation (A2.4). Ultimately this ambiguity will be resolved by the elasticity of "between period consumption." An elasticity of substitution greater than one results in the substitution effect dominating the income effect and first period saving rising in response to an increase in interest rates. The converse applies for an elasticity of substitution less than one.

A2.3 Equilibrium for the producer

For equilibrium in the production side of the economy (which includes the production of financial information) quite simply the demand for goods must equate to the supply of goods (or equivalently, investment must equate to savings). This is represented as follows

$$\acute{K}_{t+1} - \acute{K}_t = Z_\emptyset(w_t, r_{t+1}) - \acute{K}_t, \tag{A2.5}$$

where $\acute{K}$ is the capital stock, so the LHS of Equation (A2.5) effectively represents net investment while the RHS represents net saving (the first term represents the saving by the young, the second term represents the dissaving by the old).

Eliminating $\acute{K}_t$ from both sides of Equation (A2.5) and dividing by Z_t gives the following equilibrium condition for producers

$$(1 + \breve{y})\acute{K}L_{t+1} = \emptyset(w, r_{t+1}), \tag{A2.6}$$

where $\acute{K}L_t$ is the firm's capital to labor ratio.

A2.4   Factor market equilibrium

Assuming firms act competitively and hire labor and employ capital in their attempts to meet the equilibrium condition espoused in Equation (A2.6), they do so up until a point where the marginal product of labor equates to the prevailing wage rate (w_t) and acquire capital up until a point where the marginal product of that capital equates to the interest rate

$$f(\acute{K}L_t) - \acute{K}L_t f'(\acute{K}L_t) = w_t \tag{A2.7}$$

and

$$f'(\acute{K}L_t) = r_t. \tag{A2.8}$$

Where, as stated earlier, we assume that labor is supplied inelastically and the capital available in period t is predetermined by the saving decision made by the "young" in period $t - 1$. Factor market equilibrium therefore arises when these predetermined labor supply and capital supply conditions are met by identical demand conditions by producers at each and every moment in time – which leads us nicely into the general equilibrium conditions of the model.

A2.5   General equilibrium

We now have set up a system of equations where general equilibrium revolves around the intertemporal decision of saving and investment by our incumbent population. The correct decision here will not only maximize consumption (which includes the maximization of consumption of financial information) for these individuals but also clears the labor market and the goods market. Combining our conditions for factor market equilibrium – Equations (A2.7) and (A2.8) – with the equilibrium condition decision of the producer – Equation (A2.6) – generates the following dynamic path for the capital stock

$$\acute{K}L_t = \frac{\emptyset[w(\acute{K}L_t), r(\acute{K}L_{t+1})]}{1+\breve{y}} = \frac{\emptyset[f(\acute{K}L_t) - \acute{K}L_t f'(\acute{K}L_t), f'(\acute{K}L_{t+1})]}{1+\breve{y}}. \tag{A2.9}$$

Taking the derivative of $\acute{K}L_t$ with respect to $\acute{K}L_{t+1}$ (which is the crucial decision that will ultimately determine equilibrium) gives

$$\frac{d\acute{K}L_{t+1}}{d\acute{K}L_t} = \frac{-\emptyset_w(\acute{K}L_t)\acute{K}L_t f''(\acute{K}L_t)}{1+\breve{y} - \emptyset_r(\acute{K}L_{t+1})f''(\acute{K}L_{t+1})}. \tag{A2.10}$$

But does our general equilibrium "solution" Equation (A2.10) necessarily represent a steady-state equilibrium? To determine this, we use Figure A2.1.

The 45° ray from the origin in Figure A2.1 effectively represents all steady-states at which $\acute{K}L_{t+1} = \acute{K}L_t$. Subsequently, for Equation (A2.10) to represent a steady-state equilibrium in our model of – among other things – intertemporal information consumption and production, the consumption decision of our finite-lived (young to old) population must be such that the intertemporal path of savings – as represented in Equation (A2.10) – will effectively cross this 45° line. By way of example, we illustrate two

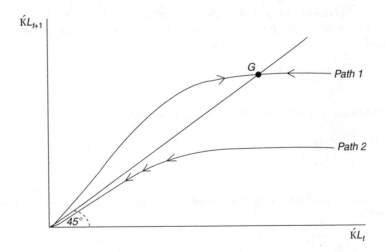

Figure A2.1   Steady-state equilibrium for intertemporal saving/investment

such paths in Figure A2.1 – Path 1 illustrates a steady-state equilibrium point G, Path 2 implies no steady-state capital stock exists for the associated values attached to the consumption/saving decision. In short, our model as presently specified does not *guarantee* the existence of a steady-state equilibrium, rather it only specifies the criteria necessary for a steady-state equilibrium to exist. Subsequently, it is the relative orders of magnitude attached to the consumption/savings decisions of our two-stage "young to old" finite population that ultimately determines whether a steady-state equilibrium is actually achieved.[4]

A2.6   Introducing money and prices

So far our model has relied upon a system of barter for exchange. What happens if we introduce money (and by default, prices) to facilitate intertemporal consumption smoothing? To illustrate this, we restate our consumers maximization condition

$$\text{Max}\quad U(C_{1t}, C_{2t+1}),\tag{A2.11}$$

but now add the ancillary conditions

$$P_t(1 - C_{1t}) = Þ_t^d,\tag{A2.12}$$

and

$$P_{t+1}C_{2t+1} = Þ_t^d,\tag{A2.13}$$

where $Þ_t^d$ is the individual's demand for money.

Subsequently, the first order maximization condition for solving our intertemporal consumption problem is now as follows

$$\frac{-U_1(C_{1t}, C_{2t+1})}{P_t} + \frac{U_2(C_{1t}, C_{2t+1})}{P_{t+1}} = 0.\tag{A2.14}$$

This implies a demand/supply function for money of

$$\frac{Þ_t^d}{P_t} = Æ\left(\frac{P_t}{P_{t+1}}\right),\tag{A2.15}$$

---

[4] Likewise, the stability of this equilibrium is only guaranteed when such decisions result in

$$0 < \frac{-ø_w \acute{K}L^* f''(\acute{K}L^*)}{1 + \check{y} - ø_r f''(\acute{K}L^*)} < 1$$

where $\acute{K}L^*$ represents the steady-state capital to labor ratio.

where the LHS of Equation (A2.15) represents the supply of money while the RHS of Equation (A2.15) represents money demand. In expanding upon the money demand determinants, Æ(.) represents the savings function (hopefully consistent with a steady-state equilibrium in the real economy as represented by point G in Figure A2.1) and P_t/P_{t+1} is the rate of return on money (which happens to be the inverse of the interest rate since by holding money you are no longer saving). Subsequently, we define the inverse of the interest rate (g) as the opportunity cost of holding money (or, from another perspective "the rate of return" on money)

$$g = \frac{1}{r_t} = \frac{P_t}{P_{t+1}}. \tag{A2.16}$$

To equate the monetary sector of our model we must therefore equate *intertemporal* money supply (which for convenience we will denote by Đ) with *intertemporal* money demand. Assuming the "young" buy money in accordance with Equation (A2.15), this implies an equilibrium condition of

$$(1 + \check{y})^t P_t^d = Đ. \tag{A2.17}$$

Thus combining our equilibrium money supply/money demand condition – Equation (A2.17) – with the relative price condition – Equation (A2.15) – and the opportunity cost of holding money under the real economy equilibrium – Equation (A2.16) – gives the following solution for the monetary sector of our model

$$(1 + g)^{-1}(1 + \check{y}) = \frac{Æ(1 + g_t)}{Æ(1 + g_{t+1})}. \tag{A2.18}$$

This implies the inverse of the interest rate $r_t(g)$ must equal the growth of the population, $\check{y}$. In short, prices must decrease at a rate such that real money supply grows at the same rate as money demand – which, in turn, is growing at the rate of population growth, $\check{y}$. It is only then that the monetary sector of our model of intertemporal information consumption and production is equilibrated.

As was the case with the barter-based system, these equilibrium conditions for money supply equating to money demand (as dictated by the relative price levels between the two periods within our model) do not necessarily imply an equilibrium in the real economy. It is only when there is an alignment between the savings function Æ(.) – representing a real economy equilibrium (as consistent with point G in Figure A2.1) – and interest rates being at a level such that real money demand equates to real money supply, that the full general equilibrium conditions of our monetized OLG model of intertemporal information consumption/production are satisfied. Which ultimately brings us to the issue of how expectations are actually formed. Does the informational component of intertemporal consumption/production model have any impact in forcing an equilibrating solution?

A2.7   Adding analyst driven expectations

Assuming, as stated earlier, information production represents a share of output within our model and that – at least as a starting point assumption – the general equilibrium conditions are initially satisfied, the next question to ask is what are the ramifications of this information production in either maintaining this equilibrium solution or alternately, forcing the system away from equilibrium?

To start with, let us simplify matters by using y_t to denote that set of conditions where general equilibrium under a monetized OLG model of intertemporal information consumption/production are initially met. This includes intertemporal saving at a level consistent with equilibrium in the real economy by clearing factor markets and which equates real money supply to real money demand. Subsequently, we can conveniently summarize the expectational difference equation as

$$y_t = a\text{Ex}(y_{t+1}|t) + c(\text{er})_t, \tag{A2.19}$$

where $\text{Ex}(y_{t+1}|t)$ denotes the expectation of y_{t+1} held at time t and $(\text{er})_t$ represents an "error" term in the sense that it can dictate whether prices depart from their fundamental value.

It is in the formation of these expectations about this "error" term that the analysts role is crucial. To formalize this, we state expectations are conditioned upon the information-set provided by analysts

$$\text{Ex}(y_{t+1}|t) = \text{Ex}(y_{t+1}|\ddot{\text{I}}_t), \tag{A2.20}$$

where

$$\ddot{\text{I}}_t = \{y_{t-i}, \text{er}_{t-i}, {}'\text{H}_{t-i}, i = 0, \dots, \infty\}.$$

Conveniently, as was highlighted with our interdependence between information assumption outlined in Chapter 4, the information-set contains current and lagged variables of both y and (er) (past and present expectations and their associated error terms). Also, importantly, the vector $'\text{H}$ is identified in Equation (A2.20). This represents analyst advice regarding both y and (er).

By endogenizing analyst advice into our equilibrium conditions for a monetized OLG model of an economy containing explicit information consumption and production as a fraction of its output, we can now represent the money supply/money demand function previously represented in

Equation (A2.15) as

$$\frac{\mathrm{P}_t}{P_t} = \mathbb{E}\left[\frac{\mathrm{Ex}(P_{t+1}|\ddot{\mathrm{I}}_t) - P_t}{P_t}\right], \tag{A2.21}$$

where the LHS of Equation (A2.21) represents the real money supply that is supplied inelastically by the old, while the RHS of Equation (A2.21) represents real money demand by the young – taking into account analyst supplied information acting as either an equilibrating force or alternately, driving our system of equations away from equilibrium. Taking logs (as denoted by the use of lowercase letters) yields

$$\mathrm{P}_t - p_t = -[\mathrm{Ex}(p_{t+1}|\ddot{\mathrm{I}}_t) - p_t]. \tag{A2.22}$$

Reorganizing gives

$$p_t = a\mathrm{Ex}(p_{t+1}|\ddot{\mathrm{I}}_t) + (1-a)\mathrm{P}_t, \tag{A2.23}$$

where

$$a \equiv \frac{}{1+}.$$

Note the similarities between Equations (A2.23) and (A2.19).[5] For our model, it is the vector of analyst advice ('H) with regard to prices – including asset prices – that will determine the intertemporal money supply conditions within our system of equations and through this either equilibrate the model or force a departure of prices from their fundamental value. Indeed, it is the characteristics of the savings function Æ(.) that determine both and a – which happen to be derivative outcomes of this "analyst advice as to future prices" function.

Analyst advice forcing a general equilibrium solution

So what are the conditions under which expectations are met and general equilibrium obtained? If one assumes rational expectations, then the solution procedure is relatively simple to ascertain for our expectation drive monetized OLG model of intertemporal information consumption/production. Using the process of iterated expectations, the solution for Equation (A2.19) under rational expectations is

$$y_t = c\sum_{i=0}^{\infty} \mathrm{Ex}(er_{i+1}|\ddot{\mathrm{I}}_t), \tag{A2.24}$$

---

[5] Indeed, both equations bear a striking resemblance to the standard "evolution of beliefs" framework as mapped out by Chamley (2004).

which satisfies

$$\lim_{T \to \infty} a^{T+1} \mathrm{Ex}(y_{t+T+1}|\ddot{\mathrm{I}}_t) = 0. \tag{A2.25}$$

Mapping this to Equation (A2.23) implies the price level depends upon the *sequence* of expected future money stocks – which since the error term converges to zero, implies prices are set at their "fundamental" value. Thus, as stated earlier, prices act as an intertemporal equilibrating force on our system of equations through the intertemporal money supply conditions. Assuming condition (A2.25) is met, general equilibrium is indeed achieved and our system of equations is equilibrated.

Analyst advice promoting the appearance of speculative bubbles

But what happens when analyst projections as to the future *fundamental-driven* direction of prices is less than perfect? This is where the infamous case of speculative "bubbles" arise. In particular, the presence of speculative bubbles is determined by the relativity of g with respect to $\breve{y}$, and it is in the determining of g in particular that analysts play a role.

Using y_t^* to denote the "rational expectations" fundamental solution given by Equation (A2.24) and $Ł_t$ to denote the bubble effect, we can express various forms of analyst driven bubble solutions as

$$y_t = y_t^* + Ł_t. \tag{A2.26}$$

For example, the case of an ever-expanding bubble

$$Ł_t = Ł_0 a^{-t} \quad \text{for arbitrary } Ł_0, \tag{A2.27}$$

implies prices – including financial asset prices – are expanding exponentially through time. However, since individuals are *expecting* this outcome (thanks to the presence of analyst research) the model is still in "equilibrium" – it's just that asset prices are accelerating ad-infinitum thanks to analysts continuously being too bullish. Under such circumstances, information consumers readily believe the analysts and set their savings levels accordingly (and by default, this affects price levels).

However, more realistically, we can also introduce the prospect of bursting bubbles – with a given probability each period – by stating

$$Ł_{t+1} = (a\mathrm{Pr})^{-1}Ł_t + \mathrm{er}_{t+1} \quad \text{with probability } Pr \tag{A2.28}$$

and

$$Ł_{t+1} = \mathrm{er}_{t+1} \quad \text{with probability } 1 - Pr,$$

where $\mathrm{Ex}(\mathrm{er}_{t+1}|\ddot{\mathrm{I}}_t) = 0$.

Again, the long term "zero error" arbitrage conditions will ultimately be satisfied and our model is in a general form of intertemporal "equilibrium" as analyst behavior not only drives the information-set that ensures the bubble/bust result in asset prices but also by default equilibrates the system as consumers are effectively informed of such probabilities and these outcomes are built into their intertemporal decision-making. This illustrates the self-fulfilling nature of the analyst provided price information advice in an environ where decision-makers are ultimately informationally constrained.

Without doubt, the number of various permutations and combinations of alternatives as to analyst behavior is near endless.[6] We could for example explore the byte to meme, meme to theme, theme to overall market sentiment informational building consequences of analyst research output in similar fashion as we have done in Chapter 5. But that is not the purpose of the model presented here. Rather, what we have done in the model presented in this Appendix is to provide the reader with a framework that actually endogenizes the provision of financial information within an intertemporal model where *all* goods and services (along with financial "assets" in the form of money) are considered. Future extensions to this model could consider varying types of analyst behavior (and their associated calibrations) as we detailed extensively in our alternate framework presented in Chapter 5. An explicit stock market sector – along with bonds (to facilitate intertemporal consumption smoothing) could also be appended to the model. As mentioned earlier, the scope for expansion is near endless. Still, we hope we have provided the reader with some insight within this Appendix that an intertemporal model of information consumption and production can be developed where analysts do on occasion get it both "right" and "wrong" and there are economic (and associated financial market) consequences stemming from both outcomes. We leave the extensions to this skeletal framework to the labors of future researchers and wish them well in their endeavors.

---

[6] For an outline of some of these permutations, refer to Azariadis (1993).

Appendix 3
Some Background on Evolutionary Finance™ Ltd

Like any good scientist who feels the only way to prove the efficacy of their new vaccine is to inject themselves with their own formula (and to thereby expose themselves to the worst of all possible circumstances), we ourselves believe that the only way to prove the efficacy of our revolutionary/evolutionary perspective as to the informational "building blocks" of asset prices is to put our exciting new thought paradigm into action in the *applied* finance sphere. To do this, we established our company – Evolutionary Finance Ltd – so that a wider array of investors can benefit from our unique perspective upon markets.

Founded at Cambridge in the UK, Evolutionary Finance Ltd is a specialist absolute return manager and boutique research firm specializing in the implementation of our revolutionary Evolutionary Finance principles across an array of absolute return mandates. Our investment philosophy is simple – markets are in a constant state of evolution as market participants respond to the unrelenting flow of information that besieges the market each and every day. Using our unique "adaptively predatory" investment style, Evolutionary Finance Ltd develops an array of sophisticated evolutionary investment strategies that are designed specifically to capitalize upon our structured approach toward information and by doing so, capitalize upon the fallibilities of others.

Using our patented software platform – Natural Selection™ – Evolutionary Finance Ltd screens the market at the close of each business day looking for telltale "memetic footprints" as to winning strategy theme formation. Our unique structured information investment approach can then tell us whether a particular investing thematic is in its early stages of formation (our potential "longs") or nearing full maturity – and is thus vulnerable to thematic "strategy failure" (our potential "shorts"). We then position ourselves accordingly to benefit from the "greater herd's" investment response.

Given the fact that our predatory strategy investment triggers are calibrated at the close of each business day, we have subsequently found it convenient to establish our main office in the Pacific time-zone – so as to benefit from the main European and North American market closing. Doing so enables us to better manage risk when markets are at a heightened state of turmoil. For further information on how to contact us and for our latest investment performance results please visit our website at: www.Evolutionary-Finance.co.uk.

References

Aharony, J. and Swary, I. (1980) "Quarterly Dividend and Earnings Announcements and Stockholders Returns: An Empirical Analysis" *Journal of Finance*, 35, 1, (March), 1–12.

Aitchison, J. (2000) *The Seeds of Speech: Language Origin an Evolution*, Cambridge University Press, Cambridge.

Alexander, C. (2001) *Market Models: A Guide to Financial Data Analysis*, John Wiley & Sons, Chichester.

Allouche, J.P. and Shallit, J. (2003) *Automatic Sequences: Theory, Applications, Generalizations*, Cambridge University Press, Cambridge.

Aoki, M. (2000) "Cluster Size Distributions of Economic Agents of Many Types in a Market" *Journal of Mathematical Applied Analysis*, 249, 32–52.

Aoki, M. (2002) *Modeling Aggregate Behavior and Fluctuations in Economics: Stochastic Views of Interacting Agents*, Cambridge University Press, Cambridge.

Arthur, W.B. (1989) "Competing Technologies Increasing Returns and Lock-in by Historical Events" *The Economic Journal*, 45, 538–52.

Arthur, W.B. (1990) "A Learning Algorithm that Mimics Human Learning" *Santa Fe Institute Working Paper* 90-026.

Arthur, W.B. (1994) *Increasing Returns and Path Dependence in the Economy*, University of Michigan Press, Ann Arbor, MI.

Arthur, W.B., Holland, J.H., LeBaron, B., Palmer, R. and Taylor, P. (1997) "Asset Pricing Under Endogenous Expectations in an Artificial Stock Market" in Arthur, W.B., Durlauf, S.N. and Lane, D.A. (eds) *The Economy as an Evolving Complex System II: Proceedings Volume XXVII Santa Fe Institute Studies in the Sciences of Complexity*, Perseus Books, Reading, MA.

Ariel, R.A. (1987) "A Monthly Effect in Stock Returns" *Journal of Financial Economics*, 18, 1 (March), 161–74.

Asquith, P., Mikhail, M.B. and Au, A.S. (2002) "Information Content of Equity Analyst Reports" *NBER Working Paper No. 9246*, National Bureau of Economic Research, Cambridge, MA.

Attenborough, D. (1980) *Life on Earth*, Book Club Associates, London.

Aumann, R.J. (1959) "Acceptable Points in General Cooperative N-Person Games" in Tucker, A.W. and Luce, R.D. (eds) *Contributions to the Theory of Games, Volume IV Annals of Mathematics Studies*, 40, pp. 287–324, Princeton University Press, Princeton, NJ.

Aumann, R.J. and Maschler, M. (1966) "Game – Theoretic Aspects of Gradual Disarmament" Chapter V in *Report to the U.S. Arms Control and Disarmament Agency ST-80*, Princeton, NJ.

Aumann, R.J. and Maschler, M. (1985) "Game Theoretic Analysis of a Bankruptcy Problem from the Talmud" *Journal of Economic Theory*, 36, 195–213.

Aumann, R.J. and Peleg, B. (1960) "Von Neumann-Morgenstern Solutions to Cooperative Games without Side Payments" *Bulletin of the American Mathematical Society*, 66, 173–79.

Auyang, S.Y. (1998) *Foundations for Complex System Theories in Economics, Evolutionary Biology and Statistical Physics*, Cambridge University Press, Cambridge.

Avellaneda, M., Levy, A. and Parás, A. (1995) "Pricing and Hedging Derivative Securities in Markets with Uncertain Volatilities" *Applied Mathematical Finance*, 2, 73–88.

Axelrod, R. and Hamilton, W.D. (1981) "The Evolution of Cooperation" *Science*, 211 (March), 1390–96.

Ayres, R.V. (1994) *Information, Entropy and Progress: A New Evolutionary Paradigm*, American Institute of Physics Press, New York.

Azariadis, C. (1993) *Intertemporal Macroeconomics*, Blackwell, Oxford.

Bak, P. (1997) *How Nature Works: The Science of Self-Organized Criticality*, Oxford University Press, Oxford.

Bak, P. and Chen, K. (1991) "Self Organized Criticality" *Scientific American*, II (3), January.

Ball, R., Kothari, S.P. and Wasley, C. (1995) "Can We Implement Research on Stock Trading Rules?" *Journal of Portfolio Management*, 21, 2 (Winter), 54–63.

Balvers, R.J., Cosimano, T.F. and McDonald, B. (1990) "Predicting Stock Returns in an Efficient Market" *Journal of Finance*, 45, 4 (September), 1109–28.

Banerjee, A.V. (1992) "A Simple Model of Herd Behavior" *Quarterly Journal of Economics*, 107, 3, 797–817.

Banerjee, A.V. (1993) "The Economics of Rumors" *The Review of Economic Studies*, 60, 2, 309–27.

Banz, R.W. (1981) "The Relationship Between Return and Market Value of Common Stocks" *Journal of Financial Economics*, 9, 1 (March), 3–18.

Barber, B., Lehavy, R., McNichols, M. and Trueman, B. (2001) "Can Investors Profit from the Prophets? Security Analyst Recommendations and Stock Returns" *Journal of Finance*, 56, 2, 531–63.

Barberis, N., and Shleifer, A. (2000) "Style Investing" *Journal of Behavioral Finance*, 16 (November), 103–41.

Barberis, N., Shleifer, A. and Vishny, R. (1997) "A Model of Investor Sentiment" *Journal of Behavioral Finance*, 1, 4(April), 213–24.

Barkow, J., Cosmides, L. and Tooby, J. (1992) *The Adapted Mind: Evolutionary Psychology and the Generation of Culture*, Oxford University Press, New York.

Basu, S. (1983) "The Relationship Between Earnings Yield, Market Value and Return for NYSE Common Stocks" *Journal of Financial Economics*, 12, 1 (June), 129–56.

Beale, R. and Jackson, T. (1990) *Neural Computing: An Introduction*, The Institute of Physics, London.

Beltrami, E. (1999) *What is Random? Chance and Order in Mathematics and Life*, Springer-Verlag, New York.

Beltratti, A., Margarita, S. and Terna, P. (1996) *Neural Networks for Economic and Financial Modelling*, International Thomson Computing Press, London.

Benink, H. and Bossaerts, P. (2001) "An Exploration in Neo-Austrian Theory Applied to Financial Markets" *Journal of Finance*, 56, 3, 1011–27.

Benjamin, S.C., Johnson, N.F. and Hui, P.M. (1996) "Cellular Automata Models of Traffic Flow along a Highway Containing a Junction" *Journal of Physics*, 29, 1053–76.

Berges, A., McConnell, J.J. and Schlarbaum, G.G. (1984) "The Turn of the Year in Canada" *Journal of Finance*, 39, 1 (March), 185–92.

Bernheim, B.D. (1994) "A Theory of Conformity" *The Journal of Political Economy*, 102, 5, 841–77.

Bikhchandani, S. and Sharma, S. (2001) "Herd Behavior in Financial Markets" *IMF Staff Papers*, 47, 3, 279–310.

Black, F. (1973) "Yes, Virginia. There is Hope: Tests of the Value Line Ranking System" *Financial Analysts Journal*, 29, 5, 10–14.

Black, F. (1986) "Noise" *Journal of Finance*, 41 (July), 31–52.

Black, F. and Scholes, M. (1973) "The Pricing of Options and Corporate Liabilities" *Journal of Political Economy*, 81, 637–59.

Blackmore, S. (1999) *The Meme Machine*, Oxford University Press, Oxford.

Blanchard, O.J. and Fischer, S. (1993) *Lectures on Macroeconomics*, MIT, Cambridge, MA.

Bonabeau, E., Dorigo, M. and Theraulaz, G. (1999) *Swarm Intelligence: From Natural to Artificial Systems*, Sante Fe Institute Studies in the Sciences of Complexity, New York.

Bouchaud, J.P. and Cont, R. (1998) "A Langevin Approach to Stock Market Fluctuations and Crashes" *The European Physical Journal*, 6, 543–50.

Bouchaud, J.P. and Potters, M. (2003) *Theory of Financial Risk and Derivative Pricing: From Statistical Physics to Risk Management*, Cambridge University Press, Cambridge.

Bouchaud, J.P. and Sornette, D. (1994) "The Black-Scholes Option Pricing Problem in Mathematical Finance: Generalization and Extensions for a Large Class of Stochastic Processes" *Journal of Physics*, 4, 863–81.

Bouchaud, J.P., Iori, G. and Sornette, D. (1996) "Real World Options: Smile and Residual Risk" *Risk*, 9 (March), 61–5.

Boyle, P.P. and Emanuel (1980) "Discretely Adjusted Option Hedges" *Journal of Financial Economics*, 8, 241–50.

Branch, B. (1977) "A Tax Loss Trading Rule" *Journal of Business*, 50, 2 (April), 198–207.

Brock, W.A. (1997) "Time and Money" in Arthur, W.B., Durlauf, S.N. and Lane, D.A. (eds) *The Economy as an Evolving Complex System II: Proceedings Volume XXVII Santa Fe Institute Studies in the Sciences of Complexity*, Perseus Books, Reading, MA.

Brock, W.A. and Hommes, C.H. (1997a) "A Rational Route to Randomness" *Econometrica*, 65, 1059–95.

Brock, W.A. and Hommes, C.H. (1997b) "Models of Complexity in Economics and Finance" in Hey C. *et al.* (eds) *System Dynamics in Economic and Financial Models*, John Wiley & Sons, London.

Brock, W.A. and Hommes, C.H. (1998) "Heterogenous Beliefs and Routes to Chaos in a Simple Asset Pricing Model" *Journal of Economic Dynamics and Control*, 22, 1235–74.

Brock, W.A. and Hommes, C.H. (1999) "Rational Animal Spirits" in Herings, P.J.J., van der Lan, G. and Talman, A.J.J. (eds) *The Theory of Markets*, North-Holland, Amsterdam.

Brock, W.A., Lakonishok, J. and LeBaron, B. (1992) "Simple Technical Trading Rules and the Stochastic Properties of Stock Returns" *Journal of Finance*, 47, 1731–64.

Brown, P., Kleidon, A.W. and Marsh, T.A. (1983) "New Evidence on the Nature of Size Related Anomalies in Stock Prices" *Journal of Financial Economics*, 12, 1 (June), 33–56.

Brown, P., Keim, D.B., Kleidon, A.W. and Marsh, T.A. (1983) "Stock Return Seasonalities and the Tax-Loss Selling Hypothesis" *Journal of Financial Economics*, 12, 1 (June), 105–27.

Brunnermeier, M.K. (2001) *Asset Pricing under Asymmetric Information: Bubbles, Crashes, Technical Analysis, and Herding*, Oxford University Press, Oxford.

Brush, J.S. (1986) "Eight Relative Strength Models Compared" *Journal of Portfolio Management*, 13, 1 (Fall), 21–8.

Byrne, R. and Whiten, A. (1988) *Machiavellian Intelligence: Social Expertise and the Evolution of Intellect in Monkeys, Apes and Humans*, Clarendon Press, Oxford.

Caldwell, B.J. (1986) "Economic Methodology and Behavioral Economics: An Interpretive History" in B. Gilad and S. Kaish (eds) *Handbook of Behavioral Economics*, Jai Press Inc., London.

Camerer, C. *et al.* (1997) "Labor Supply of New York City Cabdrivers: One Day at a Time" *Quarterly Journal of Economics*, 112, 2 (May), 407–41.

Campbell, H. (1991) "The World Price of Covariance Risk" *Journal of Finance*, 46, 1, 111–57.

Campbell, J., Grossman, S. and Wang, J. (1993) "Trading Volume and Serial Correlation in Stock Returns" *Quarterly Journal of Economics*, 108, 3, 905–34.

Cartwright, J. (2000) *Evolution and Human Behavior*, Palgrave, Hampshire.

Caudill, M. and Butler, C. (1992) *Understanding Neural Networks: Computer Explorations: Vol 1 & 2*, MIT Press, Cambridge, MA.

Chakravarty, S., Gulen, H. and Mayhew, S. (2004) "Informed Trading in Stock and Option Markets," *Journal of Finance*, 59, 3, 1235–58.

Chamley, C.P. (2004), *Rational Herds: Economic Models of Social Learning*, Cambridge University Press, Cambridge.

Chan, L.K.C., Hamao, Y. and Lakonishok, J. (1991) "Fundamentals and Stock Returns in Japan" *Journal of Finance*, 46, 5, 1739–64.

Chang, E.C. and Pinegar, J.M. (1986) "Return Seasonality and Tax-Loss Selling in the Market for Long Term Government and Corporate Bonds" *Journal of Financial Economics*, 17, 2 (December), 391–415.

Charest, G. (1978) "Dividend Information, Stock Returns and Market Efficiency II" *Journal of Financial Economics*, 6, 2/3 (June/September), 297–330.

Chiang, A.C. (1984), *Fundamental Methods of Mathematical Economics*, McGraw-Hill, New York.

Chiang, A.C. (1992), *Elements of Dynamic Optimization*, McGraw-Hill, New York.

Cohen, B. (1997) *The Edge of Chaos: Financial Dooms, Bubbles, Crashes and Chaos*, John Wiley & Sons, Chichester.

Coley, D.A. (1999) *An Introduction to Genetic Algorithms for Scientists and Engineers*, World Scientific Publishing, London.

Conrad, J. and Kaul, G. (1988) "Time Variation in Expected Returns" *Journal of Business*, 61, 4 (October), 409–25.

Cootner, P. (1974) *The Random Character of Stock Market Prices*, MIT Press, Cambridge, MA.

Copeland, T.E. and Mayers, D. (1982) "The Value Line Enigma (1965–1978): A Case Study of Performance Evaluation Issues" *Journal of Financial Economics*, 10, 3 (November), 289–321.

Cournot, A.A. (1838) *Recherches sur les Principes Mathematiquesde la Theorie des Richesses*, Hachette, Paris.

Cox, J.C., Ross, S.A. and Rubinstein, M. (1979) "Option Pricing: A Simplified Approach" *Journal of Financial Economics*, 7, 229–63.

Crank, J. (1975) *The Mathematics of Diffusion* (2nd Edn), Oxford University Press, Oxford.

Daley, M. and Wilson, M. (1983) *Sex, Evolution and Behavior*, Wadsworth Publishing Company, Belmont, CA.

Darley, V.M. and Kauffman, S.A. (1997) "Natural Rationality" in Arthur, W.B. Durlauf, S.N. and Lane, D.A. (eds) *The Economy as an Evolving Complex System II: Proceedings Volume XXVII Santa Fe Institute Studies in the Sciences of Complexity*, Perseus Books, Reading, MA.

Darwin, C. (1859) *On the Origin of Species by Means of Natural Selection*, Murray, London.

Dawkins, R. (1976) *The Selfish Gene*, Oxford University Press, Oxford.

DeLong, J.B., Shleifer, A., Summers, L.H. and Waldmann, R.J. (1990) "Noise Traders Risk in Financial Markets" *The Journal of Financial Economics*, 98, 103–41.

Dempster, M.A.H. and Jones, C.M. (1999a) "Can Technical Pattern Trading be Profitably Automated?" *Cambridge University Judge Institute of Management Working Paper* 11/99.

Dempster, M.A.H. and Jones, C.M. (1999b) "Can Technical Pattern Trading be Profitably Automated: The Head and Shoulders?" *Cambridge University Judge Institute of Management Working Paper* 12/99.

Dempster, M.A.H. and Jones, C.M. (2000) "The Profitability of Intra-Day FX Trading using Technical Indicators" *Cambridge University Judge Institute of Management Working Paper* 35/00.

Dempster, M.A.H. and Jones, C.M. (2001) "A Real-Time Adaptive Trading System using Genetic Programming" *Journal of Quantitative Finance*, 1, 397–413.

Derman, E. and Kani, I. (1994) "Riding on a Smile" *Risk Magazine*, 7, 2, 17–23.

Devaney, R.L. (1989) *An Introduction to Chaotic Dynamical Systems*, Addison-Wesley, Menlo Park California, CA.

Diamond, P.A. (1965) "National Debt in a Neoclassical Growth Model" *American Economic Review*, 55, 5, 1126–50.

Dimson, E. (1979) "Risk Measurement When Shares are Subject to Infrequent Trading" *Journal of Financial Economics*, 7, 2 (June), 197–226.

Dimson, E. and Marsh, P. (1984) "An Analysis of Brokers and Analysts Unpublished Forecasts of UK Stock Returns" *Journal of Finance*, 39, 5 (December), 1257–92.

Dixit, A.K. and Stiglitz, J.E. (1977) "Monopolistic Competition and Optimum Product Diversity" *American Economic Review*, 67, 297–308.

Dodd, P. and Ruback, R. (1977) "Tender Offers and Stockholders Returns" *Journal of Financial Economics*, 5, 3 (December), 351–75.

Doyle, J. (2001) "Robustness and Complexity" *Santa Fe Institute Colloquium Talk*, Santa Fe Institute, New Mexico, March.

Dupire, B. (1994) "Pricing with a Smile" *Risk Magazine*, 7, 1, 24–6.

Dyl, E.A. (1977) "Capital Gains Taxation and Year-end Stock Market Behavior" *Journal of Finance*, 32, 1 (March), 165–75.

Easley, D., Hvidkjoer, S. and O'Hara, M. (2002) "Is Information Risk a Determinant of Asset Returns?" *The Journal of Finance*, 57, 5, 2185–221.

Edwards, W. (1968) "Conservatism in Human Information Processing" in B. Kleinmutz (ed.) *Formal Representation of Human Judgement*, John Wiley & Sons, New York.

Eldredge, N. (1995) *Reinventing Darwin*, John Wiley & Sons, New York.

Elton, E.J. and Gruber, M.J. (1995) *Modern Portfolio Theory and Investment Analysis*, John Wiley & Sons, New York.

Elton, E.J., Gruber, M.J. and Grossman, S. (1986) "Discreet Expectational Data and Portfolio Performance" *Journal of Finance*, 41, 3 (July), 699–712.

Erlich, D., Guttman, P., Schpenbach, P. and Mills, J. (1957) "Postdecision Exposure to Relevant Information" *Journal of Abnormal and Social Psychology*, 54, 98–102.

Ewens, W.J. (1972) "The Sampling Theory of Selectively Neutral Alleles" *Theoretical Population Biology*, 3, 87–112.

Ewens, W.J. (1979) *Mathematical Population Genetics*, Springer-Verlag, Berlin.

Fama, E. (1965) "The Behavior of Stock Prices" *Journal of Business*, 38 (January), 34–105.

Fama, E. (1970) "Efficient Capital Markets: A Review of Theory and Empirical Work" *Journal of Finance* (March), 383–417.

Fama, E. (1991) "Efficient Capital Markets II" *Journal of Finance*, 26, 5 (December), 1575–617.

Fama, E. (1998) "Market Efficiency, Long-Term Returns, and Behavioral Finance" *Journal of Financial Economics*, 49, 3, 283–306.

Fama, E. and Blume, M. (1966) "Filter Rules and Stock Market Trading" *Journal of Business*, 39 (January), 226–41.

Fama, E. and French, K.R. (1988) "Permanent and Temporary Components of Stock Prices" *Journal of Political Economy*, 96 (April), 246–73.

Fama, E. and MacBeth, J. (1973) "Risk, Return and Equilibrium Empirical Tests" *Journal of Political Economy*, 81, 3 (May/June) 607–36.

Farmer, J.D. (1998) "Market Force, Ecology, and Evolution" *Santa-Fe Institute Working Paper, 98-12-117*.

Farmer, J.D. (1999) "Physicists Attempt to Scale the Ivory Towers of Finance" *Computing in Science and Engineering*, (November/December), 26–39.

Farmer, J.D. and Lo, A.W. (1999) "Frontiers of Finance: Evolution and Efficient Markets" *Santa Fe Institute Working Paper* 99-06-039.

Favre, A., Guitton, H.J., Lichnerowicz, A. and Wolff, E. (1995) *Chaos and Determinism: Turbulence as a Paradigm for Complex Systems Converging toward Final States*, The Johns Hopkins University Press, Baltimore, MA.

Fernandez-Rodriguez, F., Garcia-Arties, M.D. and Martin-Gonzalez, F.M. (2002) "A Model of Speculative Behavior with a Strange Attractor" *Applied Mathematical Finance*, 9, 3, 143–61.

Fogel, L.J., Owens, A.J. and Walsh, M.J. (1996) *Artificial Intelligence through Simulated Evolution*, John Wiley & Sons, New York.

Foster, G., Olsen, C. and Shevlin, T. (1984) "Earnings Releases, Anomalies and the Behavior of Security Returns" *Accounting Review*, 4 (October), 574–603.

Firth, M. (1975) "The Information Content of Large Investment Holdings" *Journal of Finance*, 30, 5 (December), 1265–81.

French, K.R. (1980) "Stock Returns and the Weekend Effect" *Journal of Financial Economics*, 8, 1 (March), 55–70.

Friedman, J.W. (1991) *Game Theory with Applications to Economics*, Oxford University Press, Oxford.

Friedman, M. (1953) "The Case for Flexible Exchange Rates" in *Essays in Positive Economics*, University of Chicago Press, Chicago, IL.

Ghashghaie, S., Breymann, W., Peinke, J., Talkner, P. and Dodge, Y. (1996) "Turbulent Cascades in Foreign Exchange Markets" *Nature*, 381, 767–70.

Gibbons, M.R. and Hess, P.J. (1981) "Day of the Week Effects and Asset Returns" *Journal of Business*, 54, 579–96.

Gintis, H. (2000) *Game Theory Evolving: A Problem-Centered Introduction to Modeling Strategic Interaction*, Princeton University Press, Princeton, NJ.

Gladwell, M. (2000) *The Tipping Point: How Little Things Can Make a Big Difference*, Little Brown and Company, Boston, MA.

Glaser, R. (2001) *Biophysics*, Springer, Berlin.

Gleick, J. (1998) *Chaos: The Amazing Science of the Unpredictable*, Vintage, London.

Glosten, L. (1989) "Insider Trading, Liquidity and the Role of the Monopolist Specialist" *Journal of Business*, 54, 2, 579–96.

Goldberg, D.E. (1989) *Genetic Algorithms in Search, Optimization and Machine Learning*, Addison-Wesley, Redwood City California, CA.

Granger, C.W. (1975) "A Survey on Empirical Studies on Capital Markets" in Elton, E.J. and Gruber, M.J. (eds) *International Capital Markets*, North Holland, Amsterdam.

Green, T.C. (2004) "Economic News and the Impact of Trading on Bond Prices" *Journal of Finance*, 59, 3, 1201–34.

Greene, B. (2000) *The Elegant Universe: Superstrings, Hidden Dimensions and the Quest for the Ultimate Theory*, Vintage, London.

Grier, P. and Albin, P. (1973) "Non-Random Price Changes in Association with Trading in Large Blocks" *Journal of Business*, 46, 3 (July), 425–33.

Grilliches, Z. (1957) "Hybrid Corn: An Exploration in the Economics of Technological Change" *Econometrica*, 25, 501–22.

Grilliches, Z. (1960) "Hybrid Corn and the Economics of Innovation" *Science*, 70, 275–80.

Grossman, G. and Helpman, E. (1991) *Innovation and Growth in the Global Economy*, MIT Press, Cambridge, MA.

Grossman, S.J. and Stiglitz, J.E. (1980) "On the Impossibility of Informationally Efficient Markets" *American Economic Review* (6), 4, 103–47.

Gruen D.W.R. and Gizycki, M. (1993) "Explaining Forward Discount Bias: Is it Anchoring?" *Reserve Bank of Australia Discussion Paper* RDP 9307, Reserve Bank of Australia.

Gultekin, M. and Gultekin, N.B. (1983) "Stock Market Seasonality: International Evidence" *Journal of Financial Economics*, 12, 469–81.

Hald, A. (1990) *A History of Probability and Statistics and Their Approaches Before 1750*, John Wiley & Sons, New York.

Hamilton, J.D. (1994) *Time Series Analysis*, Princeton University Press, Princeton, NJ.

Harris, L. (1986) "A Transaction Data Study of Weekly and Intradaily Patterns in Stock Returns" *Journal of Financial Economics*, 14 (May), 99–117.

Harsanyi, J.C. (1966) "A General Theory of Rational Behavior in Game Situations" *Econometrica*, 34, 613–34.

Harsanyi, J.C. (1967–8) "Games with Incomplete Information Played by "Bayesian" Players" Parts I, II and III, *Management Science*, 14, 159–82, 320–34 and 486–502.

Hart, W.E. (2003) "Locally Adaptive and Memetic Evolutionary Pattern Search Algorithms" *Evolutionary Computation*, 11, 1, 29–51.

Haug, E.G. (1997) *The Complete Guide to Option Pricing Formulas*, McGraw-Hill, New York.

Hayek, F. (1945) "The Use of Knowledge in Society" *American Economic Review*, 35, 519–30.

Hens, T., Schenk-Hoppe, K.R. and Stalker, M. (2002). "An Application of Evolutionary Finance to Firms Listed in the Swiss Market Index" *Swiss Journal of Economics and Statistics*, 138, 4, 465–87.

Hirschfeld, L. and Gelman, S. (1994) *Mapping the Mind: Domain Specificity in Cognition and Culture*, Cambridge University Press, New York.

Hirshleifer, J. and Riley, J.G. (1992) *The Analytics of Uncertainty and Information*, Cambridge University Press, Cambridge.

Hodgson, G.M. (1996) *Evolution and Institutions: On Evolutionary Economics and the Evolution of Economics*, Edward Elgar, Cheltenham, UK.

Holland, J.H. (1974) "Genetic Algorithms and the Optimal Allocation of Trials" *SIAM Journal of Computing*, 2, 88–105.

Holland, J.H. (1992) *Adaptation in Natural and Artificial Systems*, MIT Press, Cambridge, MA.

Holland, J.H. (1995) *Hidden Order: How Adaptation Builds Complexity*, Perseus Books, Reading, MA.

Holloway, C. (1981) "A Note on Testing an Aggressive Investment Strategy Using Value Line Ranks" *Journal of Finance*, 36, 3 (June), 711–19.

Hoggard, T., Whalley, A.E. and Wilmott, P. (1994) "Hedging Option Portfolios in the Presence of Transaction Costs" *Advances in Futures and Options Research*, 7, 21–35.

Hull, J.C. (1997) *Options, Futures, and Other Derivatives*, Prentice Hall International, London.

Hull, J.C. and White, A. (1987) "The Pricing of Options on Assets with Stochastic Volatilities" *Journal of Finance*, 42, 281–300.

Ibbtoson, R.G. (1975) "Price Performance of Common Stock New Issues" *Journal of Financial Economics*, 2, 3 (September), 235–72.

Jain, P.C. (1988) "Response of Hourly Stock Prices and Trading Volume to Economic News" *Journal of Business*, 61, 2 (April), 219–31.

Jefferies, P., Lamper, D. and Johnson, N.F. (2002) "Anatomy of Extreme Events in a Complex Adaptive System" *Condensed Matter Archive Paper* 0201510.

Jensen, M. and Bennington, G. (1970) "Random Walks and Technical Theories: Some Additional Evidence" *Journal of Finance*, 25, 2, 469–82.

Jha, V., Lichtblau, D. and Mozes, H. (2003) "The Usefulness of Analysts' Recommendations" *The Journal of Investing* (Summer), Vol no 7–18.

Joesphs, R., Larrick, R.P., Steele, C.M. and Nisbett, R.E. (1996) "Protecting the Self from the Negative Consequences of Risky Decisions" *Journal of Personality and Social Psychology*, 62, 26–37.

Jones, C.D., Pearce, O.K. and Wilson, J.W. (1987) "Can Tax-Loss Selling Explain the January Effect? A Note" *Journal of Finance*, 42, 2 (June), 453–61.

Johnson, S. (2001) *Emergence: The Connected Lives of Ants, Brains, Cities and Software*, Penguin Press, New York.

Jorion, P. (1997) *Value-at-Risk: The New Benchmark for Controlling Derivative Risk*, Irwin Publishing, Chicago, IL.

Joy, O.M., Litzenberger, R.H. and McEnally, R.W. (1977) "The Adjustment of Stock Prices to Announcements of Unanticipated Changes in Quarterly Earnings" *Journal of Accounting Research*, 15, 2, 207–25.

Kaheman, D. and Tversky, A. (1979) "Prospect Theory of Decisions under Risk" *Econometrica*, 47, 2, 1156–67.

Kato, K. and Shallheim, J. (1985) "Seasonal and Size Anomalies in the Japanese Stock Market" *Journal of Financial and Quantitative Analysis*, 20, 2 (June), 243–60.

Kauffman, S. (1993) *The Origins of Order*, Oxford University Press, New York.

Kauffman, S. (1995) *At Home in the Universe*, Oxford University Press, Oxford.

Kaufman, P. (1998) *Trading Systems and Methods* (3rd Edn), John Wiley & Sons, New York.

Kawazynski, A.L. and Legowiec, B. (2001) "Two Dimensional Model of a Reaction-Diffusion System as a Typewriter" *Physical Review*, 64, 056202.

Keim, D.B. (1983) "Size Related Anomalies and Stock Return Seasonaility" *Journal of Financial Economics*, 12, 1 (June), 13–32.

Keim, D.B. (1985) "Dividend Yields and Stock Returns: Implications of Abnormal January Returns" *Journal of Financial Economics*, 14, 3 (September), 473–89.

Keim, D.B. (1986) "Dividend Yields and the January Effect" *Journal of Portfolio Management*, 12, 2, 54–60.

Keim, D.B. and Stambaugh, R.F. (1984) "A Further Investigation of the Weekend Effect in Stock Returns" *Journal of Finance*, 39, 3 (July), 819–40.

Keim, D.B. and Stambaugh, R.F. (1986) "Predicting Returns in Stock and Bond Markets" *Journal of Financial Economics*, 17, 2 (December), 357–90.

Keynes, J.M. (1936) *The General Theory of Employment, Interest and Money*, Macmillan, London.

Kindleberger, C.P. (1996) *Manias, Panics and Crashes*, John Wiley & Sons, New York.

Kirman, A. (1993) "Ants, Rationality and Recruitment" *Quarterly Journal of Economics*, 108, 1, 137–56.

Kirzner, I.N. (1979) *Perception, Opportunity and Profit*, University of Chicago Press, Chicago, IL.

Kraus, A. and Stoll, H. (1972) "Price Impacts of Block Trading on the New York Stock Exchange" *Journal of Finance*, 27, 3 (June), 435–55.

Kremyanskiy, V.I. (1960) "Certain Peculiarities of Organisms as a 'System' from the Point of View of Physics, Cybernetics and Biology" *General Systems*, 5, 221–30.

Kreps, D.M., Milgrom, P., Roberts, J. and Wilson, R. (1982) "Rational Cooperation in the Finitely Repeated Prisoners' Dilemma" *Journal of Economic Theory*, 27, 245–52.

Krische, S. and Lee, C. (2001) "Analyzing the Analysts: Are Stock Recommendations Informative?" *Cornell University Working Paper*.

Kroll, Y., Levy, H. and Rapoport, A. (1988) "Experimental Tests of the Separation Theorem and the Capital Asset Pricing Model" *American Economic Review*, 52, 103–47.

Kuhn, H.W. (1953) "Extensive Games and the Problem of Information" in Kuhn, H.W. and Tucker, A.W. (eds) *Contributions to the Theory of Games: Volume II Annals of Mathematics Studies*, pp. 193–216, 28 Princeton University Press, Princeton, NJ.

Kuhn, H.W. (1968) "Preface to Waldegrave's Comments: Excerpt from Montmort's Letter to Nicholas Bernoulli" in Baumol W.J. and Goldfeld S.M. (eds) *Precursors in Mathematical Economics: An Anthology: Series of Reprints of Scarce Works on Political Economy* Baumol, London School of Economics and Political Science, London.

Lakonishok, J. and Smidt, S. (1984) "Volume and Turn of the Year Behavior" *Journal of Financial Economics*, 13, 3 (September), 435–55.

Latane, H.A., Joy, O.M. and Jones, C.P. (1970) "Quarterly Data, Sort-Rank Routines and Security Evaluation" *Journal of Business*, 43, 4 (October), 427–38.

Leland, H.E. (1985) "Option Pricing and Replication with Transaction Costs" *Journal of Finance*, 40, 1283–1301.

LeRoy, S. (1973) "Risk Aversion and the Martingale Property of Stock Returns" *International Economic Review*, 14, 436–46.

Levy, P. (1925) *Calcul des Probabilities*, Gauthier-Villas, Paris.

Levy, R. (1967) "Relative Strength as a Criterion for Investment Selection" *Journal of Finance*, 22 (December), 595–610.

Levy, M., Levy, H. and Solomon, S. (2000) *Microscopic Simulation of Financial Markets: From Investor Behavior to Market Phenomena*, Academic Press, San Diego, CA.

Lewin, R. (1999) *Complexity: Life at the Edge of Chaos*, University of Chicago Press, Chicago, IL.

Lewontin, R. C. (1961) "Evolution and the Theory of Games" *Journal of Theoretical Biology*, 1, 382–403.

Litterman, B. (2003) *Modern Investment Management: An Equilibrium Approach*, John Wiley & Sons, New York.

Lotka, A.J (1925) *Elements of Physical Biology*, Williams and Wilkins, Baltimore, MA.

Lo, A.W. and MacKinlay, A.C. (1999) *A Non-Random Walk Down Wall Street*, Princeton University Press, Princeton, NJ.

Lo, A.W. and Repin, D.V. (2002) "The Psychophysiology of Real-Time Financial Risk Processing" *Journal of Cognitive Neuroscience*, 14, 323–39.

Lucas, R.E. (1978) "Asset Prices in an Exchange Economy" *Econometrica*, 46, 1429–46.

MacKay, D.J.C. (2003) *Information Theory, Inference and Learning Algorithms*, Cambridge University Press, Cambridge.

Maheu, J.M. and McCurdy, T.H. (2004) "News Arrival, Jump Dynamics and Volatility Components for Individual Stock Returns" *Journal of Finance*, 59, 2, 755–93.

Mainzer, K. (1997) *Thinking in Complexity: The Complex Dynamics of Matter, Mind, and Mankind*, (3rd Edn), Springer-Verlag, Berlin.

Majorana, E. (1942) "Il volore delle leggi statistiche nella fiscal e nelle scienze sociali" *Scientia*, 36, 58–66.

Malkiel, B.G. (1996) *A Random Walk Down Wall Street*, W.W. Norton & Company, New York.

Mandelbrot, B.B. (1963) "The Variation of Certain Speculative Prices" *Journal of Business*, 36, 394–419.

Mandelbrot, B.B. (1997) *Fractals and Scaling in Finance: Discontinuity, Concentration, Risk*, Springer, New York.

Mantegna, R.N. and Stanley, E.H. (1996) "Turbulence and Financial Markets" *Nature*, 383, 588.

Mantegna, R.N. and Stanley, E.H. (2000) *An Introduction to Econophysics: Correlations and Complexity in Finance*, Cambridge University Press, Cambridge.

Markowitz, H. (1952) "Portfolio Selection" *Journal of Political Economy*, XI, 2, 151–58.

Markowitz, H. (1991) *Portfolio Selection: Efficient Diversification of Investments*, Blackwell Publishers, Oxford.

Marx, K. (1954) *Capital*, Foreign Languages Publishing House, Moscow.

Maynard-Smith, J. (1972) "Game Theory and the Evolution of Fighting" in *On Evolution (John Maynard Smith)*, pp. 8–28, Edinburgh University Press, Edinburgh.

Maynard Smith, J. (1982) *Evolution and the Theory of Games*, Cambridge University Press, Cambridge.

McClelland, J.L. and Rogers, T.T. (2003) "The Parallel Distributed Processing Approach to Semantic Cognition" *Nature*, 4, 310–22.

McFadden, D. (1998) "Rationality for Economists?" *Santa Fe Institute Working Paper* N. 98-09-086.

Mertens, J.F. and Zamir, S. (1985) "Formulations of Bayesian Analysis for Games with Incomplete Information" *International Journal of Game Theory*, 14, 1–29.

Merton, R.C. (1992) *Continuous-Time Finance*, Blackwell, MA.

McKenna, C.J. (1986) *The Economics of Uncertainty*, Wheatsheaf Books/St. Martin's Press, New York.

Michaud, R.O. (1998) *Efficient Asset Management: A Practical Guide to Stock Portfolio Optimization and Asset Allocation*, Harvard Business School Press, Boston, MA.

Mills, T.C. (1993) *The Econometric Modelling of Financial Time Series*, Cambridge University Press, Cambridge.

Minsky, H.P. (1982) "The Financial Instability Hypothesis: Capitalistic Processes and the Behavior of the Economy" in Kindleberger, C.P. and Laffargue, J.-P. (eds) *Financial Crises: Theory, History and Policy*, Cambridge University Press, Cambridge.

Mitchell, M. (1998) *An Introduction to Genetic Algorithms*, MIT Press, Cambridge, MA.

Montier, J. (2002) *Behavioural Finance: Insights into Irrational Minds and Markets*, John Wiley & Sons, Chichester.

Morris, R. (2001) *The Evolutionists: The Struggle for Darwin's Soul*, Henry Holt and Company, New York.

Morrison, A.D. and Vulkan, N. (2003) "Making Money out of Publicly Available Information" *Oxford University Working Paper*.

Mullainathan, S. and Thaler, R.H. (2000) "Behavioral Economics" *National Bureau of Economic Research Working Paper* No. 7948, October.

Muller, B., Reinhardt, J. and Strickland, M. (1995) *Neural Networks: An Introduction*, Springer-Verlag, Berlin.

Muth, J.F. (1961) "Rational Expectations and the Theory of Price Movements" *Econometrica*, 29, 315–35.

Nash, J.F. (1950a) "Equilibrium Points in N-Person Games" *Proceedings of the National Academy of Sciences of the United States of America*, 36, 48–9.

Nash, J.F. (1950b) "The Bargaining Problem" *Econometrica*, 18, 155–62.

Nash, J.F. (1951) "Non-Cooperative Games" *Annals of Mathematics*, 54, 286–95.

Nash, J.F. (1953) "Two Person Cooperative Games" *Econometrica*, 21, 128–40.

Nasur, S. (1994) *A Beautiful Mind: A Biography of John Forbes Nash Jr, Winner of the Nobel Prize for Economics 1994*, Simon & Schuster, New York.

Nelson, R.R. and Winter, S.G. (1982) *An Evolutionary Theory of Economic Change*, Harvard University Press, Cambridge, MA.

North, D. (1990) *Institutions, Institutional Change and Economic Performance*, Cambridge University Press, Cambridge.

Osborne, M.F.M. (1964) "Brownian Motion in the Stock Market" in P. Cootner (ed.) *The Random Character of Stock Prices*, MIT Press, Cambridge.

Osborne, M.J. and Rubenstein, A. (1994) *A Course in Game Theory*, MIT Press, Cambridge, MA.

Olson, M. (1982) *The Rise and Decline of Nations*, Yale University Press, New Haven, CT.

Pareto, V. (1897) *Cours d'Economie Politique*, Lausanne, Switzerland.

Pascale, R.T., Millemann, M. and Gioja, L. (2000) *Surfing the Edge of Chaos: The Laws of Nature and the New Laws of Business*, Texere, New York.

Peretto, P. (1992) *An Introduction to the Modeling of Neural Networks*, Cambridge University Press, Cambridge.

Pesaran, M.H. and Timmermann, A. (1995) "Predictability of Stock Returns: Robustness and Economic Significance" *Journal of Finance*, 50, 4 (September), 1201–28.

Peters, E. (1996) *Chaos and Order in Capital Markets; A New View of Cycles, Prices and Market Volatility*, John Wiley & Sons, New York.

Pierce, D. and Roley, V. (1985) "Stock Prices and Economic News" *Journal of Business*, 59, 1, 49–67.

Pettit, R.R. (1972) "Dividend Announcements, Security Performance and Capital Market Efficiency" *Journal of Finance*, 27, 5 (December), 993–1007.

Pinches, G. (1970) "The Random Walk Hypothesis and Technical Analysis" *Financial Analysts Journal*, 26, 2, 104–10.

Pinker, S. (1994) *The Language of Instinct: How the Mind Creates Language*, William Morrow, New York.

Poterba, J. and Summers, L.H. (1988) "Mean Reversion in Stock Prices: Evidence and Implications" *Journal of Financial Economics*, 22, 1 (October), 22–59.

Poteshaman, A.M., Serbin, V. (2003) "Cleary Irrational Financial Market Behavior: Evidence from the Early Exercise of Exchange Traded Stock Options" *Journal of Finance*, 58, 1, 37–70.

Potters, M., Cont, R. and Bouchaud, J.P. (1998) "Financial Markets as Adaptive Systems" *Europhysics Letters*, 41, 3, 239–44.

Praetz, P. (1972) "The Distribution of Share Price Changes" *Journal of Business*, 45, 1 (January), 49–55.

Pruitt, S.W. and White, R.E. (1988) "Who Says Technical Analysis Can't Beat the Market?" *Journal of Portfolio Management*, 14, 3, 55–58.

Ramsey, F.P. (1928) "A Contribution to the Theory of Taxation" *Economic Journal*, 37, 47–61.

Rao, V. and Rao, H. (1995) *C++ Neural Networks and Fuzzy Logic* (2nd Edn), World Scientific, London.

Rebonato, R. (1999) *Volatility and Correlation in the Pricing of Equity, FX and Interest Rate Options*, John Wiley & Sons, Chichester.

Refenes, P.A. (1995) *Neural Networks in the Capital Market*, John Wiley and Sons, London.

Reilly, F.K. and Brown, K.C. (1997) *Investment Analysis and Portfolio Management*, Dryden Press, Fort Worth.

Reilly, F.K. and Hatfield, K. (1969) "Investor Experience with New Stock Issues" *Financial Analysts Journal*, 25, 5, 73–80.

Reinganum, M.R. (1981) "Abnormal Returns in Small Firm Portfolios" *Financial Analysts Journal*, 37, 52–57.

Reinganum, M.R. (1983a) "The Anomalous Stock Market Behavior of Small Firms in January" *Journal of Financial Economics*, 12 (June), 89–104.

Reinganum, M.R. (1983b) "Portfolio Strategies Based on Market Capitalization" *Journal of Portfolio Management*, 9, 29–36.

Reinganum, M.R. (1992) "A Revival of the Small Firm Effect" *Journal of Portfolio Management*, 18, 3, 29–36.

Rendleman, R.J. and Bartter, B.J. (1979) "Two-State Option Pricing" *Journal of Finance*, 34, 1093–110.

Rendleman, R.J., Jones, C.P. and Latane, H.A. (1982) "Empirical Anomalies Based on Unexpected Earnings and the Importance of Risk Adjustments" *Journal of Financial Economics*, 10, 3 (November) 269–87.

Ritov, I. (1996) "Probability of Regret: Anticipation of Uncertainty Resolution in Choice" *Organizational Behavior and Human Decision Processes*, 66, 228–36.

Roberts, H.V. (1967) "Statistical versus Clinical Prediction of the Stock Market" Unpublished Paper presented to the *Seminar on the Analysis of Security Prices*, University of Chicago, IL.

Rodgers, E.M. (1995) *Diffusion of Innovations* (4th Edn), The Free Press, New York.

Roll, R. (1981) "A Possible Explanation of the Small Firm Effect" *Journal of Finance*, 36, 4 (September), 879–88.

Romer, P.M. (1986) "Increasing Returns and Long-Run Growth" *Journal of Political Economy*, 94, 5, 123–76.

Romer, P.M. (1989) "Human Capital Growth: Theory and Evidence" *National Bureau of Economic Research*, Working Paper No. 3173, Cambridge, MA.

Romer, P.M. (1990) "Endogenous Technological Change" *Journal of Political Economy*, 98, Pt 2, S71–102.

Rozeff, M. (1984) "Dividend Yields as Equity Risk Premiums" *Journal of Portfolio Management*, 11, 1 (Fall), 68–75.

Rozeff, M. and Kinney, W.R. (1976) "Capital Market Seasonality: The Case of Stock Returns" *Journal of Financial Economics*, 3, 4 (December), 379–402.

Rubinstein, M. (1994) "Implied Binomial Trees" *Journal of Finance*, 49, 771–818.

Ryan, B. and Gross, N.C. (1943) "The Diffusion of seed Corn in Two Iowa Communities" *Rural Sociology*, 8, 15–24.

Samuelson, P. (1938) "A Note on the Pure Theory of Consumers Behavior" *Economica*, 5, 61–71.

Samuelson, P. (1958) "An Exact Consumption-Loan Model of Interest with or without the Social Contrivance of Money" *Journal of Political Economy*, 66, 6, 467–82.

Samuelson, P. (1965) "Proof that Properly Anticipated Prices Fluctuate Randomly" *Industrial Management Review*, 6, 41–9.

Sargent, T.J. (1993) *Bounded Rationality in Macroeconomics*, Oxford University Press, Oxford.

Scheinkman, J.A. and LeBaron, B. (1989) "Nonlinear Dynamics and Stock Returns" *Journal of Business*, 62, 311–38.

Schrödinger, E. (1967) *What is Life?*, Cambridge University Press, Cambridge.

Schumpeter, J.A. (1939) *Business Cycles: A Theoretical and Statistical Analysis of the Capitalist Process*, McGraw-Hill, New York.

Shafer, G. and Vovk, V. (2001) *Probability and Finance: It's Only a Game!*, John Wiley & Sons, New York.

Shannon, C.E. and Weaver, W. (1963) *The Mathematical Theory of Communication*, Urbana, University of Illinois.

Shapley, L.S. and Shubik, M. (1954) "A Method for Evaluating the Distribution of Power in a Committee System" *American Political Science Review*, 48, 787–92.

Sharpe, W. (1963) "A Simplified Model for Portfolio Analysis" *Management Science* (January), 2, 277–93.

Sharpe, W. (2000) *Portfolio Theory and Capital Markets*, McGraw-Hill, New York.

Shefrin, H. (2000) *Beyond Greed and Fear: Understanding Behavioral Finance and the Psychology of Investing*, Harvard Business School Press, Boston, MA.

Shefrin, H. and Statman, M. (1985) "The Disposition to Sell Winners too Early and Ride Losers too Long" *Journal of Finance*, 40, 777–90.

Shiller, R. (1984) "Stock Prices and Social Dynamics" *Brookings Papers on Economic Activity*, 2, 457–510.

Shiller, R. (1999) "Human Behavior and the Efficiency of Financial Markets" in Taylor, J.B. and Woodford, M. (eds), *Handbook of Macroeconomics*, Vol 1C, pp. 1305–40, North Holland, New York.

Shiller, R. (1995) "Conversation, Information and Herd Behavior" *American Economic Review*, 85, 181–85.

Shubik, M. (1997) "Time and Money" in Arthur, W.B. Durlauf, S.N. and Lane, D.A. (eds) *The Economy as an Evolving Complex System II: Proceedings Volume XXVII Santa Fe Institute Studies in the Sciences of Complexity*, Perseus Books, Reading, MA.

Simon, H.A. (1956) "Rational Choice and the Structure of Environments" *Psychological Review*, 69, 129–38.

Simon, H. (1982) *Models of Bounded Rationality*, MIT Press, Cambridge, MA.

Skinner, B.F. (1948) "Superstition in the Pigeon" *Journal of Experimental Psychology*, 38, 168–72.

Sornette, D. (2001) "Predicatability of Catastrophic Events: Material Rupture, Earthquakes, Turbulence, Financial Crashes and Human Birth" *Archives of Condensed Matter Working Paper No 0107173*.

Steeb, W.H. (1999) *The Nonlinear Workbook: Chaos, Fractals, Cellular Automatic, Neural Networks, Genetic Algorithms, Fuzzy Logic with C++, Java, Symbolic C++ and Reduce P*, World Scientific, London.

Sterman, I.D. (2000) *Business Dynamics: Systems Thinking and Modeling for a Complex World*, McGraw-Hill, New York.

Stoll, H.R. and Whaley, R.E. (1983) "Transactions Costs and the Small Firm Effect" *Journal of Financial Economics*, 12, 1 (June), 57–80.

Stickel, S.E. (1985) "The Effect of Value Line Investment Survey Changes on Common Stock Prices" *Journal of Financial Economics*, 14, 1 (March), 121–43.

Taleb, N.N. (2004) *Fooled by Randomness: The Hidden Role of Chance in Life and the Markets*, Texere Publishing, New York.

Tay, N.S.P. and Linn, S.C. (2001) "Fuzzy Inductive Reasoning, Expectation Formation and the Behavior of Security Prices" *Journal of Economic Dynamics and Control*, 25, 321–61.

Thaler, R. (1994) *Quasi Rational Economics*, Russel Sage Foundation, New York.

Thompson, D.W. (1942) *On Growth and Form*, Cambridge University Press, Cambridge.

Tinic, S.M. and West, R.R. (1984) "Risk and Return: January vs the Rest of the Year" *Journal of Financial Economics*, 13, 4 (December), 561–74.

Tirole, J. (1982) "On the Possibility of Speculation under Rational Expectations" *Econometrica*, 50 (September), 1163–81.

Trivers, R.L. (1971) "The Evolution of Reciprocal Altruism" *Quarterly Review of Biology*, 46, 35–57.

Tucker A.W. (1950) "On Jargon: The Prisoner's Dilemma" *UMAP Journal*, 1, 101.

Tversky, A. and Kahneman, D. (1974) "Judgment Under Uncertainty: Heuristics and Biases" *Science*, 30, 211, 453–58.

Urbach, R.M. (2000) *Footprints of Chaos in the Markets: Analyzing Non-linear Time series in Financial Markets and Other Real Systems*, Financial Times Prentice Hall, London.

Vaga, T. (1991) "The Coherent Market Hypothesis" *Financial Analysts Journal*, (December/January) 41–56.

Vaga, T. (1994) *Profiting from Chaos*, McGraw-Hill, New York.

Voit, J. (2000) *The Statistical Mechanics of Financial Markets*, Springer, Berlin.

Volteraa, V. (1931) *Lecons sur la theorie mathematique de la lutte pour la vie*, Gauthier-Villars, Paris.

Volteraa, V. (1937) "Principes de biologie mathematique" *Acta Biother*, 3, 1–36.

von Bertalanffy, L. (1950) "The Theory of Open Systems in Physics and Biology" *Science*, 111, 23–9.

von Neuman, J. (1966) *Theory of Self-Reproducing Automata*, edited and completed by Burks, A.W. (ed.) University of Illinois Press, Champaign-Urbana.

von Neuman, J. and Morgenstern, O. (1944) *Theory of Games and Economic Behavior*, Princeton University Press, Princeton, NJ.

Waldrop, M.W. (1992) *Complexity: The Emerging in Science at the Edge of Chaos*, Touchstone, New York.

Wärneryd, K.E. (2001) *Stock-Market Psychology*, Edward Elgar, Cheltenham, UK.

Watson, J. (2004) *DNA: The Secret of Life*, Arrow Books, London.

Watts, D. J. (1999) *Small Worlds: The Dynamics of Networks between Order and Randomness*, Princeton University Press, Princeton, NJ.

Watts, R. (1973) "The Information Content of Dividends" *Journal of Business*, 45, 2 (April), 191–211.

Watts, R. (1978) "Systematic Abnormal Returns after Quarterly Earnings Announcements" *Journal of Financial Economics*, 6, 2/3, 127–50.

Whiting, M.F., Bradler, S. and Taylor, M. (2003) "Loss and Recovery of Wings in Stick Insects" *Nature*, 241, 264–67.

Wilmott, P. (1998) *Derivatives: The Theory and Practice of Financial Engineering*, John Wiley & Sons, Chichester.

Wilmott, P., Howison, S. and Dewynne, J. (1997) *The Mathematics of Financial Derivatives: A Student Introduction*, Cambridge University Press, Cambridge.

Winston, R. (2002) *Human Instinct: How our Primeval Impulses Shape our Modern Lives*, Bantam Press, London.

Winston, R. (2003) *The Human Mind: And How we Can Make the Most of It*, Bantam Press, London.

Wolfram, S. (1986) *Theory and Applications of Cellular Automata*, World Scientific, Singapore.

Womack, K.L. (1996) "Do Brokerage Analysts Recommendations Have Investment Value?" *Journal of Finance*, 51, 1 (March), 137–67.

Yamada, Y. and Primbs, J. (2002) "Distribution-Based Option Pricing on Latic Asset Dynamics Models" *International Journal of Theoretical and Applied Finance*, 5, 6, 599–618.

Index